Bonding Energetics
in Organometallic Compounds

ACS SYMPOSIUM SERIES **428**

Bonding Energetics in Organometallic Compounds

Tobin J. Marks, EDITOR

Northwestern University

Developed from a symposium sponsored
by the Division of Inorganic Chemistry
at the 198th National Meeting
of the American Chemical Society,
Miami Beach, Florida
September 10–15, 1989

American Chemical Society, Washington, DC 1990

Library of Congress Cataloging-in-Publication Data

Bonding energetics in organometallic compounds
Tobin J. Marks, editor

p.   cm.—(ACS Symposium Series, ISSN 0065–6156; 428).

"Developed from a symposium sponsored by the Division of
Inorganic Chemistry at the 198th National Meeting of the
American Chemical Society, Miami Beach, Florida, September
10–15, 1989."

Includes bibliographical references and indexes

ISBN 0–8412–1791–2

1. Organometallic chemistry—Congresses. 2. Chemical
bonds—Congresses.

I. Marks, Tobin J., 1944–   . II. American Chemical Society.
Division of Inorganic Chemistry. III. American Chemical Society.
Meeting (198th: 1989: Miami Beach, Fla.). V. Series.

QD411.B65   1990
547′.05—dc20   65655

                                          90–36268
                                              CIP

The paper used in this publication meets the minimum requirements of American
National Standard for Information Sciences—Permanence of Paper for Printed Library
Materials, ANSI Z39.48–1984.

∞

PRINTED IN THE UNITED STATES OF AMERICA

ACS Symposium Series

M. Joan Comstock, *Series Editor*

1990 ACS Books Advisory Board

Foreword

The ACS SYMPOSIUM SERIES was founded in 1974 to provide a medium for publishing symposia quickly in book form. The format of the Series parallels that of the continuing ADVANCES IN CHEMISTRY SERIES except that, in order to save time, the papers are not typeset but are reproduced as they are submitted by the authors in camera-ready form. Papers are reviewed under the supervision of the Editors with the assistance of the Series Advisory Board and are selected to maintain the integrity of the symposia; however, verbatim reproductions of previously published papers are not accepted. Both reviews and reports of research are acceptable, because symposia may embrace both types of presentation.

Contents

INDEXES

Preface

As CONTEMPORARY ORGANOMETALLIC CHEMISTRY HAS GROWN in sophistication and as we have learned more about reactivity, reaction mechanisms, molecular architecture, and electronic structure, it is only natural that we should inquire more deeply about the strengths of the bonds holding organometallic molecules together and how these are altered in various chemical transformations. At present, the acquisition and understanding of metal–ligand bond energy information for organometallic molecules is an active and important area of chemical research. It impacts directly upon central issues in contemporary organometallic, inorganic, organic, physical, enzymatic, and catalytic chemistry.

This volume and the symposium that served as its basis represent an effort to bring together leading experimental and theoretical researchers concerned with organometallic bonding energetics in the gas phase, in solution, and on well-defined surfaces. There has traditionally been minimal interaction between these diverse communities of activity, and the present volume attempts to convey the essence of the oral presentations and active dialogue that took place at the symposium. In addition, an overview chapter has been added to introduce basic concepts and experimental methodologies, as well as to provide a bibliography of relevant review articles, textbooks, compilations of thermodynamic data, and other source materials. This book is intended for all chemical researchers who are interested in a broad, in-depth survey of bonding energetics in organometallic molecules.

I am grateful to the Division of Inorganic Chemistry for generous financial support of this symposium and to the participants for their enthusiastic dedication. I also thank the ACS Books Department staff for their diligence and excellent advice.

TOBIN J. MARKS
Department of Chemistry
Northwestern University
Evanston, IL 60208

March 14, 1990

Chapter 1

Importance of Metal–Ligand Bond Energies in Organometallic Chemistry: An Overview

Tobin J. Marks

Department of Chemistry, Northwestern University, Evanston, IL 60208

The acquisition and analysis of metal-ligand bond energy information in organometallic molecules represents an active and important research area in modern chemistry. This overview begins with a brief historical introduction to the subject, followed by a discussion of basic principles, experimental methodology, and issues, and concludes with an overview of the Symposium Series volume organization and contents. Finally, a bibliography of thermodynamic data compilations and other source materials is provided.

One need only browse through any general chemistry or introductory organic chemistry text to appreciate just how fundamental the notions of bonding energetics are to modern chemistry. It is these compilations of bond energy data for simple organic and inorganic molecules that afford students their first quantitative ideas about the strengths of chemical bonds as well as the possibility of understanding the course of chemical transformations in terms of the strengths of bonds being made and broken. Likewise, the genesis of valence ideas as basic as the electronegativities of atoms can be traced back to perceived irregularities in bond energy trends (1). Over the past several decades, major advances have occurred in the accurate measurement and systematization of thermochemical data for organic and relatively simple (binary and ternary) inorganic molecules. The former represent a cornerstone of modern physical organic chemistry while the latter provide a useful tool for understanding large segments of main group, transition element, and f-element reaction chemistry. All such information is of obvious technological importance for process design and predicting product characteristics.

The past several decades have also witnessed the phenomenal development of contemporary organometallic chemistry. This field has had a major impact on our understanding of structure/bonding/reactivity relationships in metal-centered molecules, in practicing and/or modelling homogeneous and heterogeneous catalysis, in stoichi-

0097–6156/90/0428–0001$06.00/0

ometric synthetic organic chemistry, in metal ion biochemistry, and in the synthesis of important electronic and ceramic materials. While one can only be dazzled by the plethora of unprecedented reactions and equally beautiful molecular structures, and while our understanding of bonding and reaction mechanisms has advanced considerably, it is fair to concede that a parallel understanding of bonding energetics and the thermodynamics of reactions involving most organometallic compounds does not yet exist. Indeed, in most cases, we do not know the strengths of the bonds holding these fascinating molecules together nor do we even know whether these molecules are kinetic or thermodynamic products of the reactions that produce them.

Although the thermodynamics of organometallic substances is justifiably a topic of considerable current interest, activity in this area is by no means new. Thus, Guntz reported the heat of formation of dimethyl zinc (determined by combustion calorimetric methods) in 1887 (2) and Berthelot the heats of formation of several mercury alkyls (by similar techniques) in 1899 (3). In 1928, Mittasch reported the heat of formation of iron pentacarbonyl, again determined by combustion calorimetry (4). The 1930's and 1940's saw important developments in instrumentation which allowed far more accurate calorimetry (5) and a rapidly expanding data base of information on organic (6) and simple inorganic (1) compounds. Nevertheless, only a handful of organometallic compounds had been studied prior to 1940, and the 1939 "The Nature of the Chemical Bond" contains no bond energy information on organometallic compounds except for organosilanes (1).

The period after the Second World War saw greatly accelerated activity in the study of organometallic compounds by calorimetry and by a growing number of gas phase techniques. By 1964, Skinner was able to publish the first substantial review article on the strengths of metal-to-carbon bonds (7). Further advances were evident in the 1970 "Thermochemistry of Organic and Organometallic Compounds," by Cox and Pilcher (8), as well as in the key 1977 review article by Connor on the thermochemistry of transition metal carbonyls and related compounds (9). The post-1980 period has been one of heightened interest in bond energy information for a variety of important reasons. The developing sophistication of structural and mechanistic organometallic chemistry has raised increasing numbers of thermodynamic questions, sometimes as fundamental as why a particular reaction does or does not occur (10-12). The power of contemporary quantum chemistry to map out the energies and spatial characteristics of molecular orbitals in complex organometallic systems in turn raises quantitative questions about the strengths of the bonds being portrayed. Finally, the impressive experimental advances in gas phase, solution phase, and surface chemical physics have allowed studies of metal-ligand interactions in heretofore inaccessible environments and on heretofore inaccessible timescales. Conceptual bridges to more traditional organometallic chemistry are only just emerging and should have a major impact.

<u>Basic Concepts, Measurement Approaches, and Issues.</u>

In a strict spectroscopic sense, the bond dissociation energy, D_0, for a diatomic molecule AB can be defined as the change in internal energy accompanying homolytic bond dissociation (Equation 1) at T = 0

K in the gas phase (13). The equilibrium bond dissociation energy, D_e, measures the depth of the Morse potential well describing AB, and

$$AB(g,0\ K) \longrightarrow A(g,0\ K) + B(g,0\ K) \qquad (1)$$

differs from D_0 by the zero-point energy (Equation 2). Here x_e is the anharmonicity constant and ω_0 is the harmonic stretching frequen-

$$D_e = D_0 + \frac{1}{2}[1 - \frac{x_e}{2}]\hbar\omega_0 \qquad (2)$$

cy. Such parameters are commonly derived from a Birg-Sponer analysis of spectroscopic data (13). For thermochemical purposes, the enthalpy required to homolytically dissociate AB at 298 K in the gas phase (Equation 3) is commonly referred to as the "bond dissociation energy," the "bond disruption enthalpy," the "bond energy," the "bond

$$AB(g,298\ K) \longrightarrow A(g,298\ K) + B(g,298\ K) \qquad (3)$$

enthalpy," and the "bond strength (7, 14)." As given in Equation 4, it refers to fragments which are in relaxed (equilibrium) states, and

$$D_{AB} = \Delta H_f^0(A,g,298\ K) + \Delta H_f^0(B,g,298\ K) - \Delta H_f^0(AB,g,298\ K) \qquad (4)$$

is related to D_0 via Equation 5. Here N_A is Avagadro's number. Some authors use abbreviations of the form $\Delta H(AB)$ rather than D_{AB} (13).

$$D = N_A D_0 + RT \qquad (5)$$

The description of bonding energetics for polyatomic molecules is more complex. For homoleptic systems such as MX_n (X = an atom), the enthalpy of atomization can be defined as in Equation 6, where a temperature of 298 K is normally assumed. From this quantity, it is

$$\Delta H_{atom} = \Delta H_f^0(M,g) + n\Delta H_f^0(X,g) - \Delta H_f^0(MX_n,g) \qquad (6)$$

also possible to define a mean bond dissociation energy (or enthal-py), $\bar{D}$, which describes the average M-X bond enthalpy (Equation 7), as well as first, second, etc. stepwise bond dissociation energies

$$\bar{D}_{MX} = \Delta H_{atom}/n \qquad (7)$$

(or enthalpies) (Equations 8, 9, etc). It is not in general correct

$$D_1 = \Delta H_f^0(MX_{n-1},g) + \Delta H_f^0(X,g) - \Delta H_f^0(MX_n,g) \qquad (8)$$

$$D_2 = \Delta H_f^0(MX_{n-2},g) + \Delta H_f^0(X,g) - \Delta H_f(MX_{n-1},g) \qquad (9)$$

to assume that $\bar{D} = D_1$, nor that either $\bar{D}$ or D_1 will be the same in all MX_m and MY_mX_{n-m} molecules. The latter issue of whether D_{MX} values are transferable among different environments is of great interest not only for developing bond energy parameters of broad applicability but also for understanding ancillary ligand steric and electronic effects. The adherence of redistribution processes (e.g., Equation 10) to a purely statistical model is one straightforward

$$MY_n + MX_n \rightleftharpoons MY_{n-x} X_x + etc. \tag{10}$$

test of transferability.

For elaborate polyatomic organometallic molecules, the atomization enthalpy of taking the entire molecule to gaseous atoms is not a particularly informative quantity, and it is more meaningful to use the terminology of disruption enthalpies which describe dissociation into identifiable fragments. For a homoleptic compound, MR_n, this term is defined as in Equation 11 for the process depicted in Equation 12 ($\underline{7}$, $\underline{14}$). There is now a considerable body of

$$\Delta H_{disrupt} = \Delta H_f^0(M,g) + n\Delta H_f^0(R\cdot,g) - \Delta H_f^0(MR_n,g) \tag{11}$$

$$MR_n(g) \longrightarrow M(g) + nR\cdot(g) \tag{12}$$

$\Delta H_f^0(M,g)$ and $\Delta H_f^0(R\cdot,g)$ data available ($\underline{15}$-$\underline{17}$). Because of uncertainties in transferability surrounding $\Delta H_{disrupt}$ values and because chemical situations are more commonly encountered in which the enthalpies of single dissociation/disruption processes as in Equation 13 are more useful, the author suggests defining the M-R bond disrup-

$$L_nMR \longrightarrow L_nM + R\cdot \tag{13}$$

tion enthalpy (or dissociation energy) as in Equation 14 ($\underline{12}$). Although problems of transferability are still important, it should be

$$D(L_nM-R) = \Delta H_f^0(L_nM) + \Delta H_f^0(R\cdot) - \Delta H_f^0(L_nMR) \tag{14}$$

recognized that many processes of greatest chemical interest will take place within the same coordination sphere (e.g., Equation 15) and that $D(L_nM-R)$ implicitly contains the information needed to understand/predict such processes. Ways to test parameter transfer-

$$L_nMR \longrightarrow L_nMR' \tag{15}$$

ability in such systems and to account for the lack of useful volatility in many of the molecules of interest will be discussed shortly (vide infra).

Returning to Equations 11 and 12 and similar situations, problems arise when ΔH_f^0 values of disruption/atomization fragments are not available. Various schemes exist to estimate such quantities by apportioning the energetics among the constituent bonds ($\underline{7}$, $\underline{14}$). The resulting, estimated M-R energies are generally referred to as bond enthalpy contributions and denoted for an MR_n molecule by $\bar{E}(M-R)$. Importantly, such values do not incorporate the enthalpies of reorganization involved in relaxation of the newly formed fragments to their equilibrium configurations. As such, $\bar{E}$ values are less readily applied to transformations of the type depicted in Equation 15.

The classical approach to deducing the thermodynamic properties of a substance has traditionally been combustion calorimetry ($\underline{18}$). Here the substance of interest is completely burned, and the heat of

combustion apportioned among the known/assumed and unknown constituent bond energies. While such approaches are well-suited for
simple organic molecules, the apportioning problem becomes exceeding-
ly difficult for complicated organometallic molecules. In addition,
organometallic molecules are notoriously difficult to burn cleanly,
so that major errors can be introduced in data analysis (despite many
advances in calorimeter design and combustion procedures). A useful
alternative approach is sometimes high temperature calorimetry with
other oxidants (e.g., halogens) ($\underline{19}$).
    Various types of solution reaction calorimetry have proven
especially useful in thermochemical studies of organometallic
molecules. Provided the reactions employed are clean, rapid, and
quantitative, it is possible to employ processes as in Equation 16 to
express $\underline{relative}$ M-R bond disruption enthalpies in terms of the meas-

$$L_nMR + X_2 \longrightarrow L_nMX + RX \qquad (16)$$

ured heat of reaction and generally available or readily estimated
dissociation energy parameters (Equation 17). Particularly useful
reactions for this purpose are halogenolysis and protonolysis ($\underline{12}$,

$$D(L_nM\text{-}R)_{soln} = \Delta H_{rxn} + D(L_nM\text{-}X)_{soln} + D(RX)_{soln} - D(X_2)_{soln} \qquad (17)$$

$\underline{20}$). When this procedure is carried out in a vacuum-tight isoperibol
(quasi-adiabatic) batch titration calorimeter ($\underline{21}$), there is the
added advantage of anaerobic security for highly sensitive samples, a
readily incorporated protocol for measuring heats of solution, and a
means (via titration) of sequentially investigating a series of bonds
in a molecule as well as providing a built-in check on reaction
stoichiometry and calorimeter system integrity. As should be evident
in Equation 16, derived relative $D(L_nM\text{-}R)$ parameters in such a
determination are "anchored" to a particular $D(L_nM\text{-}X)$ value.
Conversion from a series of relative bond enthalpies to a scale
approximating absolute can be achieved if $D(L_nM\text{-}X)$ can be estimated.
Use of the corresponding $D_1(MX_n)$ value appears to be realistic for M
= an early transition metal, lanthanide, or actinide in the same
formal oxidation state, and X = halogen ($\underline{22}$). $\underline{Absolute}$ $D(M\text{-}R)$ values
can be obtained via one- and two-electron redox sequences as shown in
Equations 18-23 and 24-28 ($\underline{23\text{-}27}$). The availability of lower-valent,
less coordinatively saturated L_nM species is of course essential for

$$L_nMR + X_2 \longrightarrow L_nMX + RX \qquad (18)$$

$$L_nM\text{-}X \longrightarrow L_nM + 1/2\ X_2 \qquad (19)$$

$$X\cdot \longrightarrow 1/2\ X_2 \qquad (20)$$

$$R\text{-}X \longrightarrow R\cdot + X\cdot \qquad (21)$$

$$L_nMR \longrightarrow L_nM + R\cdot \qquad (22)$$

$$D(L_nM\text{-}R) = \Delta H_{rx(16)} + \Delta H_{rx(19)} - 1/2\ D(X_2) + D(R\text{-}X) \qquad (23)$$

this approach. In addition, note that Equations 24-29 as written

$$L_nMR_2 + 2 X_2 \longrightarrow L_nMX_2 + 2 RX \qquad (24)$$

$$L_nMX_2 \longrightarrow L_nM + X_2 \qquad (25)$$

$$2 X\cdot \longrightarrow X_2 \qquad (26)$$

$$2R-X \longrightarrow 2 R\cdot + X\cdot \qquad (27)$$

$$L_nMR_2 \longrightarrow L_nM + 2 R\cdot \qquad (28)$$

$$D(L_nM-R_2) = \Delta H_{rx(24)} + \Delta H_{rx(25)} - D(X_2) + 2 D(RX) \qquad (29)$$

afford a $\bar{D}$(M-R) value, while stepwise cleavage of M-R groups will yield D($L_nM(R)$-R) and D($L_nM(X)$-R) parameters referenced to a $\bar{D}$(M-X).

Bond enthalpy values derived from the aforementioned calorimetric approaches and all other D(M-R) quantities determined in solution (vide infra) are not strictly gas phase quantities as specified in the formal definitions (Equations 1,3,4). However, since most chemistry of interest to organometallic chemists occurs in the solution phase, it can be argued that this is not an undesirable state of affairs. There is also good evidence that D(R-H) values for simple organic molecules are the same in nonpolar solvents as in the gas phase (27), and that the heats of solution (ΔH_{soln}) of uncharged organometallic molecules are generally rather small in nonpolar solvents (~2-4 kcal/mol) (12, 21, 24). Conversion of solution phase D(M-R) data to the gas phase requires ΔH_{soln} and heat of sublimation (ΔH_{sub}) data for the species involved (e.g., on both sides of Equation 18). Such parameters are readily available for small inorganic and organic molecules, while ΔH_{soln} values can be easily measured for the organometallic molecules of interest. In contrast, ΔH_{sub} data are not easily measured for organometallic compounds and may be inaccessible for compounds of low volatility and/or thermal stability. Hence, it is usually assumed that ΔH_{sub} is the same for all organometallic molecules involved. Where this assumption has been checked by experiment (19, 28), it has been found to be reasonable, but not extremely accurate.

Other solution calorimetric techniques which have been applied to the study of organometallic molecules have included batch, flow, heat flux (Calvet), and pulsed, time-resolved photoacoustic calorimetry. The former method is a simpler isoperibol variant of the aforementioned batch titration technique and involves breaking ampoules of the compound of interest into a reaction dewar containing an excess of reagent. It is not chemically as selective as a method in which the reagent is added in a titration mode, and in some systems the excess reagent is likely to cause side reactions. The heat flux or Calvet technique is an isothermal approach, and has the advantage that reactions can be studied at elevated temperatures and pressures (19, 29, 30). It has also been employed in a solventless mode using halogen reagents as a substitute for combustion calorimetry in studying metal carbonyls (19, 30). As in the batch technique, the reagent cannot be added incrementally. It is, in principle, adaptable to the study of extremely air-sensitive compounds.

Pulsed, time-resolved photoacoustic calorimetry employs a pulsed uv laser source to induce a photochemical reaction in the molecule of interest (e.g., Equation 30). The absorbed photochemical energy which is subsequently released as thermal energy is measured by

$$ML_n \xrightarrow{h\nu} ML_{n-1} + L \tag{30}$$

acoustic techniques (from the resulting shock waves), and combined with quantum yield information on nonradiative processes, yields the M-L bond enthalpy as well as kinetic information on recombination processes (31-33). Comparison of photoacoustic D(M-L) data with gas phase data for the same molecules obtained by other techniques (vide infra) has been particularly informative in quantifying the energetics of weak M-solvent interactions (32). Disadvantages of organometallic systems for photoacoustic calorimetry may include the lack of clean, well-defined photochemistry, low nonradiative quantum yields, and the pitfalls of multiphoton processes at high light fluences (32).

Equilibrium and kinetic techniques have also been successfully applied to studying metal-ligand bonding energetics in solution. In the former approach, relative metal-ligand bond enthalpies are obtained by studying (usually via NMR, IR, or optical spectroscopy) the positions of various equilibria (e.g., Equation 31). Study of the equilibrium constant as a function of temperature combined with

$$L_nM-R + R'H \rightleftarrows L_nM-R' + RH \tag{31}$$

a van't Hoff analysis, yields $\Delta H°$ and $\Delta S°$ values for the process (34). Of course, data must be acquired over a sufficiently wide temperature range, which may be prohibited by thermal instability. Alternatively, it can be assumed that $\Delta S \approx 0$ for the equilibration process, so that $\Delta H \approx \Delta G$ (34-36). Major disadvantages of the equilibration technique for organometallic systems include the lack of suitably clean or rapid equilibria in many systems, the inherent inaccuracy in measuring large equilibrium constants (only small steps in ΔH values can be used in constructing scales of relative bond enthalpies), and the inability to measure absolute bond enthalpies.

Kinetic techniques for measuring metal-ligand bond enthalpies focus on measuring the activation enthalpies for homolytic processes as in Equation 32. The resulting organic radical is usually trapped.

$$L_nMR \longrightarrow L_nM + R\cdot \tag{32}$$

If it is reasonably assumed that $\Delta H^{\ddagger}$ for the reverse, recombination reaction is small (ca. 2-3 kcal/mol), then the absolute L_nM-R bond enthalpy can be directly calculated from $\Delta H^{\ddagger}$ for the forward reaction (37-42). Radical cage diffusion effects cannot be ignored in such kinetic analyses, however they appear only to be important in more viscous media. Good agreement has recently been noted between kinetically-derived Co-R bond enthalpies and those obtained using one-electron oxidative (cf., Equations 18-23) calorimetric techniques (26). These kinetic techniques are of course only applicable to systems with relative low M-R bond enthalpies and clean, homolytic dissociation chemistry.

Differential scanning calorimetry has been applied to thermodynamic studies of organometallic compounds in the solid state (43).

Thus, the enthalpies of various decomposition reactions (e.g., gaseous olefin loss) can be readily measured for rather small samples (a few mg.). Drawbacks of this technique include the necessity of correcting data to 298 K, the uncertain influence that solid state effects may have on measured ΔH values, and the possibility of side reactions at elevated temperatures. The latter effects can be assayed by thermogravimetric and volatile product analysis.

Electrochemical techniques have also been applied to determining metal-metal bond energies in dinuclear organometallic molecules (44). Using a combination of redox equilibration experiments with reagents of known potentials and fast scan cyclic voltammetry, it is possible to obtain E° data for the kinds of reactions shown in Equations 33 and 34, respectively. The result is the free energy (but not the

$$Mn_2(CO)_{10} \longrightarrow 2\ Mn(CO)_5^- \qquad (33)$$

$$2Mn(CO)_5^- \longrightarrow 2\ Mn(CO)_5 \qquad (34)$$

$$Mn_2(CO)_{10} \longrightarrow 2\ Mn(CO)_5 \qquad (35)$$

enthalpy) of Equation 35. In Equations 36-39, an approach to measuring metal-hydrogen bond enthalpies in acetonitrile solution is shown which uses pKa, electrochemical, and tabulated gas phase electron affinity data (45). When approximate corrections for solvation and entropy of solvation effects are made, good agreement

$$L_nMH \longrightarrow L_nM^- + H^+ \qquad (36)$$

$$L_nM^- \longrightarrow L_nM \qquad (37)$$

$$H^+ \longrightarrow H\cdot \qquad (38)$$

$$L_nMH \longrightarrow L_nM + H\cdot \qquad (39)$$

is obtained with $D(L_nM\text{-}H)$ values measured by other techniques. Crucial to this method is the amenability of the subject compounds to both the pK_a and electrochemical (Equation 37) measurements. Solvation effects are expected to be important in coordinatively unsaturated systems.

Gas phase approaches to the study of organometallic metal-ligand bonding energetics can be roughly divided into those employing ion-molecule reactions and those which employ neutral molecular precursors. In the most common embodiment of the former approach, bare gaseous metal ions are created by evaporation/ionization or pulsed laser desorption/ionization of bulk metal sources or by dissociation/ionization of volatile molecular precursors (e.g., by electron impact). These ions are then brought to well-defined kinetic energies either by mass filtering in a guided ion beam (tandem mass spectrometer) apparatus (46, 47) or by trapping in circular orbits in an ion cyclotron resonance (ICR) spectrometer (48, 49). The latter device can be readily coupled to a highly sensitive Fourier transform mass spectrometry detection system. A wide variety of informative

experiments can then be performed in these two experimental configurations.
In a common type of ion-molecule experiment (46), a guided beam
of monoenergetic metal ions is impinged upon gaseous neutral molecules under single collision conditions, and the reaction products
analyzed by mass spectrometry as the kinetic energy of the metal ions
is varied. For the relatively straightforward case of endothermic
reactions (e.g., Equation 40), the MH^+ yield is analyzed as a
function of M^+ kinetic energy to obtain the threshold E_T for the

$$M^+ + RH \longrightarrow MH^+ + R\cdot \qquad (40)$$

reaction. Making the reasonable assumption (which has been verified
in several simple cases) that the endothermic reaction has a negligible kinetic activation barrier, then Equation 41 applies. For
processes as in Equation 42, the relationship of Equation 43 holds,
where IP is the ionization potential. In addition to the assumption

$$D(M\text{-}H^+) = D(R\text{-}H) - E_T \qquad (41)$$

$$M^+ + RH \longrightarrow MH + R^+ \qquad (42)$$

$$D(M\text{-}H) = D(R\text{-}H) + IP(R) - IP(M) - E_T \qquad (43)$$

of negligible activation barriers, other uncertainties in the above
analyses include the cross-section model used to calculate E_T and
internal energy uncertainties in the reactants. The latter issue
includes the possibility that M^+ is formed and used in an electron-
ically excited state. In addition, large molecular ions will have
many degrees of freedom so that cross-section versus kinetic energy
plots will be rather flat and E_T values difficult to accurately
determine in such cases. Although the nature of the cross-section
model and the data analysis are quite different, bond enthalpy
information can also be derived from exothermic ion-molecule reactions by studying product kinetic energy release distributions (50).
ICR-FTMS experiments have been employed to derive thermodynamic
information using several approaches (48). These include studies of
exothermic and endothermic ion-molecule reactions, equilibration
studies, competitive collision-induced dissociation reactions, and
photodissociation studies. Exothermic reactions involving thermalyzed (cooled by an inert buffer gas) ions provide brackets on
metal-ligand bond enthalpies, as illustrated by Equation 44 which
implies that $D(Fe\text{-}C_6H_4^+) \geq 66$ kcal/mol. Endothermic reactions can
also be studied in the manner described above for ion beam experi-

$$Fe^+ + C_6H_5Cl \longrightarrow FeC_6H_4^+ + HCl \qquad (44)$$

ments (vide supra). However, the greater uncertainty in M^+ energies
in ICR-FTMS experiments renders this approach somewhat less accurate
than that described above. ICR-FTMS experiments can yield relative
metal-ligand bond enthalpy information by measuring equilibrium
constants for ligand equilibration processes (e.g., Equation 45) and
examining other reactions where ancillary thermodynamic information
exists (e.g., Equation 46). These approaches assume that $\Delta S \approx 0$ and

$$\overset{+}{ML} + L' \rightleftarrows \overset{+}{ML'} + L \qquad (45)$$

yield relative metal-ligand bond enthalpies unless an absolute anchor

$$FeNH_2^+ + /\!\!/\!\!\diagdown \longrightarrow Fe \overset{+}{\diagup\!\!\!\!\diagdown} + NH_3 \qquad (46)$$

point has been established. The gaseous metal-ligand ion reactants
for such studies are normally prepared by ion-molecule reactions or
by the decomposition of neutral organometallic precursors. Competitive collision-induced dissociation studies (e.g., Equation 47) can
also be employed to estimate metal-ligand bond enthalpies by observing the distribution of fragments following collisions with neutral
molecules. Finally, photo-dissociation techniques can be used to
obtain metal-ligand bond enthalpies by determining the photon

$$
\begin{array}{l}
AM^+B \longrightarrow AM^+ + B \\
\qquad or \\
\qquad \longrightarrow MB^+ + A
\end{array} \qquad (47)
$$

energetic threshold for dissociation (Equation 48). Such an approach
is only workable if the ion absorbs the photon, if the metal-ligand

$$ML^+ \overset{h\nu}{\longrightarrow} M^+ + L \qquad (48)$$

ion undergoes clean photodissociation (it is in some cases possible
to disentangle side reactions), and the metal-ligand ion has a high
density of low-lying excited states so that dissociation is thermo-
dynamically- rather than spectroscopically-determined.

In all of the above ion-molecule experiments, it should be
evident that structural ambiguities can arise in products which are
assigned purely on the basis of an M/e ratio. In some cases,
additional product structural information can be obtained by isotopic
labelling, collision-induced dissociation, and photodissociation
experiments. The latter techniques rely upon understanding the
nature of the fragments produced when the metal-ligand ion is
dissociated. It should also be noted that ICR-FTMS experiments are
by no means limited to small metal-ligand ions, and that bracketing
studies leading to relative metal-ligand bond enthalpies (cf.,
Equations 45, 46) can be successfully performed on ions as elaborate
as $(C_5H_5)_2ZrR^+$ (50).

Flowing afterglow techniques are yet another approach to
studying ion-molecule reactions of organometallic species (51, 52).
Here ion-molecule reactions are conducted in a flow of inert buffer
gas (usually He), so that all reagents and products are completely
thermalyzed. Representative experiments have employed bracket-
ing/competition experiments (cf., Equations 45-47) to obtain information such as hydride affinities of metal carbonyls and transition
metal formyl complex stabilities (51, 52).

Another class of gas phase experiments begins with the molecular
organometallic compounds of interest and measures the added energy
required for ligand dissociation. The mass spectrometric appearance
potential method relates the metal-ligand bond energy to the mass

spectrometer electron accelerating potential needed to observe the metal-ligand fragment of interest and the fragment ionization potential (e.g., Equation 49). Major uncertainties in this approach

$$D_1(ML_n) = AP(ML_{n-1}^+) - IP(ML_{n-1}) \tag{49}$$

concern the precision with which the threshold can be measured and the possible formation of thermally "hot" products. Photodissociation approaches were discussed in the previous section.

Very low pressure pyrolysis kinetic techniques measure (via mass spectrometry) the activation energy for ligand dissociation under collisionless conditions (53, 54). As in the solution kinetic approach (vide supra), it is assumed that the kinetic barrier for recombination is small. This approach suffers from problems associated with catalytic sample decomposition on the hot walls of the apparatus and radical chain reactions. A recent improvement in this technique (53, 54), the LPHP (laser powered homogeneous pyrolysis) method, employs a pulsed infrared (CO_2) laser to heat an SF_6 "sensitizer" in the reaction mixture containing a low pressure of the compound of interest. An N_2 thermalizing bath and an internal $M(CO)_n$ "thermometer" of known thermolytic characteristics are also present so that sample pyrolysis is rapid and homogeneous, reactor walls remain cool, and the internal temperature is known accurately. Sample decomposition is allowed to proceed for 10-15 μs before rapid expansion cooling and mass spectrometric analysis. This technique has been successfully applied to a wide variety of organotransition metal compounds.

Measurements of metal-ligand bonding energetics on clean metal surfaces are generally carried out by equilibrium or kinetic methods (55). In the former approach, adsorbate coverage, θ, is measured as a function of equilibrium pressure at varying temperatures. The enthalpy of adsorption is related to these isotherm data via Equation 50. In kinetic methods, desorption rates are measured as a function

$$\left(\frac{\partial \ln p}{\partial T}\right)_{\theta=constant} = -\frac{\Delta H_{ads}}{RT^2} \tag{50}$$

of temperature and ΔH_{ads} calculated assuming a negligible activation energy for the (reverse) adsorption process. These approaches are complicated by the realistic possibilities that ΔH_{ads} is dependent upon θ, that the adsorbate bonding mode may be temperature-dependent, and that the surface itself may be heterogeneous. Surface thermochemical measurements are clearly most informative when analyzed in conjunction with detailed adsorbate surface structural information (56, 57).

Symposium Volume Motivation and Organization.

The present Symposium Series volume and the September 1989 symposium from which it derives are part of an effort to bring together leading investigators active in the diverse subdisciplines concerned with bond energies in organometallic compounds. There has traditionally been little interaction between such researchers, and the present work attempts to convey the essence of the oral presentations and active dialogue which occurred at the symposium. All contributions have been carefully reviewed by experts in the respective fields, and all efforts have been made to expedite the editorial process so that coverage is as up-to-date as possible. This volume begins with a discussion of relatively simple systems in the gas phase. Thus, Armentrout, Beauchamp, Bowers, and Freiser focus upon the course and thermodynamics of both endothermic and exothermic gas phase reactions between metal ions and simple organic or inorganic molecules. Metal-ligand bond enthalpy data are reported for a wide range of ions, neutrals, and metal-metal bonded species. Important trends in bonding energetics are revealed and convincingly explained. Richardson next describes gas phase organo-zirconium chemistry of the molecular ion $(C_5H_5)_2ZrCH_3^+$ and draws useful connections to known solution phase chemistry. He also compares the gas and solution phase redox properties of metallocenes and metal acetylacetonates. Lichtenberger reports a new way to estimate metal-ligand bond energies based upon ionization potentials and simple molecular orbital concepts. The approach is successfully applied to several classes of metal carbonyl complexes. The focus then shifts to solution phase studies, and Halpern presents a detailed study of metal-alkyl bond energies derived from kinetic measurements. Close attention is given to supporting ligand effects on D(M-alkyl) as well as to the possible errors incurred in such analyses. A complementary contribution by Koenig further discusses the determination of bond energies by kinetic methods and the importance of radical pair cage effects in such determinations. Hoff and Kubas report calorimetric and kinetic studies of metal-ligand binding energetics and kinetics in a fascinating series of molecular N_2 and H_2 complexes. Wayland next describes the CO and H_2 chemistry of several series of rhodium porphyrin complexes, elucidating the thermodynamics and kinetics of several unprecedented migratory CO insertion and reduction processes. Metal-ligand bond energies, ancillary ligand effects, and their consequences for unusual reactivity are then surveyed by Marks for a broad series of organolanthanides. Drago next discusses the application of E and C parameters to accurately predicting ligation thermodynamics in several classes of organometallic/coordination compounds. Photoacoustic calorimetry is an attractive and relatively new technique for studying metal-ligand bonding energetics in solution, and Yang applies this method to Mn-L and Cr-olefin bond energies in a series of $(C_5H_5)Mn(CO)_2L$ and $Cr(CO)_5(olefin)$ complexes. Finally, Simoes presents a "classical" (calorimetric) study of metal-ligand bonding energetics in a series of $(C_5H_5)_2MR_2$ and $(C_5H_5)_2ML$ complexes, as well as a "non-classical" (photoacoustic calorimetric) study of Si-H bond energies in a series of silanes.

In the area of surface phenomena, Somorjai discusses the restructuring processes which occur when clean metal surfaces undergo chemisorption, and the major consequences that such structural reorganizations have for the measured thermodynamics of chemisorption. The binding of small molecules to chemically modified surfaces is next surveyed by Stair. Dramatic effects on the chemisorption thermodynamics of Lewis bases and on heterogeneous catalytic activity are observed when clean molybdenum surfaces are modified with carbon, oxygen, or sulfur. The theoretical section begins with a presentation by Pearson on the use of absolute electronegativity and hard/-

soft concepts for understanding bond energies and chemical reactivity in organometallic systems. These rather simple concepts are shown to have considerable predictive power. Harrison and Allison next focus on transition metal-N bond energy systematics in a series of gas phase ions. Correlations are drawn between experimental data and the results of high-level electronic structure calculations. Finally, Ziegler discusses the use of local density functional calculational methods to probe $M-CH_3$, M-H, and M-CO bonding trends as a function of the position of M in the Periodic Table. Important and convincing explanations of trends in solution experimental data can be made.

Acknowledgments

The author thanks all of the "Bond Energies and the Thermodynamics of Organometallic Reactions" symposium participants for their enthusiasm in this enterprise and for the prompt preparation of outstanding Symposium Series chapters. He likewise thanks the manuscript reviewers and the ACS Books staff for their conscientious efforts in ensuring a high-quality symposium volume, and Ms. Rachel Harris for expert secretarial assistance. The support of NSF under grant CHE8800813 is also gratefully acknowledged.

Literature Cited

1. Pauling, L. "The Nature of the Chemical Bond," Cornell University Press: Ithaca, NY, 1939.
2. Guntz, A. Compt. Rend., 1887, 105, 673.
3. Berthelot, M. P. E. Compt. Rend., 1899, 129, 918.
4. Mittasch, A. Z. Angew. Chem., 1928, 41, 827-833.
5. See, for example: Rossini, F. D., ed., "Experimental Thermochemistry," Wiley: New York, Vol. 1, 1956, and references therein.
6. See, for example: Parks, G. S.; Huffman, H. M. "The Free Energies of Some Organic Compounds," American Chemical Society Monograph Series, Chemical Catalog Co.: New York, 1932.
7. Skinner, H. A. Advan. Organometal. Chem., 1964, 2, 49-114.
8. Cox, J. D.; Pilcher, G. "Thermochemistry of Organic and Organometallic Compounds," Academic Press: London, 1970.
9. Connor, J. A. Topics Curr. Chem., 1977, 71, 72-109.
10. Halpern, J. Acc. Chem. Res. 1982, 15, 238-244.
11. Ibers, J. A.; DiCosimo, R.; Whitesides, G. M. Organometallics, 1982, 1, 13-20.
12. Bruno, J. W.; Marks, T. J.; Morss, L. R. J. Am. Chem. Soc. 1983, 105, 6824-6832.
13. See, for example: Atkins, P. W. "Physical Chemistry," 3rd ed., Freeman: New York, 1986, Chapts. 4, 17.
14. Pilcher, G.; Skinner, H. A. in "The Chemistry of the Metal-Carbon Bond"; Hartley, F. R.; Patai, S., Eds.; Wiley: New York, 1982; pp. 43-90.
15. Wagman, D. D., et al, "The NBS Tables of Chemical Thermodynamic Properties. Selected Values for Inorganic and C_1 and C_2 Organic Substances in SI Units," J. Phys. Chem. Ref. Data, 1982, 11, Supplement No. 2.

16. Pedley, J. B.; Naylor, R. D.; Kirby, S. P. "Thermochemical
     Data of Organic Compounds," Second Ed., Chapman and Hall:
     London, 1986.
17. McMillan, D. F.; Golden, D. M. Ann. Rev. Phys. Chem. 1982,
     33, 493-532.
18. Sunner, S.; Mansson, M., Eds. "Combustion Calorimetry";
     Pergamon Press: Oxford, 1979.
19. Connor, J. A.; Zafarani-Moattar, M. T.; Bickerton, J.;
     Saied, N. I.; Suradi, S.; Carson, R.; Takhin, G. A.;
     Skinner, H. A. Organometallics 1982, 1, 1166-1174, and
     references therein.
20. Marks, T. J.: Gagné, M. R.; Nolan, S. P.; Schock, L. E.;
     Seyam, A. M.; Stern, D. Pure Appl. Chem., 1989, 16, 1665-
     1672.
21. Schock, L. E.; Marks, T. J. J. Am. Chem. Soc. 1988, 110,
     7701-7715.
22. Nolan, S. P.; Stern, D.; Hedden, D.; Marks, T. J., this
     volume.
23. Schock, L. E.; Seyam, A. M.; Marks, T. J., Polyhedron,
     1988, 7, 1517-1530.
24. Nolan, S. P.; Stern, D.; Marks, T. J. J. Am. Chem. Soc.,
     1989, 111, 7844-7853.
25. Dias, A. R.; Galena, M. S.; Simoes, J. A. M.; Pattiasina,
     J. W.; Teuben, J. J. Organometal. Chem., 1988, 346, C4-C6.
26. Toscano, P. J.; Seligson, A. L.; Curran, M. T.; Skrobutt,
     A. T.; Sonnenberger, D. C. Inorg. Chem., 1989, 28, 166-168.
27. Castelhano, A. L.; Griller, D. J. Am. Chem. Soc., 1982,
     104, 3655-3659.
28. Yoneda, G.; Lin, S.-M.; Wang, L.-P.; Blake, D. M. J. Am.
     Chem. Soc., 1981, 103, 5768-5771.
29. Nolan, S. P.; Lopez de la Vega, R.; Hoff, C. D. J. Am.
     Chem. Soc., 1986, 108, 7852-7853, and references therein.
30. Connor, J. A.; Skinner, H. A.; Virmani, Y. J. Chem. Soc.,
     Faraday Trans I, 1972, 1754-1763.
31. Rudzki, J. E.; Goodman, J. L.; Peters, K. S. J. Am. Chem.
     Soc., 1985, 107, 7849-7854.
32. Yang, G. K.; Peters, K. S.; Vaida, V. Chem. Phys. Lett.,
     1986, 125, 566-568.
33. Yang, G. K.; Vaida, V.; Peters, K. S. Polyhedron, 1988, 7,
     1619-1622.
34. Bulls, A. R.; Bercaw, J. E.; Manriquez, J. M.; Thompson, M.
     E. Polyhedron, 1989, 7, 1409-1428.
35. Bryndza, H. E.; Fong, L. K.; Paciello, R. A.; Tam, W.;
     Bercaw, J. E. J. Am. Chem. Soc. 1987, 109, 1444-1456.
36. Stoutland, P. O.; Bergman, R. G.; Nolan, S. P.; Hoff, C. D.
     Polyhedron, 1988, 7, 1429-1440.
37. Halpern, J. Polyhedron, 1988, 7, 1483-1490.
38. Hay, B. P.; Finke, R. G. Polyhedron, 1988, 7, 1469-1481.
39. Koenig, T. W.; Hay, B. P.; Finke, R. G. Polyhedron, 1988,
     7, 1499-1516.
40. Wayland, B. B. Polyhedron, 1988, 7, 1545-1555.
41. Collman, J. P.; McElwee-White, L.; Brothers, P. J.; Rose,
     E. J. Am. Chem. Soc., 1986, 108, 1332-1333.
42. Bakac, A.; Espenson, J. H. J. Am. Chem. Soc., 1984, 106,
     5197-5202.

43. Puddephatt, R. J. Coord. Chem. Rev., 1980, 33, 149-194, and references therein.
44. Pugh, J. R.; Meyer, T. J. J. Am. Chem. Soc., 1988, 110, 8245-8246.
45. Tilset, M.; Parker, V. D. J. Am. Chem. Soc., 1989, 111, 6711-6717.
46. Armentrout, P. B., this volume.
47. Ervin, K. M.; Armentrout, P. B. J. Chem. Phys., 1985, 83, 166-189.
48. Freiser, B. S., this volume.
49. Buckner, S. W.; Freiser, B. S. Polyhedron, 1988, 7, 1583-1603.
50. van Koppen, P. A. M.; Bowers, M. T.; Beauchamp, J. L.; Dearden, D. V., this volume.
51. Lane, K. R.; Squires, R. R. Polyhedron, 1988, 7, 1609-1618.
52. Lane, K. R.; Lee, R. E.; Sallans, L.; Squires, R. R. J. Am. Chem. Soc., 1984, 106, 5767-5772.
53. Lewis, K. E.; Golden, D. M.; Smith, G. P. J. Am. Chem. Soc., 1984, 106, 3905-3912, and references therein.
54. Smith, G. P. Polyhedron, 1988, 7, 1605-1608.
55. Somorjai, G. A. "Chemistry in Two Dimensions: Surfaces," Cornell University Press: Ithaca, 1981, Chapts. 2, 6.
56. van Hove, M. A.; Wang, S.-W.; Ogletree, D. F.; Somorjai, G. A. Advan. Quantum Chem., 1989, 20, 1-184.
57. Somorjai, G. A., this volume.

Bibliography

The following is a listing by subject area of recent data compilations, review articles, and monographs dealing with thermodynamic properties, bond energies, and associated measurement techniques.

Data Compilations, Organic Compounds
1. Pihlaja, K. in "Molecular Structure and Energetics," Liebman, J. F.; Greenberg, A., Eds., VCH Publishers: New York, 1987, Vol. 2, Chapt. 5. (Thermodynamic properties; bond energies).
2. Chickos, J. S. in "Molecular Structure and Energetics," Liebman, J. F.; Greenberg, A., Eds., VCH Publishers: New York, 1987, Vol. 2, Chapt. 3. (Heats of sublimation).
3. Pedley, J. B.; Naylor, R. D.; Kirby, S. P. "Thermochemical Data of Organic Compounds," Second Ed., Chapman and Hall: London, 1986. (Thermodynamic properties; bond energies).
4. Smith, B. D.; Srivastava, R. "Thermodynamic Data for Pure Compounds, Part A. Hydrocarbons and Ketones," Elsevier: Amsterdam, 1986. (Thermodynamic properties).
5. Smith, B. D.; Srivastava, R. "Thermodynamic Data for Pure Compounds, Part B. Halogenated Hydrocarbons and Alcohols," Elsevier: Amsterdam, 1986. (Thermodynamic properties).
6. McMillan, D. F.; Golden, D. M. Ann. Rev. Phys. Chem. 1982, 33, 493-532. (Bond energies).
7. Benson, S. W. "Thermochemical Kinetics," 2nd ed.; John Wiley and Sons: New York, 1976. (Thermodynamic properties; bond energies).

8.   Cox, J. D.; Pilcher, G. "Thermochemistry of Organic and
     Organometallic Compounds," Academic Press: London, 1970.
     (Thermodynamic properties).
9.   Stull, D. R.; Westrum, E. F.; Sinke, G. C. "The Chemical
     Thermodynamics of Organic Compounds"; Wiley: New York,
     1969. (Thermodynamic properties; bond energies).

Data Compilations. Inorganic Compounds.
1.   Morss, L. R. in "The Chemistry of the Actinide Elements,"
     2nd ed.; Katz, J. J.; Seaborg, G. T.; Morss, L. R., Eds.,
     Chapman and Hall: London, 1986, Chapt. 17. (Thermodynamic
     properties of actinide compounds).
2.   Chase, M. W., Jr., et al, "JANAF Thermochemical Tables,"
     3rd ed., Part I, Al-Co J. Phys. Chem. Ref. Data, 1985, 14,
     Supplement No. 1. (Thermodynamic properties).
3.   Chase, M. W., Jr., et al, "JANAF Thermochemical Tables,"
     3rd ed., Part II, Cr-Zr, J. Phys. Chem. Ref. Data, 1985,
     14, Supplement No. 1. (Thermodynamic properties).
4.   Christensen, J. J.; Izatt, R. M. "Handbook of Metal Ligand
     Heats," 3rd ed., Marcel Dekker: New York, 1983. (Thermo-
     dynamic properties of coordination compounds).
5.   Wagman, D. D., et al, "The NBS Tables of Chemical Thermo-
     dynamic Properties. Selected Values for Inorganic and C_1
     and C_2 Organic Substances in SI Units," J. Phys. Chem. Ref.
     Data, 1982, 11, Supplement No. 2. (Thermodynamic proper-
     ties).
6.   Drobot, D. V.; Pisarev, E. A. Russ. J. Inorg. Chem., 1981,
     26, 3-16. (Metal-halogen, metal-oxygen, and metal-metal
     bond energies).
7.   Glidewell, C. Inorg. Chim. Acta, 1977, 24, 149-157. (Bond
     energies in oxides and oxo-anions).
8.   Huheey, J. E. "Inorganic Chemistry," 2nd ed.; Harper and
     Row: New York, 1978; pp. 824-850. (Bond energies).

Data Compilations and Reviews. Organometallic Compounds.
1.   Simoes, J. A. M.; Beauchamp, J. L. Chem. Rev., in press.
     (Large compilation of bond energies).
2.   Marks, T. J., Ed. "Metal-Ligand Bonding Energetics in
     Organotransition Metal Compounds," Polyhedron Symposium-in-
     Print, 1988, 7. (Short review articles containing much
     data).
3.   Skinner, H. A.; Connor, J. A. in "Molecular Structure and
     Energetics," Liebman, J. F.; Greenberg, A., Eds., VCH
     Publishers: New York, 1987, Vol. 2, Chapt. 6. (Compila-
     tions of bond energies).
4.   Pilcher, G.; Skinner, H. A. in "The Chemistry of the Metal-
     Carbon Bond"; Hartley, F. R.; Patai, S., Eds.; Wiley: New
     York, 1982, pp. 43-90. (Extensive compilation of bond
     energies).
5.   Connor, J. A. Top. Curr. Chem. 1977, 71, 71-110. (Exten-
     sive compilation of bond energies).
6.   Halpern, J. Acc. Chem. Res. 1982, 15, 238-244. (Review
     article).
7.   Mondal, J. U.; Blake, D. M. Coord. Chem. Rev. 1983, 47,
     204-238. (Extensive compilation of bond energies).

8.   Skinner, H. A.; Connor, J. A. Pure Appl. Chem. 1985, 57, 79-88. (Compilation of bond energies).
9.   Pearson, R. G. Chem. Rev. 1985, 85, 41-59. (Review article).
10. See also Cox and Pilcher above.

Data Compilations. Gas Phase Basicities and Proton Affinities.
1.   Lias, S. G.; Bartmess, J. E.; Liebman, J. F.; Holmes, J. L.; Levin, R. D.; Mallard, W. G. J. Phys. Chem. Ref. Data 1988, 17, Suppl. 1.
2.   Lias, S. G.; Liebman, J. F.; Levin, R. D. J. Phys. Chem. Ref. Data, 1984, 13, 695-808.

Gas Phase Studies.
1.   Russell, D. H., Ed. "Gas Phase Inorganic Chemistry," Plenum Press: New York, 1989. (Much data on organometallic systems).
2.   Allison, J. Prog. Inorg. Chem., 1986, 34, 627-676. (Extensive compilation of data on organometallic systems).
3.   Bowers, M. T., Ed. "Gas Phase Ion Chemistry," Academic Press: New York, 1979, Vols. 1, 2. (Much information on gas phase techniques and results).
4.   Lehman, T. A.; Bursey, M. M. "Ion Cyclotron Resonance Spectrometry," Wiley: New York, 1976. (Introductory text).

Calorimetry.
1.   Wiberg, K. B. in "Molecular Structure and Energetics," Liebman, J. F.; Greenberg, A., Eds., VCH Publishers: New York, 1987, Vol. 2, Chapt. 4. (Review of experimental methods).
2.   Grime, J. K., Ed. "Analytical Solution Calorimetry," Wiley: New York, 1985.
3.   Hemminger, W.; Höhne, G. "Calorimetry," Verlag-Chemie: Weinheim, 1984.
4.   Ribeiro da Silva, M. A. V. "Thermochemistry and Its Applications to Chemical and Biochemical Systems," Reidel: Dordrecht, 1982. (NATO Advanced Study Institute monograph).
5.   Sunner, S.; Mansson, M., Eds. "Combustion Calorimetry"; Pergamon Press: Oxford, 1979.
6.   Barthell, J. "Thermometric Titrations," Wiley: New York, 1975.
7.   Vaughn, G. A. "Thermometric and Enthalpimetric Titrimetry," Van Nostrand-Reinhold: New York, 1973.
8.   Tyrrell, H. J. V.; Beezer, A. E. "Thermometric Titrimetry," Chapman and Hall: London, 1968.

Surface Studies
1.   Van Hove, M. A.; Wang, S.-W.; Ogletree, D. F.; Somarjai, G. A. Advan. Quantum Chem. 1989, 20, 1-184.
2.   Somorjai, G. A. "Chemistry in Two Dimensions: Surfaces," Cornell University Press: Ithaca, 1981, Chapts. 2-6.

RECEIVED January 22, 1990

Chapter 2

Periodic Trends in Transition Metal Bonds to Hydrogen, Carbon, and Nitrogen

Peter B. Armentrout

Department of Chemistry, University of Utah, Salt Lake City, UT 84112

Guided ion beam mass spectrometry has been used to obtain a variety of metal-hydrogen, metal-carbon, and metal-nitrogen bond strengths. The experimental method is reviewed with an emphasis on the requirements for obtaining good thermochemical information. The gas-phase values which have been obtained are listed. Trends in the values are evaluated for changes in charge, periodic variations, and bond-energy bond-order relationships. These comparisons result in the determination of several "intrinsic" bond energies which may be of use in estimating metal-ligand bonds for condensed-phase organometallic systems.

A comprehensive characterization of organometallic thermochemistry is a difficult task because of the multitude of metals, ligands, and environments. Nevertheless, this task has seen much progress over the past decade in part because of increasing communication between scientists working in the areas of gas-phase and condensed-phase organometallics, surface science, and theoretical chemistry. Our efforts in this area center on gas-phase determinations and involve several distinct parts: develop a novel, experimental tool for thermochemical measurements; provide a substantial data base of organometallic bond energies; and analyze the data to relate this thermochemistry to the broader organometallic perspective. Progress in these areas is reviewed here. Discussions that emphasize different aspects of this information have been published elsewhere (1-3).

Guided Ion Beam Mass Spectrometry

The Apparatus. Experimental methods used in our laboratory to measure gas phase bond dissociation energies have been detailed before (4,5,6,7), and involve the use of a "guided" ion beam tandem mass spectrometer. Ion beam instruments are two mass spectrometers

0097–6156/90/0428–0018$06.00/0
© 1990 American Chemical Society

back-to-back with a reaction zone in between, and an ion source and
an ion detector at either end. An ion beam experiment consists of
creating ions, mass selecting them, accelerating the ions to a
particular kinetic energy, reacting the ion with a neutral reagent,
and finally, mass separating and detecting the reactant and various
product ions. The reaction zone is designed so that reactions occur
over a well-defined path length and at a pressure low enough that all
products are the result of *single* ion-neutral encounters. (This is
easily verified by pressure dependence studies.) In our apparatus,
the interaction region is surrounded by an rf octopole ion beam
"guide" (8) which ensures efficient collection of all ions. The
sensitivity of the detector (a secondary electron scintillation ion
detector (9)) is sufficiently high that *individual ions* are detected
with near 100% efficiency.

The raw data of an ion beam experiment are intensities of the
reactant and product ions as a function of the ion kinetic energy in
the laboratory frame. Before presentation of the data, we convert
this information into an absolute reaction cross section as a
function of the kinetic energy in the center-of-mass frame, $\sigma(E)$. A
cross section is the effective area that the reactants present to one
another and is a direct measure of the probability of the reaction at
a given kinetic energy. It is easily related to a rate constant, $k =
\sigma v$ where v is the relative velocity of the reactants. The center-of-
mass energy takes into account the fact that a certain fraction of
the laboratory ion energy is tied up in motion of the entire reaction
system through the laboratory. This energy must be conserved and is
unavailable to induce chemical reactions. Conversion of intensities
to cross sections is readily performed by using a Beer's law type of
formula (4). Conversion of the laboratory ion energy to the center-
of-mass frame energy involves a simple mass factor (except at very
low energies where truncation of the ion beam must be accounted for)
(4). For accurate thermochemistry, particular attention must also be
paid to a determination of the absolute zero of energy. In our
laboratories, this is routinely measured by a retarding potential
analysis which has been verified by time-of-flight measurements (4)
and comparisons with theoretical cross sections (10).

Chemical Systems. To measure metal-ligand bond energies of a species
like ML^+ or ML, we measure the cross section for reaction 1 or 2
while varying the kinetic energy available to the reactants,

$$M^+ + RL \rightarrow ML^+ + R \qquad (1)$$
$$\rightarrow ML + R^+ \qquad (2)$$

as outlined above. The neutral reactant is chosen such that
reactions 1 and 2 are endothermic, and, therefore, require the
addition of kinetic energy to proceed. The minimum kinetic energy
required is the reaction threshold energy, E_T, and this provides the
desired thermochemical information (as described below).

Table I lists the various reactants which we have used to
measure the associated metal-ligand bond energies. In general, the
best information is obtained from the simplest possible system.
Hence, MH^+ thermochemistry can be derived from studies of alkanes but
the H_2 system provides the most definitive data since there are no
competing reaction products. However, the simplest imaginable system
does not always work. For example, studies of the reactions of M^+

Table I. Chemical Systems

Product	Neutral Reactant (RL)	Metals (M)[a]
MH^+	H_2	Ca - Zn
MCH_3^+	C_2H_6	Ca - Zn
MCH_2^+	CH_4	Sc - Cr
	$c\text{-}C_3H_6$, $c\text{-}C_2H_4O$	Cr - Cu
MCH^+	CH_4	Ti - Cr
$M(CH_3)_2^+$	2-butene	Sc, Ti
MNH_2^+	NH_3	Sc - V, Co, Ni
MNH^+, MN^+	NH_3	Sc - V
MO^+	O_2	Ca - Zn
	CO	Sc - V
MH	RH (R = $2\text{-}C_3H_7$, $t\text{-}C_4H_9$)	Sc - Cu
MCH_3	RCH_3 (R = $2\text{-}C_3H_7$, $t\text{-}C_4H_9$)	Sc - Cu
MO	NO_2, $c\text{-}C_2H_4O$	Cr, Mn, Co, Ni

[a]Second and third row transition metals can generally utilize the same systems as the first row metal in the same column of the periodic table.

with N_2 convince us that the MN^+ product formed does not appear promptly at the thermodynamic threshold. Therefore, we have used ammonia to measure the MN^+ product thermochemistry in several cases.

Data Analysis. The cross sections for reactions 1 and 2 are analyzed in detail to determine E_T. In the limit that there is no activation barrier in *excess* of the endothermicity, the bond dissociation energy (BDE) of the species produced in the reaction can be obtained from the threshold value for reactions 1 and 2 by using equations 3 and 4,

$$D°(M^+\text{-}L) = D°(R\text{-}L) - E_T \qquad (3)$$
$$D°(M\text{-}L) = D°(R\text{-}L) + IE(R) - IE(M) - E_T \qquad (4)$$

respectively, where IE is the ionization energy of the appropriate species.

There are several factors which influence the accuracy of BDEs derived using the beam method. The first is the assumption that there is no activation energy in excess of the reaction endothermicity. This is equivalent to there being a barrier to reaction in the opposite (exothermic) direction. Since the long-range ion-induced dipole potential is often sufficiently strong to overcome small barriers (11), exothermic ion-molecule reactions are often (though not always) observed to proceed without an activation energy. The converse must also be true. Endothermic ion-molecule reactions are often (though not guaranteed) to proceed once the available energy exceeds the thermodynamic threshold.

This assumption is one which we have tested directly for a number of reactions where the thermochemistry is well known (6,7,12-15). In all these cases, no activation barriers are found, although this sometimes requires very good sensitivity (6). Unfortunately, such direct checks for transition metal containing species are rare since there are few alternate experimental determinations having better accuracy. Those values which are

available from other experiments (see, for example, 16) are generally
in good agreement with our work. In addition, high quality
theoretical results on diatomic metal hydride ions are available
(17-19). These values average 2 ± 3 kcal/mol lower than the
experimental values derived in our studies, a typical error for such
calculations. Note that if activation barriers were present, the
true bond energies would be larger making the agreement with theory
even worse. Good agreement is also obtained for metal methyl ions
(theory (20,21) is 3 ± 3 kcal/mol lower than our experimental
numbers) and for metal hydride and metal methyl neutrals although
here the experimental data are less precise (20).

A second limitation on the beam results concerns the
characterization of the *internal* energy of the reactants. This
energy is available to the reaction and therefore must be accounted
for in the determination of the true threshold for reaction. While
not a particular problem for the neutral reagents (since these have
well characterized temperatures of 300 K), it is a problem for the
ionic reactant since ion production is intrinsically a very energetic
process. In studies of atomic metal ions, this problem becomes one
of controlling the electronic energy of the metal ion. The extent to
which this can change the thermochemistry derived has been discussed
before (1,22). For the thermochemistry from our laboratories given
in this article, the effects of electronic energy have been carefully
considered in all cases.

A final consideration in the accuracy of the BDEs derived using
the beam method is the determination of the threshold energy, E_T.
This is truly one of the more difficult aspects in ion beam
technology since the theory behind chemical threshold phenomena is
not well established. Our general approach is to reproduce the data
by using a versatile form for the threshold dependence of the
reaction cross section, and one that can mimic the multitude of
theoretical forms which have been derived (5). This empirical
formula is

$$\sigma(E) = \sigma_0(E - E_T)^n/E^m \qquad (5)$$

where σ_0 is a scaling factor, E is the relative kinetic energy, and n
and m are adjustable parameters (5,23). We have found that a value
of $m = 1$ is among the most useful forms (7,12-15) and one which is
theoretically justifiable (24) for kinetically driven reactions of
the type investigated in our laboratories. Before comparison with
the data, equation 5 is convoluted with the known distributions in
the ion and neutral kinetic energy (4). The values of n and E_T are
then optimized by nonlinear regression analysis to best reproduce the
data. Uncertainties in E_T come from variations in values of n,
different data sets, and the absolute energy scale uncertainty.

Periodic Trends in Metal Ligand Bond Energies

Metal ligand bond dissociation energies measured in our laboratories
are given in Table II. Values from other sources are given in some
cases for completeness. In the following sections, we evaluate these
numbers to try and relate them to condensed-phase organometallic
chemistry. We do this by examining the trends in the values with
regard to (a) periodic variations, (b) charge, and (c) bond-energy
bond-order relationships.

Table II. Transition Metal Ion–Ligand Bond Dissociation Energies (kcal/mol)[a]

M	M^+-H	M^+-CH$_3$	M^+-NH$_2$	M^+-CH$_2$	M^+-(CH$_3$)$_2$	M^+-NH	M^+-CH	M^+-N
Sc	57(2)[b]	59(3)[c]	85(2)[d]	98(5)[e]	117(2)[f]	119(2)[d]		
Ti	54(3)[g]	54(2)[h]	85(3)[d]	93(4)[h]	118(6)[h]	111(3)[d]	122(4)[h]	120(3)[d]
V	48(2)[i]	50(2)[j]	73(2)[k]	80(3)[l]	98(5)[m]	99(4)[k]	115(2)[l]	107(2)[k]
Cr	32(2)[n]	30(2)[o]		54(2)[o]			75(8)[o]	
Mn	48(3)[p]	51(2)[q]		71(3)[r]				
Fe	50(2)[s]	58(2)[t]	67(12)[u]*	83(4)[v]	>96[w]		61(5)[x]*	101(7)[y]*
Co	47(2)[z]	49(4)[aa]	62(2)[bb]	78(2)[cc]	110(5)[dd]			100(7)[y]*
Ni	40(2)[z]	45(2)[aa]	56(5)[ee]	75(2)[cc]	>96[ff]			
Cu	22(3)[z]	30(2)[aa]		64(2)[cc]				
Zn	55(3)[gg]	71(3)[hh]				98(3)[hh]		
Y	62(1)[b]	59(1)[e]		95(3)[e]				
Zr	55(3)[ii]							
Nb	54(3)[ii]			109(7)[jj]*			145(8)[jj]*	
Mo	42(3)[ii]							
Ru	41(3)[kk]	54(5)[kk]						
Rh	36(3)[ll]	47(5)[kk]		91(5)[jj]*			102(7)[jj]*	
Pd	47(3)[ll]	59(5)[kk]						
Ag	16(3)[ii]							
La	58(2)[b]	55(3)[e]		98(2)[e]			125(8)[jj]*	
Lu	49(4)[b]	45(5)[e]		>57(1)[e]				

[a]Values are at 300 K with uncertainties in parentheses. Values derived from work other than ion beam data are marked by an asterisk.
[b]Elkind, J. L.; Sunderlin, L. S.; Armentrout, P. B. _J. Phys. Chem._ 1989, _93_, 3151.
[c]Sunderlin, L.; Aristov, N.; Armentrout, P. B. _J. Am. Chem. Soc._ 1987, _109_, 78.
[d]Clemmer, D. E.; Sunderlin, L. S.; Armentrout, P. B. _J. Phys. Chem._ submitted for publication.
[e]Sunderlin, L. S.; Armentrout, P. B. _J. Am. Chem. Soc._ 1989, _111_, 3845.
[f]Sunderlin, L. S.; Armentrout, P. B. _Organometallics_, submitted for publication.
[g]Elkind, J. L.; Armentrout, P. B. _Int. J. Mass Spectrom. Ion Processes_ 1988, _83_, 259.
[h]Sunderlin, L. S.; Armentrout, P. B. _J. Phys. Chem._ 1988, _92_, 1209; _Int. J. Mass Spectrom. Ion Processes_ accepted for publication.
[i]Elkind, J. L.; Armentrout, P. B. _J. Phys. Chem._ 1985, _89_, 5626.
[j]Reference 5.
[k]Reference 34.
[l]Aristov, N.; Armentrout, P. B. _J. Phys. Chem._ 1987, _91_, 6178.
[m]Aristov, N. Thesis, University of California, Berkeley, 1986.
[n]Elkind, J. L.; Armentrout, P. B. _J. Chem. Phys._ 1987, _86_, 1868.
[o]Georgiadis, R.; Armentrout, P. B. _J. Phys. Chem._ 1988, _92_, 7067; _Int. J. Mass Spectrom. Ion Processes_ 1989, _89_, 227.
[p]Elkind, J. L.; Armentrout, P. B. _J. Chem. Phys._ 1986, _84_, 4862.

(Continued on next page)

Table II. *(Continued)* Transition Metal Ion–Ligand Bond Dissociation Energies (kcal/mol)[a]

[q]Georgiadis, R.; Armentrout, P. B. Int. J. Mass Spectrom. Ion Processes 1989, 91, 123.

[r]Sunderlin, L. S.; Armentrout, P. B. J. Phys. Chem. submitted for publication.

[s]Elkind, J. L.; Armentrout, P. B. J. Am. Chem. Soc. 1986, 108, 2765; J. Phys. Chem. 1986, 90, 5736.

[t]Reference 22.

[u]Buckner, S. W.; Freiser, B. S. J. Am. Chem. Soc. 1987, 109, 4715.

[v]Schultz, R. H.; Armentrout, P. B. work in progress.

[w]Burnier, R. C.; Byrd, G. D.; Freiser, B. S. J. Am. Chem. Soc. 1981, 103, 4360.

[x]Buckner, S. W.; Gord, J. R.; Freiser, B. S. J. Am. Chem. Soc. 1988, 110, 6606.

[y]Hettich, R. L.; Freiser, B. S. J. Am. Chem. Soc. 1986, 108, 2537.

[z]Elkind, J. L.; Armentrout, P. B. J. Phys. Chem. 1986, 90, 6576.

[aa]Georgiadis, R.; Fisher, E. R.; Armentrout, P. B. J. Am. Chem. Soc. 1989, 111, 4251.

[bb]Clemmer, D. E.; Armentrout, P. B. J. Am. Chem. Soc. 1989, 111, 8280.

[cc]Fisher, E. R.; Armentrout, P. B. J. Phys. Chem. in press.

[dd]Hanratty, M. A.; Beauchamp, J. L.; Illies, A. J.; van Koppen, P.; Bowers, M. T. J. Am. Chem. Soc. 1988, 110, 1.

[ee]Clemmer, D. E.; Armentrout, P. B. work in progress.

[ff]Halle, L. F.; Crowe, W. E.; Armentrout, P. B.; Beauchamp, J. L. Organometallics 1984, 3, 1694.

[gg]Georgiadis, R.; Armentrout, P. B. J. Phys. Chem. 1988, 92, 7060.

[hh]Reference 6.

[ii]Reference 27.

[jj]Hettich, R. L.; Freiser, B. S. J. Am. Chem. Soc. 1986, 108, 5086.

[kk]Reference 26.

[ll]Elkind, J. L.; Armentrout, P. B. unpublished work.

Single Metal Ligand Bonds. Metal Hydride Ions. The periodic trends
in transition metal ligand bonds can be understood by considering the
simplest species, MH^+. Early work (25) pointed out that Cr^+ has a
weak bond to H and a very stable $3d^5$ ground state configuration. To
form a covalent bond to Cr^+, one of the electrons must be removed
from this very stable environment and placed in a suitable bonding
orbital. This concept can be quantified by the promotion energy, E_p,
which we define as the energy necessary to take a metal atom (ion or
neutral) in its ground state to an electron configuration where there
is one electron in the 4s orbital. It is also important to include
the loss of exchange energy involved in spin decoupling this bonding
electron from the nonbonding 3d electrons (26,27). While E_p can be
easily calculated from spectroscopic data, the necessary values for
first and second row elements have now been tabulated by Carter and
Goddard (28).

A plot of the first row MH^+ BDEs versus this promotion energy,
Figure 1, shows an excellent correlation. If E_p is alternately
defined as excitation to a $4s3d^{n-1}$ configuration, but the $3d\sigma$ orbital
is the bonding orbital and is spin decoupled, then a correlation
almost as good as Figure 1 is obtained (27). This implies that the
dominant binding orbital on the metal is the 4s orbital but that
there is significant $3d\sigma$ character as well. These ideas are quite
consistent with results of ab initio calculations (17,19). Second
row transition metal hydride BDEs also show a fairly good correlation
with E_p, although PdH^+ is clearly distinct (27). This has been
postulated to be due to a change in the dominant bonding orbital used
from the 5s to the 4d in Pd. This has been confirmed by theory
(17,19) which also suggests similar bonding for RuH^+ and RhH^+.

We have previously contended (1,27) that the most important
feature of this correlation is the y-axis intercept, 56 kcal/mol for
the first row metals, 58 kcal/mol for the second row. This maximum
bond energy, or "intrinsic" BDE, is the MH^+ BDE expected when all
electronic factors have been accounted for. Further, it is probably
a reasonable estimate of the bond energy of any metal-hydrogen bond
in the absence of electronic and steric effects. This is because
fully ligating a metal radical center can also put the metal bonding
electron into an orbital suitable for bonding and decouple it from
the remaining nonbonding electrons. Thus ligation can perform the
equivalent of the atomic electronic promotion. Indeed, our intrinsic
BDE is comparable to those observed for many M-H bonds in saturated
organometallic species (see for instance, the values listed in 28 and
29). Similar discussions based on more detailed theoretical analyses
have also been forwarded (28).

Metal Hydride Neutrals. One check into the utility of this
"intrinsic" bond energy is whether it is strongly affected by charge.
If the periodic trends analysis shown in Figure 1 is truly for
covalently bonded species, then a similar analysis should hold for
the neutral metal hydride diatoms. We find that while the values of
$D^0(MH)$ and $D^0(MH^+)$, Table II and III, differ for a given metal, this
is compensated by differing values of E_p for the ionic and neutral
atoms. The end result is the correlation shown in Figure 2. The
upper line shown is the best linear regression fit to the data, which
is not yet as precise as that available for the ions. (Squires (30)

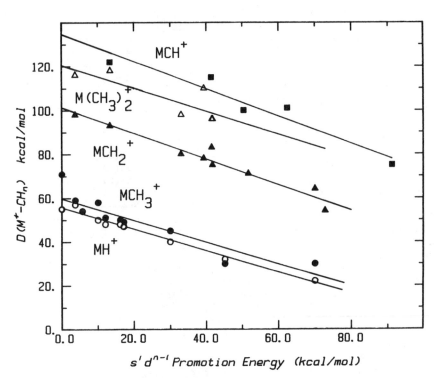

Figure 1. Metal ligand ion bond energies vs. $4s^1 3d^{n-1}$ promotion energy for first row transition metals. Data are shown for MH^+ (open circles), MCH_3^+ (closed circles), MCH_2^+ (closed triangles), $M(CH_3)_2^+$ (open triangles), and MCH^+ (closed squares). Lines are linear regression analyses of the five systems.

Table III. Transition Metal Neutral
Bond Dissociation Energies (kcal/mol)[a]

M	M-H			M-CH$_3$
Sc	48(4)[b*]			32(7)[c]
Ti	47[d*]			46(7)[c]
V	41(4),[c] 38(3)[e*]			37(9)[c]
Cr	41(3)[e*]			41(7)[c]
Mn	30(4),[f] <32[b*]			9 - 30[f]
Fe	46(3),[g] 43(6),[h] 30(3),[e*] <43[b*]			37(7)[c]
Co	46(3),[i] 45(3),[b*] 42(3)[e*]			46(3)[i]
Ni	58(3),[i] 59(2)[b*]			55(3)[i]
Cu	61(4),[i] 60(1)[b*]			58(2)[i]
Zn	20(1)[j*]			19(3)[k]

[a]Values are at 300 K with uncertainties in parentheses. Values
derived from work other than ion beam data are marked by an
asterisk.
[b]Kant, A.; Moon, K. A. High. Temp. Sci. 1981, 14, 23; 1979, 11, 55.
[c]Aristov, N.; Fisher, E. R.; Schultz, R. H.; Sunderlin, L.;
Armentrout, P. B. work in progress.
[d]Theoretical value corrected from 0 K, reference 20.
[e]Sallans, L.; Lane, K. R.; Squires, R. R.; Freiser, B. S. J. Am.
Chem. Soc. 1985, 107, 4379.
[f]Sunderlin, L. S.; Armentrout, P. B. J. Phys. Chem. submitted for
publication.
[g]Reference 22.
[h]Tolbert, M. A.; Beauchamp, J. L. J. Phys. Chem. 1986, 90, 5015.
[i]Georgiadis, R.; Fisher, E. R.; Armentrout, P. B. J. Am. Chem. Soc.
1989, 111, 4251.
[j]Huber, K. P.; Herzberg, G. Molecular Spectra and Molecular Structure
IV. Constants of Diatomic Molecules. Van Nostrand-Reinhold: New
York, 1979.
[k]Reference 6.

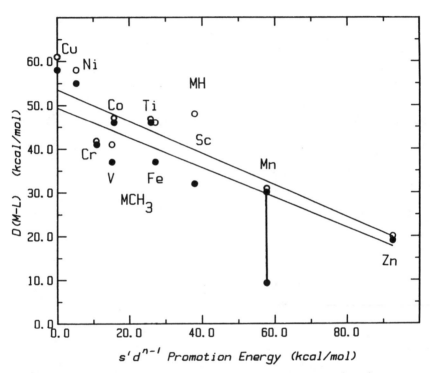

Figure 2. Metal ligand neutral bond energies vs. $4s^1 3d^{n-1}$ promotion energy for first row transition metals. Data are shown for MH (open circles) and MCH_3 (closed circles). Lines are linear regression analyses of the two systems.

has previously commented on the "crude" correlation between these quantities.) Nevertheless, the trend is clear and the intrinsic MH BDE is 54 kcal/mol, essentially the same as for the ionic hydrides. Thus, *charge is not an influential aspect of the bonding in these simple transition metal species.*

Metal Methyl Ions and Neutrals. Figure 1 also shows data for the BDEs of MCH_3^+ species vs. E_p. Note that the linear regression fit to this data yields a line which is parallel to that obtained for the MH^+ correlation. Further, the intrinsic bond energy is only slightly higher, 60 kcal/mol, probably because the methyl group is more polarizable than the H atom. In contrast, Figure 2 shows that the neutral MCH_3 BDEs have a *lower* intrinsic BDE than the MH BDEs, 49 kcal/mol, although the slope is again parallel to the MH data. These features demonstrate that the single metal-ligand bonding in M^+-H is similar to that in M^+-CH_3 (and likewise in the neutrals), but that the charge on the metal *is* influential in the *strength* of the metal-methyl bond.

Multiple Metal Ligand Bonds. Metal Methylidene Ions. Recently, we completed measurements for the first row MCH_2^+ bond energies including reevaluations of several previously measured values (Fisher, E. R.; Armentrout, P. B. J. Phys. Chem. accepted for publication. Sunderlin, L. S.; Armentrout, P. B. J. Phys. Chem. submitted for publication). When these values, Table II, are plotted vs. E_p(double) the result is shown in Figure 1 (Armentrout, P. B.; Sunderlin, L. S.; Fisher, E. R. Inorg. Chem. accepted for publication). Values for E_p(double) are taken from Carter and Goddard (28) for convenience and consist of the promotion energy necessary to form the atomic ion in an electron configuration of $4s^13d^{n-1}$ where the 4s electron and a 3d electron are spin-decoupled from the remaining nonbonding 3d electrons. The correlation is again quite remarkable considering its simplicity. We note that the use of alternate electron configurations for E_p ($3d^n$ or the lowest of the $3d^n$ and $4s^13d^{n-1}$ promotion energies) yield a significantly poorer correlation.

Also remarkable is the fact that the slope of the linear regression line is nearly parallel to that obtained for the MH^+ and MCH_3^+ correlations. This suggests that the bonding characteristics are similar between these species. The intrinsic metal-carbon double BDE is 101 kcal/mol, ~43 kcal/mol above the intrinsic single bond energies. This intrinsic bond energy is close to that for the saturated species, $D^o[(CO)_5Mn^+=CH_2] = 104 \pm 3$ kcal/mol (31).

Dimethyl Metal Ions. Another correlation between E_p(double) might be expected for metal ions with two singly bound ligands. While the data is still somewhat sparse, Table II, the correlation of the sum of the bonds in $M(CH_3)_2^+$ vs. E_p(double) is shown in Figure 1. Only lower limits are available for Fe^+ and Ni^+, but the best linear regression analysis again shows a line with a slope similar to those observed for MH^+, MCH_3^+, and MCH_2^+. Further, the intrinsic dimethyl BDE, 120 kcal/mol, is almost exactly twice that of the single methyl BDE, 58 kcal/mol. The BDE for $Zn(CH_3)_2^+$ is not included in this analysis since E_p(double) is not easily defined due to the $4s^13d^{10}$

configuration of Zn^+. Indeed, the $CH_3Zn^+-CH_3$ bond is essentially a single electron bond.

Metal Methylidyne Ions. Again the data for MCH^+ species is limited, Table II, however the correlation of these BDEs with E_p(triple), taken from Carter and Goddard (28), is quite consistent with the other correlations, Figure 1. The slope of the linear regression line is comparable to the other correlations and the intercept is 135 kcal/mol, the intrinsic metal-carbon triple bond. This agrees nicely with the theoretical value of 129 kcal/mol which can be derived (see 1) from the theoretical work of Harrison and coworkers on $TiCH^+$, VCH^+, and $CrCH^+$ bond energies (32).

Bond-Energy Bond-Order Relationships.

Comparison of the intrinsic BDEs for MH^+, MCH_3^+, MCH_2^+, and MCH^+ shows that they increase in the ratio of 0.9:1.0:1.7:2.2. This compares favorably to the ratios for the organic analogues (H-CH_3, CH_3-CH_3, CH_2=CH_2, and CH≡CH), 1.0:0.9:1.6:2.2. We first noticed this type of correlation between the single, double, and triple BDEs of metal ligand bonds vs. those of organic analogues for the case of M = V (33). This correlation is reproduced in Figure 3. Similar correlations have now been pointed out for nearly all the first row metal ions and many second and third row metals as well.

The utility of this comparison has been resurrected by the need to understand the bonding in metal-ligand BDEs involving heteroatoms. Specifically, we have recently measured the values of the BDEs for VNH_2^+, VNH^+, VN^+ (34), and VOH^+ (35, Clemmer, D. E.; Armentrout, P. B. work in progress). If these species and VO^+ (36) are compared with the organic analogues, CH_3NH_2, CH_2NH, HCN, CH_3OH, and CO, Figure 3 (correlations to NH_2NH_2, N_2H_2, and N_2 are nearly equivalent), we find that only VN^+ and VO^+ lie on the line established by the metal-carbon BDEs. The remaining values lie well above the line. A similar result is obtained for Sc^+ and Ti^+ species. We explain this deviation by participation of the lone pair electrons on the nitrogen and oxygen atoms. Harrison (Harrison, J. F. personal communication) has verified this picture theoretically for $ScNH_2^+$ and $ScNH^+$.

Figure 3 shows that the VNH_2^+ and VNH^+ values lie on a line parallel to the metal-carbon correlation but displaced upward by 28 kcal/mol. Displacements of similar magnitude are found for Sc^+ and Ti^+ species. Thus, 28 kcal/mol is the experimentally determined contribution that the nitrogen lone pair makes to the bonding energy. The two lone pairs can be seen to each make a similar 28 kcal/mol contribution each in the VOH^+ case, Figure 3. Ziegler et al. have calculated a similar result for the series of saturated species, $C\ell_3Ti$-L, finding the Ti-OH bond to be 21 kcal/mol stronger than the Ti-NH_2 bond which is 23 kcal/mol stronger than the Ti-CH_3 bond (37).

For metal ions on the right side of the periodic table, the picture is less clear-cut. One would anticipate that the lone pair effect will be reduced since there are no longer any empty orbitals on the metal to accept the lone pair donation. This is indeed found experimentally for the case of $CoNH_2^+$ and $CoOH^+$ (35,38), Figure 4, and Ziegler et al. (37) also find this for the saturated species $(CO)_4CoL$. We anticipate similar results for other metals on the right side of the periodic table.

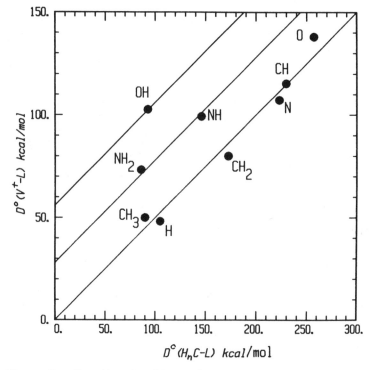

Figure 3. Vanadium ion ligand bond energies vs. analogous organic bond energies.

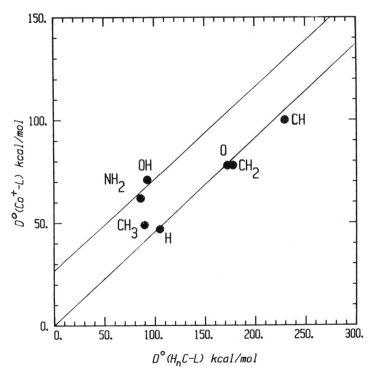

Figure 4. Cobalt ion ligand bond energies vs. analogous organic bond energies.

Future Developments

Ongoing work in our laboratory seeks to continue to extend the ideas
discussed above to additional ligands and metals. Data for second
and third row metals are being accumulated slowly because the
problems with electronic excited states are more difficult than for
the first row metals. We are also endeavoring to extend these
experimental techniques to systems of more direct interest to the
condensed-phase chemist. Our efforts in these areas involve two
primary research directions. The first is the extension to multiple
metal centers, i.e. metal clusters. We have recently published the
first measurements of the binding energies for transition metal
clusters larger than the dimers (39,40, Hales, D. A.; Armentrout, P.
B. J. Cluster Science submitted for publication). This work is being
extended to other metals and to the chemistry of such species.
    We have also recently developed a flowing afterglow source for
our guided ion beam apparatus. This source is designed to produce
ions that are thermalized in all degrees of freedom. The inability
to definitively study "cold" metal ligand complexes is a significant
reason why gas phase ion studies have infrequently included ligated
metals. Such studies are being increasingly pursued, however, and
the combination of high pressure sources with beam techniques
promises to be a particularly powerful means of studying metal ligand
complex ions which is being developed in several laboratories.
Recent work of Marinelli and Squires (41) on the determination of
metal ion bond energies to water and ammonia represents an early
example of this type of experiment.

Acknowledgments

This work is supported by the National Science Foundation. I also
thank my many research collaborators who contributed to the work
described here.

Literature Cited

1.   Armentrout, P. B.; Georgiadis, R. Polyhedron 1988, 7, 1573.
2.   Armentrout, P. B. In "Gas Phase Inorganic Chemistry," Russell,
     D. H., Ed.; Plenum: New York, 1989; 1.
3.   Armentrout, P. B.; Beauchamp, J. L. Accts. Chem. Res. 1989, 22,
     315.
4.   Ervin, K. M.; Armentrout, P. B. J. Chem. Phys. 1985, 83, 166.
5.   Aristov, N.; Armentrout, P. B. J. Am. Chem. Soc. 1986, 108,
     1806.
6.   Georgiadis, R.; Armentrout, P. B. J. Am. Chem. Soc. 1986, 108,
     2119.
7.   Ervin, K. M.; Armentrout, P. B. J. Chem. Phys. 1986, 84, 6738.
8.   Teloy, E.; Gerlich, D. Chem. Phys. 1974, 4, 417.
9.   Daly, N. R. Rev. Sci. Instrum. 1959, 31, 264.
10. Burley, J. D.; Ervin, K. M.; Armentrout, P. B. Int. J. Mass
     Spectrom. Ion Processes 1987, 80, 153.
11. Talrose, V. L.; Vinogradov, P. S.; Larin, I. K. In Gas Phase
     Ion Chemistry; Bowers, M. T., Ed.; Academic: New York, 1979;
     Vol. 1, p. 305.

12. Ervin, K. M.; Armentrout, P. B. J. Chem. Phys. 1987, 86, 2659.
13. Weber, M. E.; Elkind, J. L.; Armentrout, P. B. J. Chem. Phys. 1986, 84, 1521.
14. Elkind, J. L.; Armentrout, P. B. J. Phys. Chem. 1984, 88, 5454.
15. Boo, B. H.; Armentrout, P. B. J. Am. Chem. Soc. 1987, 109, 3549.
16. Buckner, S. W.; Freiser, B. S. Polyhedron 1988, 7, 1583.
17. Schilling, J. B.; Goddard, W. A.; Beauchamp, J. L. J. Am. Chem. Soc. 1986, 108, 582; 1986, 109, 5565; J. Phys. Chem. 1987, 91, 5616.
18. Rappe, A. K.; Upton, T. H. J. Chem. Phys. 1986, 85, 4400.
19. Pettersson, L. G. M.; Bauschlicher, C. W.; Langhoff, S. R.; Partridge, H. J. Chem. Phys. 1987, 87, 481.
20. Bauschlicher, C. W.; Langhoff, S. R.; Partridge, H.; Barnes, L. A. J. Chem. Phys. 1989, 91, 2399.
21. Schilling, J. B.; Goddard, W. A.; Beauchamp, J. L. J. Am. Chem. Soc. 1987, 109, 5573.
22. Schultz, R. H.; Elkind, J. L.; Armentrout, P. B. J. Am. Chem. Soc. 1988, 110, 411.
23. Armentrout, P. B.; Beauchamp, J. L. J. Chem. Phys. 1981, 74, 2819.
24. Chesnavich, W. J.; Bowers, M. T. J. Phys. Chem. 1979, 83, 900.
25. Armentrout, P. B.; Halle, L. F.; Beauchamp, J. L. J. Am. Chem. Soc. 1981, 102, 6501.
26. Mandich, M. L.; Halle, L. F.; Beauchamp, J. L. J. Am. Chem. Soc. 1984, 106, 4403.
27. Elkind, J. L.; Armentrout, P. B. Inorg. Chem. 1986, 25, 1078.
28. Carter, E. A.; Goddard, W. A. J. Phys. Chem. 1988, 92, 5679.
29. Stevens, A. E.; Beauchamp, J. L. J. Am. Chem. Soc. 1981, 103, 190.
30. Squires, R. R. J. Am. Chem. Soc. 1985, 107, 4385.
31. Stevens, A. E. Ph.D. Thesis, Caltech, Pasadena, California, 1981.
32. Mavridis, A.; Alvarado-Swaisgood, A. E.; Harrison, J. F. J. Phys. Chem. 1986, 90, 2584.
33. Aristov, N.; Armentrout, P. B. J. Am. Chem. Soc. 1984, 106, 4065.
34. Clemmer, D. E.; Sunderlin, L. S.; Armentrout, P. B. J. Phys. Chem. 1989, 93, in press.
35. Magnera, T. F.; David, D. E.; Michl, J. J. Am. Chem. Soc. 1989, 111, 4101.
36. Aristov, N.; Armentrout, P. B. J. Phys. Chem. 1986, 90, 5135.
37. Ziegler, T.; Tschinke, V.; Versluis, L.; Baerends, E. J.; Ravenek, W. Polyhedron 1988, 7, 1625.
38. Cassady, C. J.; Freiser, B. S. J. Am. Chem. Soc. 1984, 106, 6176.
39. Loh, S. K.; Lian, L.; Armentrout, P. B. J. Am. Chem. Soc. 1989, 111, 3167.
40. Loh, S. K.; Hales, D. A.; Lian, L.; Armentrout, P. B. J. Chem. Phys. 1989, 90, 5466.
41. Marinelli, P. J.; Squires, R. R. J. Am. Chem. Soc. 1989, 111, 4101.

RECEIVED November 15, 1989

Chapter 3

Organometallic Reaction Energetics from Product Kinetic Energy Release Distributions

Petra A. M. van Koppen[1], Michael T. Bowers[1], J. L. Beauchamp[2], and David V. Dearden[2]

[1]Department of Chemistry, University of California, Santa Barbara, CA 93106
[2]Arthur Amos Noyes Laboratory of Chemical Physics, California Institute of Technology, Pasadena, CA 91125

Product kinetic energy release distributions reveal important features of the potential energy surfaces associated with the formation and rupture of H-H, C-H and C-C bonds at transition metal centers. For processes in which there is no barrier for the reverse reaction (loose transition state), experimental distributions can be reproduced with phase space theory. The single important parameter in fitting theory and experiment is the reaction exothermicity, which in turn yields bond dissociation energies for organometallic reaction intermediates. Two distinct experiments are discussed. In the first, the systems are chemically activated with reaction intermediates having well defined internal energies. The second, potentially more general method, involves investigations of ions with a broad range of internal energies. In this case the temporal constraints of the experiment lead to preferential detection of ions with a narrow range of internal energies. The results obtained to date in these studies are summarized, and several cautions are discussed which must be exercised if accurate bond energies are to be obtained.

The recent development and application of techniques for studying the mechanisms and energetics of organometallic reactions in the gas phase stems in part from the importance of transition metal sites as active centers which provide low energy pathways for the selective catalytic transformation of small molecules into useful products. Considerable interest in the subject of C-H bond activation at transition-metal centers has developed in the past several years, stimulated by the observation that even saturated hydrocarbons can react with little or no activation energy under appropriate conditions. Interestingly, gas phase studies of the reactions of saturated hydrocarbons at transition-metal centers were reported as early as 1973([1]). More recently, ion cyclotron resonance and ion beam experiments have provided many examples of the cleavage of both C-H and C-C bonds of alkanes by transition-metal ions in the gas phase([2]). These gas phase studies have provided a plethora of highly speculative reaction mechanisms. Conventional mechanistic probes, such as isotopic labeling, have served mainly to indicate the complexity of "simple"

0097–6156/90/0428–0034$06.25/0
© 1990 American Chemical Society

processes such as the dehydrogenation of alkanes(3,4). More detailed studies using ion beam methods(5), multiphoton infrared laser activation(6) and the determination of kinetic energy release distributions(7,8), have revealed important features of the potential energy surfaces associated with the reactions of small molecules at transition metal centers.

A starting point for considering the energetics of these processes is the potential energy surface or reaction coordinate diagram shown in Figure 1. In neutral systems a significant barrier to reaction is usually associated with the activation of C-H bonds at coordinatively unsaturated transition metal centers. Ions, on the other hand, interact with neutrals via charge-induced dipole forces creating a chemically activated adduct, often with sufficient excitation energy to overcome intrinsic barriers for insertion into C-H bonds, as shown in Figure 1.

The initial oxidative addition product III in Figure 1 may undergo further rearrangement, leading to the elimination of small molecules and the *exothermic* formation of stable organometallic products. These reactions are regarded as being *facile* if they occur efficiently at thermal energies. This implies that all interconnecting transition states are lower in energy than the reactants. Typically, elimination of molecular hydrogen, alkanes, and alkenes are facile processes. Reagent electronic or translational excitation can promote *endothermic* reactions, the simplest processes resulting from cleavage of the newly formed bonds in III. These processes typically yield radical products.

The experimental methods(2) which have been used in these kinds of studies are those of the ion-molecule chemist: ion cyclotron resonance (ICR) mass spectrometry and its close cousin, Fourier transform mass spectrometry (FTMS), flowing afterglow (FA), tandem mass spectrometry (or beam techniques), and kinetic energy release measurements. The first three techniques permit the measurement of reaction rates, k(T), and branching ratios at thermal energies. The beam method allows the measurement of the absolute probability for reaction as a function of kinetic energy(5,9). This probability is given in terms of an energy dependent cross section $\sigma(E)$, the effective area which the reactants present to one another. Kinetic energy release measurements are, as the name implies, direct studies of the kinetic energy of the products of reaction(5,10,11). These studies provide insight into the potential energy surface in the exit channel of the reaction.

The main strength of the beam method is its ability to vary precisely the energy available to the reactants. This permits the determination of thresholds for endothermic reactions, which in the absence of a barrier for the reverse process, directly yield reaction thermochemistry and heats of formation of organometallic fragments. In the case of exothermic reactions (and the majority of identified reactions are in this category), ion beam experiments are no longer a useful technique for determining reaction thermochemistry. It is for just these cases, however, that the detailed analysis of kinetic energy release distributions can be a valuable source of data relating to reaction exothermicities and metal ligand bond dissociation energies. It is the purpose of this article to summarize the developments which have occurred in this area and demonstrate the potential of this newer method for the determination of unknown thermochemical properties for reactive intermediates in general. Since the instrumentation for these experiments is widely available, we anticipate an increasing number of applications of the experimental methodology described in this review.

Experimental Methods

The instrument used to determine kinetic energy release distributions for decomposing ions, a VG ZAB-2F reversed geometry double focusing mass spectrometer($\underline{12}$), is shown schematically in Figure 2. Reactant ions, typically adducts of atomic metal ions with small molecules, are extracted from a high pressure ion source and mass analyzed using the magnetic sector. Metastable peaks resulting from ion dissociation in the second field free region between the magnetic and electric sectors are recorded by scanning the electric sector. These are differentiated to yield kinetic energy release distributions($\underline{10},\underline{11}$). The energy width of the main beam is sufficiently narrow to avoid contributing to the metastable peak widths.

The ion source is a custom designed variable temperature EI/CI source. Metal containing precursor ions are formed by electron impact (150 eV) ionization and fragmentation of volatile precursors such as $Fe(CO)_5$ and $Co(CO)_3NO$. Typical source pressures are 10^{-3} torr, and source temperatures are kept below 280 K to minimize decomposition of the organometallics on insulating surfaces. Adduct formation results from reaction of an atomic metal ion or metal containing species with small molecules. The ion source is operated under nearly field free conditions to prevent translational excitation of the ions, which are accelerated to 8 kV before mass analysis.

Analysis of Kinetic Energy Release Distributions

General Considerations. Statistical product kinetic energy release distributions can be used to determine gas phase bond dissociation energies for organometallic species. The amount of energy appearing as product translation can be used to infer details of the potential energy surfaces primarily in the region of the exit channel and has implications for the ease with which the reverse association may occur.

To illustrate how the amount of energy released to product translation for a given reaction pathway may reflect specific details of the potential energy surface, consider the two hypothetical surfaces in Figure 3. The interaction of a metal ion M^+ with a neutral molecule A can result in the formation of a chemically activated adduct, MA^+. In the absence of collisions, the internal excitation may be utilized for molecular rearrangement and subsequent fragmentation. In Figure 3, the adduct MA^+ is shown to dissociate along two different potential energy surfaces, designated Type I and Type II, yielding products MB^+ and C. For a reaction occurring on a Type I surface, a smooth transition in the exit channel, without a barrier for the reverse association, allows for complete energy randomization prior to dissociation. This results in a statistical product kinetic energy release distribution. A typical example is a simple bond cleavage. Statistical phase-space theory($\underline{13},\underline{14}$) has been successful in modeling translational energy release distributions for reactions occurring on this type of potential energy surface. In a statistical kinetic energy release distribution, the relative probability for a given product kinetic energy maximizes near zero and drops off rapidly with increasing energy as shown on the right hand portion of Figure 3a. The average kinetic energy release for a large molecule will generally be much less than the total reaction exothermicity, ΔH, since the energy of the system in excess of that necessary for dissociation will be statistically divided between all the modes.

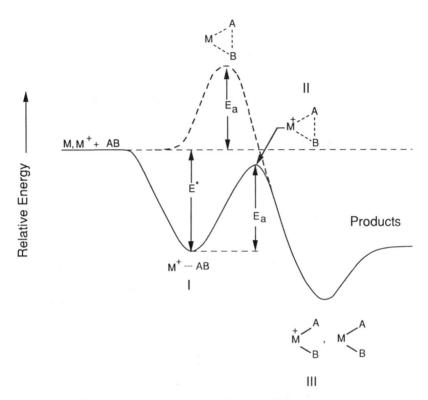

Figure 1. Schematic reaction coordinate diagram comparing the difference between metal atoms and metal ions inserting into a bond as a first step in a chemical reaction. The attractive well of the $M(AB)^+$ complex can completely offset the insertion barrier if $E^* > E_a$.

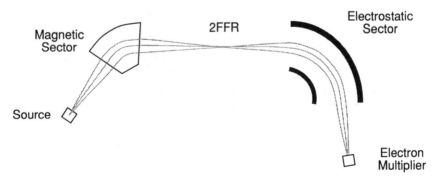

Figure 2. Schematic diagram showing the apparatus for measuring kinetic energy release distributions.

TYPE I (No Barrier for Reverse Association Reaction)

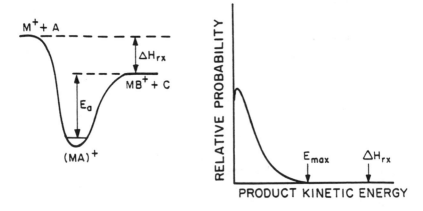

TYPE II (Large Barrier for Reverse Association Reaction)

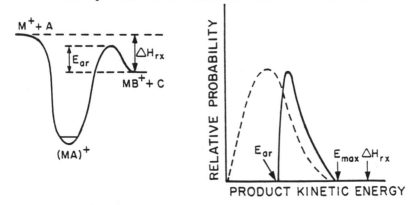

Figure 3. Hypothetical potential energy surfaces for the reaction $M^+ + A \rightarrow MB^+ + C$ and the corresponding product kinetic energy release distributions in the center-of-mass frame.

A reaction occurring on a Type II surface involves a barrier with an activation energy (E_{ar}) for the reverse association. In this case, the rate of product formation is too fast to allow for complete energy randomization, giving rise to a non-statistical kinetic energy release distribution. In the absence of coupling between the reaction coordinate and other degrees of freedom after the molecule has passed through the transition state, all of the reverse activation energy would appear as translational energy of the separating fragments. Accordingly, the translational energy release distribution would be shifted from zero by the amount E_{ar}. More typically however, some coupling does occur, yielding a broader distribution shifted to lower energy. This type of surface is often associated with complex reactions which involve the simultaneous rupture and formation of several bonds in the transition state. Statistical phase space theory is not applicable to these systems. With either potential energy surface, the maximum kinetic energy release, E_{max}, places a lower limit on the reaction exothermicity.

Complex reactions may involve the formation of several reaction intermediates which are local minima on the potential energy surfaces, separated by barriers. In such systems, the shape of the kinetic energy release distribution can be determined by the exit channel even though the rate of reaction will generally be determined by an earlier transition state. However, barriers near the asymptotic energy of the reactants can affect the kinetic energy release distribution by imposing dynamic constraints on the system, even when they are remote from the exit channel(van Koppen, P.A.M.; Brodbelt-Lustig, J.; Bowers, M.T.; Dearden, D.V.; Beauchamp, J.L.; Fisher, E.; Armentrout, P.B.; J. Am. Chem. Soc., to be published). This is discussed further below.

Studies of kinetic energy release distributions have implications for the reverse reactions. Notice that on a Type II surface, the association of ground state MB$^+$ and C to form MA$^+$ cannot occur at thermal energy. In contrast, on a Type I potential energy surface the reverse association can occur to give the adduct MA$^+$, which does not have sufficient energy to yield the reactants M$^+$ and A. Although the reaction is nonproductive, it is possible in certain cases to determine that adduct formation did occur by use of isotopic labeling (and observing isotopically mixed products) or by collisional stabilization at high pressures.

<u>Detailed Analysis of Kinetic Energy Release Distributions for Type I Surfaces using Phase Space Theory.</u> The model for the statistical phase space theory calculations(<u>8</u>) begins with Equation 1, where $F^{orb}(E,J)$ is the flux through the

$$M^+ + A \quad \overset{F^{orb}(E,J)}{\rightleftharpoons} \quad [MA^+(E,J)]^* \quad \overset{k_f(E,J)}{\rightarrow} \quad products \quad (1)$$

orbiting transition state of the formation reaction and yields the initial E,J distribution of $(MA^+)^*$. The fraction of ions decomposing in the second field-free region (2FFR) of the instrument, between the magnet and the electrostatic analyzer (ESA), is given for channel i, by

$$P_i(E,J,t_r) = \exp[-k_i(E,J)(t_1 + t_r)] - \exp[-k_i(E,J)(t_2 + t_r)] \quad (2)$$

where t_r is the time spent in the ion source after formation of the adduct, t_1 the flight time from the ion source to the exit of the magnet (entry of the 2FFR), and t_2 the flight time from the ion source to the entrance of the ESA (exit of the

2FFR). The unimolecular rate constants, $k_i(E,J)$ are given by Equation 3 where

$$k_i(E,J) = F_i^{\text{orb}}(E,J) / \rho(E,J) \qquad (3)$$

where $F_i^{\text{orb}}(E,J)$ is the total microcanonical flux through the orbiting transition state leading to products in channel i, and $\rho(E,J)$ is the microcanonical density of states of the $[MA^+(E,J)]^*$ complex. The fraction of molecules at energy E and angular momentum J decaying through the orbiting transition state to yield products i with translational energy E_t is given by Equation 4,

$$P_i(E,J;E_t) = F_i^{\text{orb}}(E,J;E_t)/F_i^{\text{orb}}(E,J) \qquad (4)$$

where $F_i^{\text{orb}}(E,J;E_t)$ is the microcanonical flux which leads to products i with energy E_t. Finally, the fraction of molecules that decay via channel i rather than some other channel is given by the expression 5. Combining these equations,

$$\gamma_i(E,J) = k_i(E,J)/\Sigma_i k_i(E,J) \qquad (5)$$

averaging over the initial Boltzman energy distributions of the reactants and the angular momentum distribution of the $[MA^+]^*$ collision complex, and normalizing yield the probability for forming products in channel i with translational energy E_t (Equation 6). For simplicity, the term $\gamma_i(E,J)$ was set

$$P_i(E_t) =$$

$$\frac{\int_0^\infty dE\exp(-E/k_BT) \int_0^{J_{max}} dJ 2J F^{orb}(E,J) P_i(E,J,t_r) P_i(E,J;E_t) \gamma_i(E,J)}{\int_0^\infty dE\exp(-E/k_BT) \int_0^{J_{max}} dJ 2J F^{orb}(E,J) P_i(E,J,t_r) \gamma_i(E,J)} \qquad (6)$$

equal to unity. In special cases where the rate constant for one reaction channel may have a very strong J dependence relative to another, this term can have an effect. However, for all systems considered here the effect will be small. The expression for $P_i(E_t)$ should also be averaged over the distribution of source residence times, $P_i(t_r)$. This term has been shown to have an effect of only a few percent in similar systems and hence is usually ignored. Further, little is generally known about the rate determining transition states along the reaction coordinate for the organometallic systems studied. Hence for simplicity it is assumed that $P_i(E,J,t_r)$ is a constant. This is a reasonable assumption since the kinetic energy distribution will not depend strongly on the detection time window for dissociating MA^+ complexes which have a narrow range of internal energies relative to the reaction exothermicity. A number of kinetic energy release distributions have been measured as a function of the accelerating voltage, over the range which maintains reasonable sensitivity and resolution, from 3 kV to 8 kV. This changes the time window by a factor of 1.6. The distributions were identical, confirming this contention (van Koppen, P. A. M.; Bowers, M. T.; Beauchamp, J. L., unpublished results). For ion complexes formed with a broad range of internal energy the time window is taken into account and will be discussed further below.

Practical Considerations in Calculations. In order to calculate kinetic energy release distributions, structures and vibrational frequencies for the various species are required. These are taken from the literature where possible, or estimated from literature values of similar species. The details of the kinetic energy release distributions are found to vary only weakly with structure or vibrational frequencies over the entire physically reasonable range for these quantities. The distributions are strongly dependent on the total energy available to the dissociating complex, and hence in our model to the ΔH^o of reaction. Often all heats of formation of product and reactants are well known except one, the organometallic product ion. This quantity can then be used as a parameter and varied until the best fit with experiment is obtained.

Results and Discussion

Chemically Activated Species with Well Characterized Internal Energies. Atomic cobalt ions react with isobutane to yield two products as indicated in reactions 7 and 8, which involve the elimination of hydrogen and methane to yield cobalt ion complexes with isobutylene and propylene, respectively(8).

$$Co^+ + i\text{-}C_4H_{10} \longrightarrow \begin{cases} Co(C_4H_8)^+ + H_2 & (7) \\ Co(C_3H_6)^+ + CH_4 & (8) \end{cases}$$

Kinetic energy release distributions for these processes are shown in Figures 4 and 5. An attempt to fit the experimental distribution for dehydrogenation of isobutane by Co^+ with phase-space theory is included in Figure 4. The observed disagreement supports a Type II surface for this process. All of the studies to date of the dehydrogenation of alkanes by group 8, 9 and 10 first row metal ions exhibit kinetic energy release distributions which are characterized by large barriers for the reverse association reactions. This is consistent with the failure to observe the reverse reaction as isotopic exchange processes when D_2 interacts with metal olefin complexes.

In contrast to the results obtained for dehydrogenation reactions, kinetic energy release distributions for alkane elimination processes can usually be fit with phase space theory. Results for the loss of methane from reaction 8 of Co^+ with isobutane are shown in Figure 5. In fitting the distribution calculated using phase space theory to the experimental distribution the single important parameter in achieving a good fit is the reaction exothermicity, which in the case of reaction 8 depends on the binding energy of propylene to the cobalt ion in the reaction products. As shown in Figure 5, a best fit is achieved with a bond dissociation energy of 1.91 eV at 0^oK (2.08 eV or 48 kcal/mol at 298 ^{o}K). An analysis of the kinetic energy release distribution for reaction 9 of Co^+ with cyclopentane(8) yields an identical value for the binding energy of propylene

$$Co^+ + cyclo\text{-}C_5H_{10} \rightarrow Co(C_3H_6)^+ + C_2H_4 \qquad (9)$$

to an atomic cobalt ion. Bond energies derived from these and other studies(15) of this type are summarized in Table I.

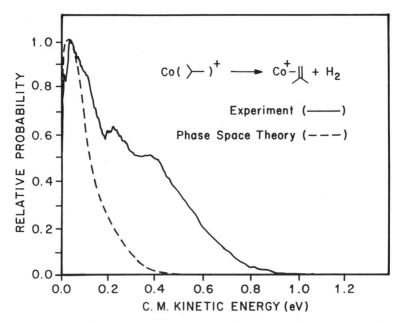

Figure 4. Experimental and theoretical kinetic energy release distributions for loss of H_2 from Co(isobutane)$^+$. The calculated distribution assumes no activation barrier for the reverse process.

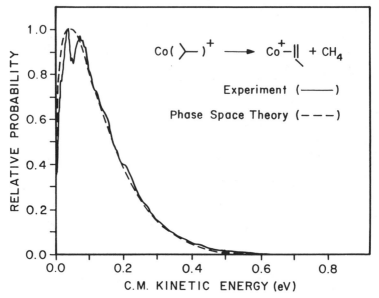

Figure 5. Experimental and theoretical kinetic energy release distributions for loss of CH_4 from Co(isobutane)$^+$. The calculated distribution assumes no activation barrier for the reverse process.

Table I. Summary of Thermochemical Data

Species	$D_0^0(D_0^{298})^a$	ΔH_{f0}^0(kcal/mol)	Reference
Co^+-CO	31(34)	224	8
Co^+-(ethene)	42(46)	255	c
Co^+-(propene)	44(48)	247	8
Co^+-$(CH_3)_2$	105^b(110)	247	8
Co^+ (cyclopropane)	81^b(86)	274	15
Fe^+-CO	26(30)	227	15
Fe^+-(ethene)	35(39)	260	c
Fe^+-(propene)	37(41)	252	c
Fe^+ (cyclopentadiene)	55(61)	260	d
Fe^+ (benzene)	70(76)	241	d
Fe^+ (cyclopropane)	85(90)	268	15

[a]The error on values of D_0^0 is on the order of $\pm$ 5 kcal/mol, reflecting the sensitivity of the fit between theory and experiment. The value in parenthesis is the estimated value for 298 K.
[b]The bond energy is the sum of the two bonds forming Co^+ and $2CH_3$ from $Co(CH_3)_2^+$ and Co^+ and trimethylene from cobaltacyclobutane ion.
[c]Values quoted in reference 15 (van Koppen, P.A.M.; Bowers, M.T.; Beauchamp, J.L., unpublished results).
[d]Dearden, D.V.; Beauchamp, J.L.; van Koppen, P.A.M.; Bowers, M.T. J. Am. Chem. Soc., submitted for publication.

As is the case for reaction 9, elimination of a π-donor or n-donor base from a coordinately unsaturated metal center would generally be expected to proceed with a Type I potential surface (no barrier for the reverse association reaction). The validity of the phase space analysis for such processes is not surprising. The statistical theory analysis for alkane elimination (e.g. reaction 8) indicates a loose transition state is operative in the exit channel. This requires that the alkane being eliminated is strongly interacting with the transition metal center, almost certainly via a significant energy well on the potential energy surface prior to product formation. This suggests an important feature of the reaction coordinate diagram indicated in Figure 6 for the reaction of Co^+ with isobutane. The kinetic energy release distribution for methane elimination is determined entirely by the dissociation of the methane adduct, and *is not useful in identifying the presence or determining the height of a reverse activation barrier which might be associated with oxidative addition.* For this reason, the features in the exit channel for methane elimination are represented by a dashed line. The

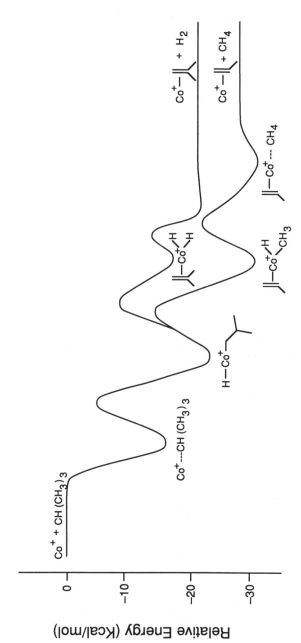

Figure 6. Qualitative potential energy diagrams for H_2 and CH_4 loss from Co(isobutane)$^+$.

substantial kinetic energy release observed for hydrogen elimination and the failure of a statistical analysis to reproduce the observed distribution is a clear signature of the barrier shown in the exit channel for this process in Figure 6.

The initial well associated with the formation of an adduct of the Co^+ with isobutane is included in Figure 6. The chemical activation associated with the formation of such an adduct is essential in overcoming the intrinsic barrier associated with insertion into a C-H bond. In comparison to larger hydrocarbons, the weaker interaction with first row group 8-10 metal ions is apparently insufficient to overcome intrinsic barriers for insertion. This would explain the failure to observe dehydrogenation of ethane by these metal ions, even though the process is known to be exothermic. The reactions of the isomeric butanes and larger hydrocarbons with cobalt ions and with other reactive metal ions as well are generally facile. Reaction rate constants and cross sections approach the Langevin or collision limit. Propane is intermediate in behavior between isobutane and ethane, and this gives rise to interesting behavior which is further discussed below.

It should be apparent that phase space fitting of kinetic energy release distributions yields important thermochemical information for exothermic reactions with no reverse activation barrier. As another example, Co^+ ions decarbonylate acetone (reaction 10) yielding a dimethyl cobalt ion as

$$Co^+ + (CH_3)_2CO \rightarrow Co(CH_3)_2^+ + CO \qquad (10)$$

the product. A phase space theory fit of the kinetic energy release distribution for this process indicates the total energy of process 11 is 4.55 eV at 0 K (4.77

$$Co(CH_3)_2^+ \rightarrow Co^+ + 2CH_3 \qquad (11)$$

eV or 110 kcal/mol at 298 K) ($\underline{8}$). Ion beam threshold measurements indicate the bond strength of Co^+-CH_3 is 49.1 kcal/mol($\underline{16}$). From our measurement we obtain, by difference, a value of 61 kcal/mol for the second bond energy. One interesting conclusion that can be drawn from this result is that insertion of Co^+ into a C-C bond in saturated alkanes should be exothermic by about 20 kcal/mol. However, the fact that the reaction is exothermic does not guarantee that the process will be observed.

Examples of Chemically Activated Systems where the Reverse Reaction is Known to be Facile. The dehydrogenation of cyclopentene by atomic iron ions (Reaction 12), is known to be reversible(Dearden, D.V.; Beauchamp, J.L.; van Koppen, P.A.M.; Bowers, M.T. J. Am. Chem. Soc., submitted for publication). In the presence of D_2, the product of reaction 12,

$$Fe^+ + cyclopentene \rightarrow Fe(C_5H_6)^+ + H_2 \qquad (12)$$

presumed to be an iron cyclopentadiene complex, rapidly undergoes isotopic hydrogen exchange. This suggests that there is no barrier for the addition of H_2 to $Fe(C_5H_6)^+$, which corresponds to the first step in the reverse of reaction 12. In accordance with the considerations outlined in the previous section, a statistical kinetic energy release distribution should be observed for this system. As shown in Figure 7, the experimental distribution for this process can be fit very closely using statistical phase space theory, which yields a bond dissociation energy $D_0{}^0(Fe^+$-$C_5H_6) = 55 \pm 5$ kcal/mol. A reaction coordinate diagram for

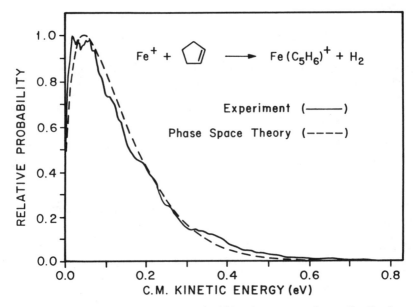

Figure 7. Experimental and theoretical kinetic energy release distributions for the dehydrogenation of cyclopentene by Fe^+.

this system is shown in Figure 8. In this case the initial interaction of the metal ion with the olefin leads to a chemically activated intermediate with around 35 kcal/mol internal energy. The postulated mechanism involves insertion of the metal into the allylic C-H bond followed by β-hydrogen transfer and reductive elimination of H_2. The reverse of the final two steps accomodates the observed isotopic exchange with D_2. As shown in Figure 8, the overall reaction is exothermic by 35 kcal/mol. The binding of cylopentadiene to the metal ion is more than enough to overcome the 20 kcal/mol required to dehydrogenate cyclopentene.

This example is particularly interesting in that dehydrogenation of alkanes by Fe^+ and other group 8 metal ions generally exhibit non-statistical kinetic energy release distributions. Apparently a full range of behavior can be observed for dehydrogenation processes, depending on the ligand environment. When bound to cyclopentadiene, oxidative addition of hydrogen to the metal center is facile. This is not generally the case when Fe^+ is ligated by a single olefin.

Another example of a hydrogen elimination process which is both reversible and exhibits a statistical kinetic energy release distribution is the dissociation of the cyclopentadienyl rhodium isopropyl ion, reaction 13(17). In this case the

$$CpRhCH(CH_3)_2^+ \quad \rightarrow \quad CpRh(C_3H_5)^+ + H_2 \qquad (13)$$

reaction of the product ion, presumed to be a π-allyl species, with D_2 leads to the incorporation of four deuterium atoms into the complex.

Quantitative Studies of Ions Formed with a Broad Range of Internal Energies. The above studies have all involved chemically activated systems in which the internal energy of the decomposing ion is defined within a narrow range. In all of our earlier investigations we had worked with the assumption that the energy content of the decomposing ion had to be precisely known in order to extract quantitative results from the application of phase space theory to reproduce kinetic energy release distributions. We now realize that if the method of ion formation produces a broad distribution of internal energies, then the experiment will select a particular internal energy (since the time window for observation is dictated by the ion flight time through the second field free region of the VG ZAB-2F reversed geometry spectrometer). Statistical theory is used to accurately calculate the internal energies of ions being observed, and this energy is then used to analyze the kinetic energy release distribution to extract the enthalpy change for the observed process. This technique has been used, for example, to determine the sequential bond energies of the metal carbonyl ions $Mn(CO)_x^+$ (Table II), and appears to be generally applicable for determination of the thermochemical properties of both ionic and neutral species(18).

The manganese carbonyl ions are of particular interest since the spin state of the system changes from septet for Mn^+ to singlet for the fully coordinated $Mn(CO)_6^+$. With the exception of the fully coordinated species, all of the $Mn(CO)_x^+$ ions undergo rapid exchange of CO with labelled ^{13}CO. Not surprisingly, then, the $Mn(CO)_x^+$ species with x = 3-6 all exhibit kinetic energy release distributions which appear statistical. These ions, however, were formed by electron impact fragmentation of various precursor ions and as a result had a broad distribution of internal energies. According to the phase space description, the statistical kinetic energy release distribution is a strong function of the amount of energy above threshold in the energized molecule, $E^{\ddagger}$, which

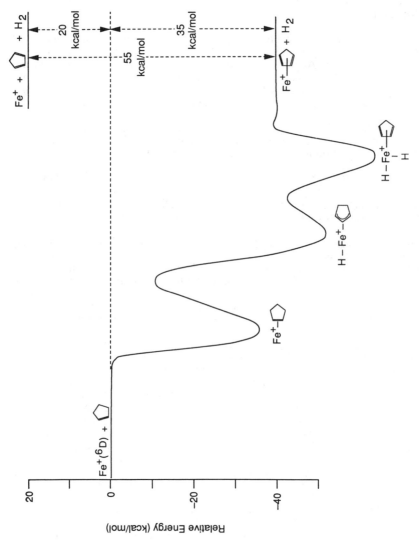

Figure 8. Schematic potential energy surface for the dehydrogenation of cyclopentene by Fe⁺.

in the dissociation of the ions is just the difference between the bond dissociation energy and E^*, the internal energy of the adduct. The lifetime requirements for the observation of metastables are stringent. They must decompose during the time they are in the second field-free region, a window of approximately 5-20 μsec after extraction from the source. The lifetime requirement in turn will serve to *select* ions with a narrow range of E^* which are detected with high sensitivity. If E^* is too high, ions will decompose while still in the source region. If E^* is too low, ions will reach the detector without decomposing. These considerations can be quantified using statistical kinetic theory to determine decomposition rates as a function of ion internal energy. Figure 9 shows calculated unimolecular decomposition rate constants for the $Mn(CO)_x^+$ ions. These calculations then yield values of E^* which can be employed in fitting statistical phase space theory to experimental kinetic energy release distributions. Typical results are shown in Figure 10 for $x = 4$. The two calculated curves correspond to logA values of 15.3 and 13.0, which in turn yield bond dissociation energies of 23 and 17 kcal/mol, respectively. These results are arrived at by iteration, with the major uncertainty in the analysis being the A factors for the reactions. The values chosen represent a range which likely includes the actual value. As a result the bond energy for the example given is taken to be 20 ± 3 kcal/mol.

Table II. Bond Dissociation Energies for Manganese
Carbonyl Ions

Species $Mn(CO)_x^+$	$D[Mn(CO)_{x-1}^+\text{-}CO]^a$
$Mn(CO)_6^+$	32 ± 5
$Mn(CO)_5^+$	16 ± 3
$Mn(CO)_4^+$	20 ± 3
$Mn(CO)_3^+$	31 ± 6
$Mn(CO)_2^+$	$<25^b$
$Mn(CO)^+$	$>7^b$

^{a}kcal/mol at 0 K
^{b}The sum of the last two bond energies is 32 kcal/mol. This is determined from the known heat of formation of $Mn(CO)_5^+$ by subtracting the first three bond energies for this species.

The two smallest ions, $Mn(CO)_2^+$, and $MnCO^+$ decompose too rapidly to be detected as metastables. This is clearly shown in Figure 9. It is of interest that this analysis works best with large ions, since they require large excess energies to decompose within the temporal constraints imposed by the experiment. This in turn gives large and easily measured kinetic energy releases.

Determination of Thermochemical Properties of Neutrals. The examples discussed above demonstrate that an analysis of kinetic energy release distributions for exothermic reactions yield accurate metal ligand bond dissociation energies. This can be extended to include neutrals as well as ions. For example, the general process 14 has been observed in the reaction

$$Co^+ + RSi(CH_3)_3 \rightarrow CoR + Si(CH_3)_3^+ \qquad (14)$$

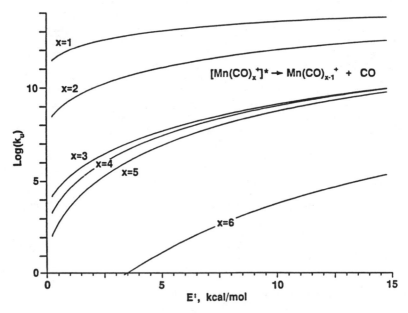

Figure 9. Unimolecular decomposition rate constants k_u for $Mn(CO)_x^+$ as a function of ion internal energy above threshold, $E^\ddagger$, calculated using RRKM theory employing Whitten-Rabinovitch state counting and bond energies listed in Table II, with $\log A = 15$.

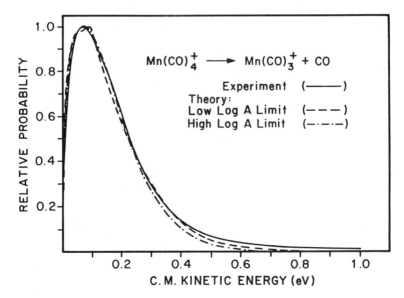

Figure 10. Experimental and theoretical kinetic energy release distributions for CO loss from $Mn(CO)_4^+$.

of atomic cobalt ions with trimethyl and tetramethyl silane, respectively, yielding Co-H and Co-CH$_3$(19). Kinetic energy release distributions for these processes and related processes are being analyzed to yield metal hydrogen and metal carbon bond dissociation energies for neutral species(Brodbelt-Lustig, J.; van Koppen, P.A.M.; Bowers, M.T., unpublished results).

Role of Excited Electronic States. Determining the electronic state of the metal ion has not been a problem for the systems studied thus far. Excited electronic states of cobalt or iron ions can be relaxed if necessary by adding a collision gas to the ion source. Co(CO)$_3$NO is known to give ~ 66% excited state Co$^+$ under electron impact conditions ($\geq$ 40 eV)(Kemper, P.R.; Bowers, M.T., unpublished results). Cobalt cyclopentadienyl dicarbonyl, however, gives 90% ground state Co$^+$ with electron impact ($\leq$ 40 eV). Greater signal intensities are obtained with the nitrosyl complex due to its higher vapor pressure and as a result it is normally employed. However, if excited states are suspected to be a problem, a comparison of reults with both compounds is made, a procedure which has been successfully applied. In the formation of a chemically activated intermediate starting with an excited state, the excess energy will most likely become available as vibrational excitation on the ground electronic surface of the system. This in turn yields higher dissociation rates (e.g. the Mn(CO)$_x^+$ system illustrated in Figure 9). A major problem arises when the dissociation rate matches the temporal requirements for detection with high sensitivity, since even a minor excited state component can then make a major contribution to the observed signal.

Coupling of Multiple Transition States by Dynamical Constraints: Effect on Kinetic Energy Release Distributions. In cases where exoergic reactions are inefficient, i.e. the reaction cross section is significantly less than the collision cross section, the rate determining transition state must be located and included in the theoretical model. Qualitatively, an initial electrostatic well followed by a tight transition state, near in energy to the reactant energy (as shown in Figure 1), can restrict the flow of reactants to products. Quantitative statistical phase space theory modeling using this simple idea has been done for a number of systems(van Koppen, P.A.M.; Bowers, M.T.; Dearden, D.V.; Beauchamp, J.L., unpublished results), including reactions of Co$^+$ with propane as noted above. Both dehydrogenation and demethanation of Co(propane)$^+$ are known to be exoergic but inefficient reactions, the sum of them occurring at 8% of the collision rate.

The reaction can be schematically written as in Equation 15.

$$M^+ + C_3H_8 \underset{k_{orb}}{\overset{k_{collision}}{\rightleftharpoons}} [M(C_3H_8)^+]^* \overset{k_\ddagger}{\rightarrow} products \quad (15)$$

Simplistically, the rate constant for product formation is given by Equation (16)

$$k_{product} = k_{collision}[k_\ddagger/(k_\ddagger + k_{orb})] \quad (16)$$

where $k_{collision}$ is the Langevin collision rate constant and k_{orb} and $k_\ddagger$ are unimolecular rate constants defined in Equation 15. The values of k_{orb} and $k_\ddagger$ are proportional to the fluxes through the orbiting and tight transition states.

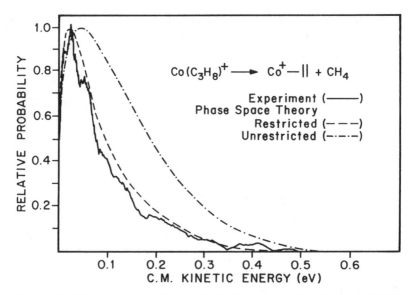

Figure 11. Kinetic energy release distribution for metastable loss of CH_4 from nascent $Co(C_3H_8)^+$ collision complexes. The "unrestricted" phase space theory curve assumes the entrance channel contains only an orbiting transition state, the exit channel has only an orbiting transition state (no reverse activation barrier), and there are no intermediate tight transition states that affect the dynamics. The "restricted" phase space theory calculation includes a tight transition state for insertion into a C-H bond located 0.08 eV below the asymptotic energy of the reactants.

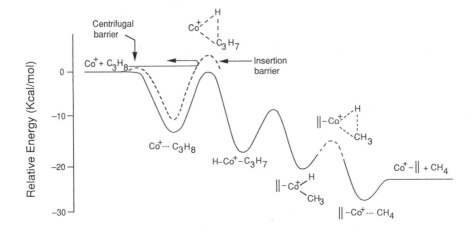

Figure 12. Schematic potential energy surface for the reaction $Co^+ + C_3H_8 \rightarrow Co(C_2H_4)^+ + CH_4$. The dashed portion schematically shows the effect of angular momentum on the initial orbiting transition state and the tight C-H insertion transition state. A collision in which the system passes over the centrifugal barrier may not result in reaction due to reflection at the second barrier.

The probability of product formation is simply $k_{products}/k_{collision}$. This ratio is calculated using statistical phase space theory as a function of the barrier height of the rate limiting transition state. A barrier 0.08 eV below the asymptotic energy of the reactants reproduces the experimental probability of product formation for Reaction 15. The kinetic energy release distribution for methane elimination is *narrower* than statistical if an orbiting transition state in the entrance channel is assumed (shown as the *unrestricted* phase space kinetic energy release distribution in Figure 11). The kinetic energy release distribution was recalculated by implementing a tight transition state for insertion into a C-H bond located 0.08 eV below the asymptotic energy of the reactants. The resulting distribution fits well with the experimental distribution and is shown as the *restricted* phase space calculations in Figure 11.

If the microcanonical (E,J) values of the $k_{products}/k_{collision}$ calculation are examined it becomes clear that the principal effect of the C-H insertion transition state is to bias against formation of products from collisions with high J values. This is a reasonable result since angular momentum increases the energy of the tight transition state more than it does the loose orbiting transition state. This is illustrated in Figure 12. Loss of the higher J portion of the (E,J) distribution results in products with less kinetic energy than would have been expected if the full (E,J) distribution were operative. This example is presented to express a cautionary note in analyzing kinetic energy release distributions. If the dynamical constraints had not been considered in fitting phase space theory to the data in Figure 11, an erroneously low bond dissociation energy would have resulted. Inefficient reactions which proceed at only a fraction of the collision rate can signal the possible importance of these factors.

Summary and Prognosis

One of the definitions which Webster gives for the word prognosis is "forecast of the course of a disease." In this case not only is the patient doing well but appears to be headed for a long and fruitful life. The VG ZAB-2F double focusing mass spectrometer and several of its cousins produced by other manufacturers are available in many laboratories around the world. Measurements of the type described in this review can be made in a relatively straightforward manner, and other groups have already started reporting kinetic energy release distributions for organometallic reactions(20). The basic computer programs for carrying out statistical phase space calculations are available from the Quantum Chemistry Program Exchange (Chesnavich, W. J., QCPE, submitted for publication). As a result, this experimental methodology can be widely applied in many laboratories to determine heats of formation and bond dissociation energies for reactive intermediates formed in the tolerant environment of the mass spectrometer. In view of the results presented in this review, it is desirable to have additional information available, such as reaction efficiencies or knowledge relating to the reverse reaction. When carefully analyzed, these experiments provide bond dissociation energies with accuracy comparable to other thermochemical kinetic methods.

Acknowledgments

This work has been funded by the National Science Foundation through Grant No. CHE87-11567 (JLB) and CHE85-12711 (MTB) and a Shell Foundation graduate fellowship (DVD). In addition we thank the donors of the Petroleum Research Fund, administered by the American Chemical Society, for additional support. This manuscript is contribution number 8069 from the Division of Chemistry and Chemical Engineering at the California Institute of Technology.

Literature Cited

1. Muller, J.; Goll, W. Chem. Ber. 1973, 106, 1129.
2. This field is well represented in the book Gas Phase Inorganic Chemistry; Russell, D. H., Ed.; Plenum: New York, 1989.
3. Houriet, R.; Halle, L. F.; Beauchamp, J.L.; Organometallics, 1983, 2, 1818.
4. Beauchamp, J. L. ACS Symp. Ser. 1987, 333, 11.
5. Armentrout, P. B.; Beauchamp, J. L. Acc. Chem. Res. 1989, 22, 315.
6. Hanratty, M. A.; Paulson, C. M.; Beauchamp, J. L. J. Am. Chem. Soc. 1985, 107, 5074.
7. Hanratty, M.; Beauchamp, J. L.; Illies, A. J.; Bowers, M. T. J. Am. Chem. Soc. 1985, 107, 1788.
8. Hanratty, M. A.; Beauchamp, J. L.; Illies, A. J.; van Kopppen, P. A. M.; Bowers, M. T. J. Am. Chem. Soc. 1988, 110, 1.
9. Ervin, K. M.; Armentrout, P. B. J. Chem. Phys. 1985, 83, 166.
10. Jarrold, M. F.; Illies, A. J.; Bowers, M. T. Chem. Phys. 1982, 91, 2573.
11. Kirchner, J. J.; Bowers, M. T. J. Phys. Chem. 1987, 91, 2573.
12. Morgan, R. P.; Beynon, J. H.; Bateman, R. H.; Green, B. N. Int. J. Mass Spectrom. Ion Phys. 1978, 28, 171.
13. Chesnavich, W. J.; Bowers, M. T. J. Am. Chem. Soc. 1977, 99, 1705.
14. Chesnavich, W. J.; Bass, L.; Su, T.; Bowers, M. T. J. Chem. Phys. 1981, 74, 2228.
15. van Koppen, P. A. M.; Jacobson, D. B.; Illies, A.; Bowers, M. T.; Hanratty, M.; Beauchamp J. L. J. Am. Chem. Soc. 1989, 111, 1991.
16. Georgiadis, R.; Fisher, E.R.; Armentrout, P.B. J. Am. Chem. Soc. 1989, 111, 4251.
17. Beauchamp, J .L.; Stevens, A. E.; Corderman, R. R. Pure and Appl. Chem. 1979, 51, 967.
18. Dearden, D. V. Hayashibara, K.; Beauchamp, J. L.; Kirchner, J. J.; van Koppen, P. A. M.; Bowers, M. T. J. Am. Chem. Soc. 1989, 111, 2401.
19. Kang, H.; Jacobson, D. B.; Shin, S. K.; Beauchamp, J. L.; Bowers, M. T. J. Am. Chem. Soc. 1986, 108, 5668.
20. Schulze, C.; Schwarz, H.; Int. J. Mass Spectrom. Ion Proc. 1989, 88, 291.

RECEIVED December 15, 1989

Chapter 4

Methods for Determining Metal–Ligand and Metal–Metal Bond Energies Using Fourier Transform Mass Spectrometry

Ben S. Freiser

Department of Chemistry, Purdue University, West Lafayette, IN 47905

In this review we discuss five techniques involving
Fourier transform mass spectrometry (FTMS) for
determining qualitative and quantitative metal ion-
ligand bond energies. These include (i) exothermic
ion-molecule reactions, (ii) equilibrium measurements,
(iii) competitive collision-induced dissociation, (iv)
endothermic ion-molecule reactions, and (v)
photodissociation. A key advantage of the FTMS
methodology is its ion and neutral manipulation
capabilities which permit the formation and study of a
limitless number of interesting metal-ion systems.

The combination of laser ionization and Fourier transform mass
spectrometry (FTMS) has proved to be ideally suited for the study of
gas-phase ion-molecule reactions involving metal ions (1-7). The
laser source permits the generation of virtually any metal ion in the
periodic table from a suitable metal target (8). The FTMS (9-14)
stores these ions in an "electro-magnetic bottle" for times typically
on the order of msec to sec (hours are possible) permitting the study
of their chemistry and photochemistry. These studies are further
facilitated by the unusual ion and neutral manipulation capabilities
of the FTMS which permit complex multistep processes to be monitored
in an MS^n fashion (1-4). These capabilities have made laser
ionization-FTMS a prominent method in what has been a rapidly growing
arsenal of techniques for studying gas-phase transition-metal ion
species.
   The major thrust of these studies has involved determining
reaction mechanisms, kinetics, and thermochemistry (15-30). In
analogy to solution studies, deriving a reaction mechanism requires
knowledge of product structure(s) and of the intermediate and overall
thermochemistry (i.e. are all of the steps proposed in the mechanism
thermodynamically feasible?). While a disadvantage of the gas phase
is that a low concentration of ions, $~10^{-17}$ $\underline{M}$, makes the use of

0097–6156/90/0428–0055$06.00/0
© 1990 American Chemical Society

conventional techniques all but impossible, a new set of mass spectrometric techniques have been developed which can readily be applied.

Among the first and most powerful of these methods used for obtaining metal-ion bond energies is the ion-beam apparatus developed by Beauchamp and Armentrout (18,19) and discussed in detail elsewhere in this book. Our laboratory has been involved in developing a complementary array of techniques for the FTMS which will be illustrated in this chapter. These include (i) exothermic ion-molecule reactions, (ii) equilibrium measurements, (iii) competitive collision-induced dissociation (CID), (iv) endothermic ion-molecule reactions, and (v) photodissociation. While much of this methodology has its roots in ion cyclotron resonance (ICR) spectroscopy, the precursor instrument to the FTMS, the Fourier method greatly enhances the number of interesting systems which can be investigated.

Exothermic Ion-Molecule Reactions.

Ion-molecule reactions generally occur with little or no activation barrier because of the long range ion-dipole and ion-induced dipole attraction. One consequence of this long range interaction is that ion-molecule reactions are several orders of magnitude faster than reactions in solution. Another is that observation of a reaction proceeding at roughly 1/10 or greater the diffusion-controlled collision rate is likely to be exothermic. Reactions which are endothermic by as much as 3 kcal/mol can be observed and recognized by their characteristically slow rates (e.g. less than about 1% of the collision rate). Thus, providing that the reactant ion is in its ground state (thermalized), the observation of a reaction yields limits on bond dissociation energies. The absence of a reaction, however, does not necessarily imply an endothermic process since other pathways may be kinetically favored. Spin restrictions may also prevent an otherwise exothermic reaction from occurring, which is the case for the lack of reaction between M^+ and N_2O to form MO^+ for M=Co, Ni, and Mn (31).

We have recently reported studies on M^+-benzyne (M = Fe and Sc) generated from reactions 1 and 2 (32,33). Observation of these

$$Fe^+ + \bigcirc\!\!-Cl \longrightarrow FeC_6H_4^+ + HCl \qquad (1)$$

$$Sc^+ + \bigcirc \longrightarrow ScC_6H_4^+ + H_2 \qquad (2)$$

facile reactions imply $D°(Fe^+-C_6H_4) \geq 66$ kcal/mol and $D°(Sc^+-C_6H_4) \geq 80$ kcal/mol. Interestingly, unlike Fe^+-benzyne, H_2 is observed to hydrogenate $ScC_6H_4^+$ to form presumably Sc^+-benzene, reaction 3. This reaction yields an upper limit on the Sc^+-benzyne bond strength of $\leq$

$$Sc^+ - \bigcirc\!\!| \quad + H_2 \longrightarrow Sc^+ \bigcirc \qquad (3)$$

133 ± 5 kcal/mol obtained using a previously determined value of $D°(Sc^+-C_6H_6) = 53 \pm 5$ kcal/mol. This type of bimolecular association reaction is extremely rare for a mononuclear metal center in the gas phase, with the only other reported example being the direct hydrogenation of $RhC_7H_6^+$ (34). As in that case, the Sc^+-benzene product ion in reaction 3 must be stabilized by an IR radiative relaxation process.

Ligand exchange (displacement) reactions are a convenient means of determining both relative and absolute metal-ligand bond energies. For example, we have reported on a wide variety of MFe^+ species including M = Co (35), Cu (36), Nb (37), Rh (38), La (38), V (39), and Sc (40). In this series the bond energies have ranged from a low of $D°(Sc^+-Fe)$ = 48 ± 5 kcal/mol to a high of $D°(V^+-Fe)$ = 75 ± 5 kcal/mol. Currently underway is a study of $MgFe^+$ where the effects of a non-transition metal is being investigated. Observation of the exchange reaction 4 implies $D°(Fe^+-C_2H_4)$ = 34 ± 5 kcal/mol (41) > $D°(Fe^+-Mg)$ which in turn yields $D°(Mg^+-Fe)$ < 29 ± 5 kcal/mol. This

$$MgFe^+ + C_2H_4 \longrightarrow FeC_2H_4^+ + Mg \qquad (4)$$

value is considerably lower than for the transition-metal dimer ions and suggests that the bonding between Mg and Fe is largely electrostatic.

Bracketing methods involving exothermic proton transfer have yielded both cationic and neutral metal-ligand bond energies. For example, $MOH^+(M$ = Fe and Co) were reacted with a series of reference bases, reaction 5 (42). For $CoOH^+$, proton transfer was observed with

$$MOH^+ + B \longrightarrow MO + BH^+ \text{ (B= various bases)} \qquad (5)$$

pyridine and stronger bases but not observed with n-propylamine or weaker bases indicating PA(pyridine) > PA(CoO) > PA(n-propylamine). Using this result in equations 6 and 7 yields $D°(Co^+-OH)$ = 71 ± 6 kcal/mol in good agreement with $D°(Co^+-OH)$ = 71 ± 3 kcal/mol obtained from photodissociation. In this case $\Delta H_f(MO)$ was known while

$$\Delta H_f(MOH^+) = \Delta H_f(MO) + \Delta H_f(H^+) - PA(MO) \qquad (6)$$

$$D°(M^+-OH) = \Delta H_f(M^+) + \Delta H_f(OH) - \Delta H_f(MOH^+) \qquad (7)$$

$\Delta H_f(MOH^+)$ was determined. In some instances ΔH_f of the ion is known and the corresponding neutral is not. Thus, proton bracketing experiments can also be used to determine neutral metal-ligand bond energies. Reaction 8, for example, was used to determine PA(4-vinyl-

$$MCH_3^+ + B \longrightarrow MCH_2 + BH^+ \qquad (8)$$

pyridine) > PA(FeCH₂) > PA(pyridine) (43). Using a previously determined value of $D°(Fe^+-CH_3)$ = 65 ± 5 kcal/mol (5), $D°(Fe-CH_2)$ = 87 ± 7 kcal/mol was obtained from equations 9 and 10.

$$\Delta H_f(FeCH_2) = PA(FeCH_2) + \Delta H_f(FeCH_3^+) - \Delta H_f(H^+) \qquad (9)$$

$$D°(Fe-CH_2) = \Delta H_f(Fe) + \Delta H_f(CH_2) - \Delta H_f(FeCH_2) \qquad (10)$$

Multiply charged metal ions have not received much attention due in part to the belief that rapid charge exchange would occur exclusively. Taking a lead from a study by Tonkyn and Weisshaar on Ti^{2+} (44), we have begun to study a series of doubly charged early transition-metal ions. Nb^{2+}(2nd IP(Nb) = 14.3 eV) is observed to react with methane (IP = 12.6 eV) not only by charge transfer, reaction 11, but by carbene and hydride abstraction as well,

$$Nb^{2+} + CH_4 \xrightarrow{\ 41\% \ } Nb^+ + CH_4^+ \tag{11}$$

$$\xrightarrow{\ 52\% \ } NbCH_2^{2+} + H_2 \tag{12}$$

$$\xrightarrow{\ 7\% \ } NbH^+ + CH_3^+ \tag{13}$$

reactions 12 and 13, respectively (45). Observation of $NbCH_2^{2+}$ from methane implies $D°(Nb^{2+}-CH_2) > 112$ kcal/mol. Reactions of $NbCH_2^{2+}$ showed this species to be a good proton donor and yielded $PA(CH_4) < PA(NbCH^+) < PA(CO)$ from which $PA(NbCH^+) = 137 \pm 7$ kcal/mol was assigned. Using this result in a thermochemical cycle yielded $D°(Nb^{2+}-CH_2) = 197 \pm 10$ kcal/mol and $D°(Nb^+-CH_2^+) = 107 \pm 10$ kcal/mol. For comparison, $D°(Nb^+-CH_2) = 109 \pm 7$ kcal/mol has recently been reported (5). Similarly, we have now observed a number of thermodynamically stable metal dication species. Dication species which are kinetically, but not thermodynamically stable can also be observed for seconds in the FTMS. A particularly fascinating example of this is $LaFe^{2+}$ which has recently been shown in our laboratory to be stable indefinitely (46).

Finally, the proton affinities of several atomic metal anions, M^- (M =V, Cr, Fe, Co, Mo, and W), have been determined by bracketing methods (47). Combining these data with measured electron affinities of the metals yielded homolytic bond energies for the neutral hydrides, $D°(M-H)$. The monohydride bond energies compare favorably with other experimental and theoretical data in the literature and were used to derive additional thermodynamic properties for metal hydride ions and neutrals.

Equilibrium Measurements

One of the early successes of ion cyclotron resonance (ICR) spectroscopy was its application to obtaining accurate proton affinity information by measuring equilibrium constants for reaction 14 (48,49). The equilibrium constant K can be determined both from the steady state ratio of BH^+/AH^+, which becomes constant beyond some trapping time (the ratio of the neutrals A/B is known from the partial pressures and remains constant), and by measuring the forward

$$AH^+ + B \; \underset{k_r}{\overset{k_f}{\rightleftharpoons}} \; BH^+ + A \tag{14}$$

and reverse rate constants with $K = k_f/k_r$. Similarly, reaction 15

$$ML^+ + L' \; \rightleftharpoons \; ML'^+ + L \tag{15}$$

has been monitored using ICR, yielding accurate relative metal-ligand binding energies for $M^+ = Li^+(50)$, $Al^+(51)$, $Mn^+(52)$, $Cu^+(53)$, and $CpNi^+$ (54). Most recently, we extended this methodology to FTMS to measure the binding energies of Mg^+ to a series of alcohols, aldehydes, ketones, and ethers (55). Some typical data are shown in Figure 1 which displays the variation of ion abundance with time for

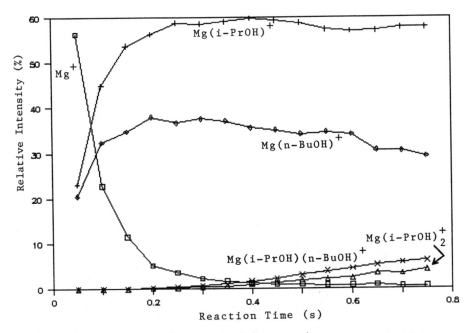

Figure 1. Variation of ion abundances with time for Mg^+ with a 3:1 mixture of i-PrOH and n-BuOH at a total pressure of 5×10^{-7} torr. (Reprinted from reference 55. Copyright 1988 American Chemical Society.)

a 3:1 mixture of i-PrOH and n-BuOH. Following initial reaction of Mg^+ with the two alcohols, the ratio of $Mg(i-PrOH)^+$ and $Mg(n-BuOH)^+$ becomes constant as reaction 15 approaches equilibrium. At longer times the abundance of secondary product ions increases appreciably, moving the system out of equilibrium. The ratio of ion abundances at equilibrium, together with the measured partial pressures of i-PrOH and n-BuOH, give an equilibrium constant in this case of K = 1.5. The average of three determinations, obtained for three different i-PrOH/n-BuOH partial pressure ratios, was K = 1.7 ± 0.2 yielding a free-energy difference of ΔG = -0.30 ± 0.08 kcal/mol. It was argued that the entropy changes were negligible and, therefore, that ΔH = -0.30 ± .08 kcal/mol = D°(Mg^+-iPrOH) - D°(Mg^+-nBuOH). Table I lists the data obtained from this study. Agreement between the various pairs of bases is generally better than 0.2 kcal/mol, and the precision of each measurement is ± 0.1 kcal/mol.

Competitive Collision-Induced Dissociation (CID)

A generalized method for determining both relative and absolute bond strengths to an ion center by CID has been developed by Cooks and co-workers. In extensive studies on X = H^+ (56) and Ag^+ (57), they demonstrated that under appropriate conditions, if AX^+ is more intense than BX^+ in the competitive CID processes 16 and 17, then D°(X^+-A) > D°(X^+-B). This relative bond energy information can be

$$AX^+B \longrightarrow \begin{cases} AX^+ + B & (16) \\ BX^+ + A & (17) \end{cases}$$

used to determine absolute bond strengths by using a series of reference Lewis bases and bracketing the value of interest. Cooks and co-workers pointed out that this method is particularly useful for obtaining information on nonvolatile compounds where insufficient vapor pressure would prevent standard ion-molecule reaction bracketing and equilibrium measurements. We have found this method to be simple and reliable and, therefore, use it extensively. As an example, we have determined several M^+-benzene bond strengths by photodissociation (58,59), as discussed below, and benzene has become a convenient reference Lewis base for competitive CID studies involving metal dimer species. A variety of MFe(benzene)$^+$ ions have been generated and their subsequent CID reactions 18 - 20 have been monitored in order to obtain information on the metal-metal bond

$$MFe(benzene)^+ \longrightarrow \begin{cases} MFe^+ + benzene & (18) \\ M(benzene)^+ + Fe & (19) \\ Fe(benzene)^+ + M & (20) \end{cases}$$

strengths (59). CID on ScFe(benzene)$^+$, for example, yields approximately equal amounts of ScFe$^+$ and Sc$^+$-benzene (Figure 2), with Sc$^+$-benzene favored at low CID energies. In addition, no production of Fe$^+$-benzene is observed. These results imply D°(Sc$^+$-Fe) ≤ D°(Sc$^+$-

Table I.   Relative and Absolute Gas Phase Ligand Binding
Energies to Mg^+ for Various Organic Molecules[a]

Ligand (L)	Measured $\Delta G_{exchange}$[b]	$D°(Mg^+\text{-}L)$ rel.[c]	abs.[d]
MeCOEt		7.64	68
	1.24		
Me_2CO		6.40	67
	1.05 1.20		
THF[e]		5.41	66
	0.23		
Et_2O		5.16	66
	0.52 1.61		
n-PrCHO		4.60	66
	0.86 1.45		
n-BuOH		3.87	65
	1.76 0.30		
EtCHO		3.67	65
	0.72 0.93		
i-PrOH		3.59	65
n-PrOH		2.95	64
	0.68 1.17		
MeCHO		2.28	63
	0.52		
EtOH		1.77	63
	1.77		
MeOH		0.00	61

[a]All data are in kcal/mol.   [b]For the equilibrium reaction 15 in text.
[c]Values (±0.1 kcal/mol) are relative to $D°(Mg^+\text{-MeOH}) = 0.00$ kcal/mol.
[d]Absolute values (±5 kcal/mol) assigned from photodissociation
results, see reference 55.   [e]Tetrahydrofuran.

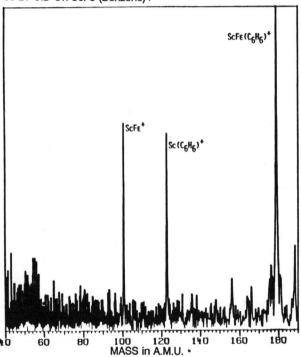

Figure 2. Collision-induced dissociation spectrum of
ScFe(benzene)$^+$ at 50 eV collision energy.

benzene) and $D°(Fe^+-Sc) > D°(Fe^+-benzene)$ which are in accordance with the photodissociation results of $D°(Sc^+-Fe) = 48 \pm 5$ kcal/mol (60), $D°(Sc^+-benzene) = 53 \pm 5$ kcal/mol (61), $D°(Fe^+-Sc) = 79 \pm 5$ kcal/mol (60), and $D°(Fe^+-benzene) = 55 \pm 5$ kcal/mol (58). Note that observation of $Sc(benzene)^+$ and not $Fe(benzene)^+$, however, does not imply $D°(Sc^+-benzene) > D°(Fe^+-benzene)$. In contrast to $ScFe(benzene)^+$, CID of $VFe(benzene)^+$ yields predominantly VFe^+ in accordance with $D°(V^+-Fe) = 75 \pm 5$ kcal/mol (39), $D°(V-Fe^+) = 101 \pm 5$ kcal/mol (39), $D°(V^+-benzene) = 55 \pm 5$ kcal/mol (58), and $D°(Fe^+-benzene) = 55 \pm 5$ kcal/mol (58). A variety of MFe^+-L (L = ligand) species can be generated in situ in the FTMS to try to bracket the values of $D°(M^+-Fe)$.

Endothermic Ion-Molecule Reactions

The kinetic energy of an ion in the trapping cell can be increased by applying a radio frequency (rf) to the transmitter plates at the cyclotron frequency of the ion. The maximum kinetic energy of an ion which can be achieved is proportional to the magnetic field squared, to the radius of the cell squared (for a cubic cell the radius would be ½ the x,y, or z dimensions), and to the inverse of the mass of the ion (62). Variation of the ion kinetic energy is achieved by adjusting the amplitude and/or the pulse duration of the rf excitation and can be estimated using a standard equation (62). As the kinetic energy of the ion increases, its radius also increases and, thus, the maximum kinetic energy is limited by a collision of the ion into one of the cell plates where it is neutralized. The radio frequency excitation is used for the collision-induced dissociation experiments described above in which an ion of interest is accelerated into a collision gas and the fragments detected. In exactly the same manner, application of rf energy is also a convenient means of adjusting the translational energy of an ion in order to study its endothermic reactions with reactive target gases. By monitoring the threshold kinetic energy required to observe the onset of an otherwise endothermic process, thermochemistry is derived which yields bond energy information. This method is analogous to the ion-beam experiment of Beauchamp and Armentrout and co-workers (18,19) which has been among the most useful in obtaining such data. In our initial study, we chose simple systems which could be directly compared to the ion-beam results (61). Thresholds for endothermic reactions of laser-generated Co^+ with cyclopropane and ethene, for example, yielded $D°(Co^+-CH_2) = 81 \pm 7$ kcal/mol and 75 ± 18 kcal/mol, respectively, compared to $D°(Co^+-CH_2) = 85 \pm 7$ kcal/mol determined from ion-beam experiments. A value of $D°(Co^+-CH_3) = 46 \pm 14$ kcal/mol by FTMS was somewhat lower than the ion-beam value of 61 ± 4 kcal/mol. Interestingly, a value of $D°(Fe_2^+-H) = 52 \pm 16$ kcal/mol was determined from the endothermic reaction of Fe_2^+, generated in situ, and ethane.

Endothermic reactions can also be used to synthesize species which are not available by exothermic reactions. For example, reactions 21 and 22 were used to generate ScH^+ and $ScCH_3^+$ in an attempt to bracket

$$Sc^+ + C_2H_6 \longrightarrow \begin{array}{l} \longrightarrow ScH^+ + C_2H_5 \quad\quad (21) \\ \longrightarrow ScCH_3^+ + CH_3 \quad\quad (22) \end{array}$$

the Sc^+-benzene bond strength (61).   In addition $CrCH_2^+$ has been generated by driving the endothermic reaction 23 for subsequent photodissociation studies (59).

$$Cr^+ + \triangle \longrightarrow CrCH_2^+ + C_2H_4 \qquad (23)$$

Kang and Beauchamp have shown that the kinetic energy of laser-desorbed metal ions increases with increasing laser power density (63).   Similarly, endothermic reactions can be observed in the FTMS without applying rf energy if the laser-desorbed metal ions are nonthermal (61).   Such reactions can mistakenly be interpreted as being exothermic yielding incorrect information.   Using laser powers just above the threshold for laser desorption, together with a collisional cooling period, however, has been found to be quite effective in circumventing this potential problem.

From these initial studies it was concluded that, when applicable, the ion-beam experiment is superior to that involving FTMS for determining accurate endothermic reaction thresholds.   The error limits of the FTMS determinations can be considerably larger due to the inability to deconvolute the effects of multiple collisions and the thermal motion of the target gas and due to the spread of ion energies arising from the phase difference between the excitation energy and the ions, an effect which increases with kinetic energy. Nevertheless, the ease of generating a wider variety of complex reactant ions whose fundamental endothermic reactivity is of interest is certainly a plus for FTMS.

Photodissociation

Photodissociation, process 24, has proven to be one of the most

$$AB^+ + h\upsilon \longrightarrow A^+ + B \qquad (24)$$

convenient and powerful methods used with FTMS to obtain both ion structure and bond energy information (47, 58, 60).   This information is a consequence of the three criteria required to observe the photodissociation process 24:   (a) the ion must first absorb a photon, (b) the photon energy must exceed the enthalpy for process 24, and (c) the quantum yield for photodissociation must be greater than zero.   If criteria b were the only consideration, the longest wavelength (lowest energy) at which photodissociation is observed would give a direct measure of the enthalpy of process 24.   Criteria a, however, makes this energy strictly a lower limit for the enthalpy since photons with sufficient energy but which are not absorbed are clearly ineffective in promoting photodissociation.

Extensive studies in our laboratory have shown, not surprisingly, that the metal center is an excellent chromophore which absorbs over a broad wavelength region due to the high density of low-lying electronic states (42, 58, 60).   Thus, in many cases the photodissociation threshold of metal-containing ions is determined by the thermodynamics and not by the absorption characteristics of the ion.   This is in contrast to most organic ions in which the first allowed excited state lies above the dissociation energy (64,65). One must still always use caution in interpreting a photodissociation threshold.   The presence of nonthermal ions and the possibility of multiphoton processes are also potential problems in obtaining meaningful thresholds. Thus, the results should preferably be supported by one or more of the other methods described.

These studies have focussed on five general categories of metal containing ions:

1. ML^+ (L = O, S, CH_2, CH_3, C_4H_6, C_4H_8, C_6H_6, etc.)
2. LM^+L' (both L = L' and L ≠ L': L = C_2H_4, C_3H_6, C_4H_8, C_6H_6, etc.)
3. MFe^+ (M = Sc, Ti, V, Cr, Fe, Co, Ni, Cu, Nb, Ta, Mg)
4. $MFeC_6H_6^+$ (M = Nb, V, Co, Sc)
5. La^{2+} - L (L = C_2H_2, C_2H_4, C_3H_6)

Table II is a representative sample of the metal-ligand bond energies obtained to date by photodissociation.

Table II. Bond Dissociation Energies Obtained From Photodissociation
Threshold Measurements

A^+ - B	Photodissociation $D°(A^+ - B)$	Literature kcal/mol
Co^+-C	90±7 (a)	98 (b)
Co^+-CH	100±7 (a)	
Co^+-CH_2	84±5 (a)	85±7 (d)
Co^+-CH_3	57±7 (c)	61±4 (d,e,f)
Cr^+-CH_2	72±5 (g)	
Fe^+-C	94±7 (a)	89 (b)
Fe^+-CH	101±7 (a)	115±20 (b)
Fe^+-CH_2	82±7 (a)	96±5 (d)
Fe^+-CH_3	65±5 (c)	69±5 (d,e)
La^+-C	102±8 (h)	
La^+-CH	125±8 (h)	
La^+-CH_2	106±5 (h)	
Nb^+-C	> 138 (h)	
Nb^+-CH	145±8 (h)	
Nb^+-CH_2	109±7 (h)	112≤ (i)
Rh^+-C	> 120 (h)	164±16 (j)
Rh^+-CH	102±7 (h)	
Rh^+-CH_2	91±5 (h)	94±5 (k)
Co^+-OH	71±3 (l)	
Co^+-S	62±5 (c)	
Fe^+-butadiene	48±5 (c)	
Fe^+-c-C_5H_6	55±5 (m)	
$FeC_5H_5^+$-H	46±5 (m)	
Fe^+-NH	61±5 (n)	
Fe^+-O	68±5 (c)	68±3 (d)
Fe^+-OH	73±3 (l)	76±5 (o)
Fe^+-S	61±6 (p)	65±5 (c)
Fe^+-S_2	48±5 (p)	
FeS^+-S_2	49±5 (p)	
FeS_2^+-S_2	49±5 (p)	
FeS_3^+-S_2	43±5 (p)	
FeS_4^+-S_2	38±5 (p)	
Ni^+-S	60±5 (c)	
Ni^+-$2C_2H_4$	80±5 (c)	90 (q)
Co^+-C_6H_6	68±5 (c)	66±7 (r)
Fe^+-C_6H_6	55±5 (c)	58±5 (s)

(Continued on next page)

Table II. (*Continued*) Bond Dissociation Energies Obtained From Photodissociation
Threshold Measurements

A^+ - B	Photodissociation $D°(A^+ - B)$	Literature kcal/mol
$La^+-C_6H_6$	50±3 (g)	
$LaC_6H_6^+-C_6H_6$	<43 (g)	
$Nb^+-C_6H_6$	64±3 (g)	
$NbC_6H_6^+-C_6H_6$	49±3 (g)	
$Sc^+-C_6H_6$	50±3 (g)	
$ScC_6H_6^+-C_6H_6$	50±3 (g)	
$V^+-C_6H_6$	62±5 (c)	
$VC_6H_6^+-C_6H_6$	57±5 (c)	
$Y^+-C_6H_6$	50±3 (g)	
$YC_6H_6^+-C_6H_6$	47±3 (g)	
Co^+-Fe	62±5 (t)	66±7 (r)
Cr^+-Fe	50±7 (t)	
Cu^+-Fe	53±7 (t)	
Fe^+-Fe	62±5 (t)	56< x <67 (u)
La^+-Fe	48±5 (v)	
Nb^+-Fe	68±5 (t)	
Ni^+-Fe	64±5 (t)	
Sc^+-Fe	48±5 (t)	
Ta^+-Fe	72±5 (t)	
Ti^+-Fe	60±6 (t)	
V^+-Fe	75±5 (w)	
$CoFe^+-C_6H_6$	48±5 (g)	
$NbFe^+-C_6H_6$	48±5 (g)	
$ScFe^+-C_6H_6$	47±4 (g)	
$VFe^+-C_6H_6$	50±7 (g)	

(a) Hettich, R.L.; Freiser, B.S. J. Am. Chem. Soc. 1986, 106, 2537.
(b) Beauchamp, J.L., private communication. (c) Reference 58. (d)
Armentrout, P.B.; Halle, L.F.; Beauchamp, J.L. J. Am. Chem. Soc.
1981, 103, 6501. (e) Halle, L.F.; Armentrout, P.B.; Beauchamp, J.L.
Organometallics 1982, 1, 963. (f) Armentrout, P.B.; Beauchamp, J.L.
J. Am. Chem. Soc. 1981, 103, 784. (g) Reference 59. (h) Hettich,
R.L.; Freiser, B.S. J. Am. Chem. Soc. 1987, 109, 3543. (i) Reference
45. (j) Jacobson, D.B.; Byrd, G.D.; Freiser, B.S. Inorg. Chem. 1984,
23, 553. (k) Jacobson, D.B.; Freiser, B.S. J. Am. Chem. Soc. 1985,
107, 5870. (l) Reference 42. (m) Huang, Y.; Freiser, B.S. J. Am.
Chem. Soc., submitted. (n) Buckner, S.W.; Gord, J.R.; Freiser, B.S.
J. Am. Chem. Soc. 1988, 110, 6606. (o) Murad, E. J. Chem. Phys.
1980, 73, 1381. (p) MacMahon, T.J.; Jackson, T.C.; Freiser, B.S. J.
Am. Chem. Soc. 1989, 111, 421. (q) Hanratty, M.A.; Beauchamp, J.L.;
Andreas, J.I.; van Koppen, P.; Bowers, M.T. J. Am. Chem. Soc. 1988,
110, 1. (r) Jacobson, D.B.; Freiser, B.S. J. Am. Chem. Soc. 1984,
106, 4623. (s) Jacobson, D.B.; Freiser, B.S. J. Am. Chem. Soc. 1984,
106, 3900. (t) Reference 60. (u) Brucat, P.J.; Zheng, L.-S.;
Pettiette, C.L.; Yang, S.; Smalley, R.E. J. Chem. Phys. 1986, 84,
3078. (v) Huang, Y.; Freiser, B.S. J. Am. Chem. Soc. 1988, 110, 387.
(w) Reference 39.

Conclusion

Less than 10 years ago, there was a dirth of information on gas-phase metal ion-ligand bond energies. Starting with the ion-beam technique, a concerted effort by many laboratories using a growing number of techniques has allowed considerable progress to be made. It is evident in the literature, however, that many of these values are not "carved in stone" and that further refinement will be forthcoming. The development of new methods is important, therefore, in checking these numbers to obtain a consensus and, therefore, confidence in these important thermochemical values. The FTMS, as shown above, combines a number of diverse methods for determining bond energies with an almost unique ability for generating and studying novel metal ion systems.

Acknowledgments are made to the Division of Chemical Sciences in the Office of Basic Energy Sciences in the United States Department of Energy (DE-FG02-87ER13766) for supporting this research and to the National Science Foundation (CHE-8612234) for providing funds for the advancement of FTMS methodology.

Literature Cited
1.   Freiser, B.S. Talanta 1985, 32, 697.
2.   Freiser, B.S. Anal. Chim. Acta 1985, 178, 137.
3.   Freiser, B.S. Techniques for the Study of Gas-Phase Ion Molecule Reactions, Farrer, J.M.; Saunders, Jr., W.H. Ed.; Wiley-Interscience, New York, 1988, p.61.
4.   Freiser, B.S. Chemtracts, 1989, 1, 65.
5.   Buckner, S.W.; Freiser, B.S. Polyhedron 1988, 7, 1583.
6.   Buckner, S.W.; Freiser, B.S. Gas Phase Inorganic Chemistry; Russell, D.H., Ed.; Plenum, New York, 1989, p. 279.
7.   Weller, R.R.; MacMahon, T.J.; Freiser, B.S. Lasers in Mass Spectrometry, Lubman, D.M. Ed.; in press.
8.   Cody, R.B.; Burnier, R.C.; Reents, Jr. W.D.; Carlin, T.J.; McCrery, D.A.; Lengel, R.K.; Freiser, B.S. Int. J. Mass Spec. Ion Phys., 1980 33, 37.
9.   Fourier Transform Mass Spectrometry: Evolution, Innovation, and Application; Buchanon, M.V., Ed.; American Chemical Society Symposium Series, V. 359, Washington, D.C., 1987.
10. Marshall, A.G. Acc. Chem. Res. 1985 18, 316.
11. Laude, Jr., D.A.; Johlman, C.L.; Brown, R.S.; Weil, D.A.; Wilkins, C.L. Mass Spec. Rev. 1986, 5, 107.
12. Gross, M.L., Rempel, D.L. Science 1984, 266, 261.
13. Russell, D.H. Mass Spec. Rev. 1986, 5, 167.
14. Two journal issues have been dedicated to FTMS: (a) Anal. Chim. Acta. 1985, 178, No. 1; (b) Int. J. Mass Spectrom. Ion Proc. 1986, 72, Nos. 1 and 2.
15. Allison, J. Progress in Inorganic Chemistry; S.J. Lippard, Ed.; Wiley-Interscience, New York, Vol. 34, 1986, p. 628.
16. Allison, J.; Freas, R.B.; Ridge, D.P. J. Am. Chem. Soc. 1979, 101, 1332.
17. Tsarbopoulos, A.; Allison, J. Organometallics 1984, 3, 86.

18. Houriet, R.; Halle, L.F.; Beauchamp, J.L. Organometallics 1983, 2, 1818.
19. Elkind, J.L.; Armentrout, P.B. J. Chem. Phys. 1986, 84, 4862.
20. Peake, D.A.; Gross, M.L. J. Am. Chem. Soc. 1987, 109, 600.
21. Weil, D.A.; Wilkins, C.L. J. Am. Chem. Soc. 1986, 108, 2765.
22. Tonkyn, R.; Weisshaar, J.C. J. Phys. Chem. 1986, 90, 2305.
23. Wang, D.; Squires, R.R. Organometallics 1985, 6, 697.
24. Shulze, C.; Schwarz, H.; Peake, D.A.; Gross, M.L. J. Am. Chem. Soc. 1987, 109, 2368.
25. Burnier, R.C.; Byrd, G.D.; Freiser, B.S. Anal. Chem., 1980, 52, 1641.
26. Jacobson, D.B.; Freiser, B.S. J. Am. Chem. Soc., 1983, 105, 736.
27. Jacobson, D.B.; Freiser, B.S. J. Am. Chem. Soc., 1983, 105, 5197.
28. Jacobson, D.B.; Freiser, B.S. J. Am. Chem. Soc., 1983, 105, 7484.
29. Carlin, T.J.; Sallans, L.; Cassady, C.J.; Jacobson, D.B.; Freiser, B.S. J. Am. Chem. Soc., 1983, 105, 6320.
30. Jacobson, D.B.; Freiser, B.S. J. Am. Chem. Soc., 1985, 107, 2605.
31. (a) Kappes, M.M.; Staley, R.H. J. Phys. Chem. 1981 85, 942; (b) Armentrout, P.B.; Halle, L.F.; Beauchamp, J.L. J. Chem. Phys. 1982, 76, 2449.
32. Huang, Y.; Freiser, B.S., J. Am. Chem. Soc., 1989, 111, 2387.
33. Huang, Y.; Sun, D.; Freiser, B.S. J. Am. Chem. Soc., submitted.
34. Jacobson, D.B.; Freiser, B.S. J. Am. Chem. Soc., 1984, 106, 1159.
35. Jacobson, D.B.; Freiser, B.S. J. Am. Chem. Soc., 1985, 107, 1581.
36. Tews, E.C.; Freiser, B.S. J. Am. Chem. Soc., 1987, 109, 4433.
37. Buckner, S.W.; Freiser, B.S. J. Phys. Chem., 1989, 93, 3667.
38. Huang, Y.; Buckner, S.; Freiser, B.S. Physics and Chemistry of Small Clusters; Jena, P.; Rao, B.K.; Khanna, S.N., Eds.; Plenum: New York, 1987; p.891.
39. Hettich, R.L.; Freiser, B.S. J. Am. Chem. Soc., 1985, 107, 6222.
40. Lech, L.M.; Freiser, B.S. J. Am. Chem. Soc., 1989, 111, 8588.
41. Jacobson, D.B.; Freiser, B.S. J. Am. Chem. Soc., 1983, 105, 7492.
42. Cassady, C.J.; Freiser, B.S. J. Am. Chem. Soc., 1984, 106, 6176.
43. Jacobson, D.B.; Gord, J.R.; Freiser, B.S., Organometallics, in press.
44. Tonkyn, R.; Weisshaar, J.C. J. Am. Chem. Soc. 1986, 108, 7128.
45. Buckner, S.W.; Freiser, B.S. J. Am. Chem. Soc., 1987, 109, 1247.
46. Huang, Y.; Freiser, B.S. J. Am. Chem. Soc., 1988, 110, 4435.
47. Sallans, L.; Lane, K.R.; Squires, R.R.; Freiser, B.S. J. Am. Chem. Soc., 1985, 107, 4379.
48. Wolf, J.F.; Staley, R.H.; Koppel, I.; Taagepera, M.; McIver, Jr.,R.T.; Beauchamp, J.L.; Taft, R.W. J. Am. Chem. Soc. 1977, 99, 5417.

49.   Aue, D.H.; Bowers, M.T. Gas Phase Ion Chemistry; Bowers, M.T., Ed.; Academic Press, New York, 1979, Vol. 2, Ch.9.
50.   Staley, R.H.; Beauchamp, J.L. J. Am. Chem. Soc. 1975, 97, 5920.
51.   Uppal, J.S.; Staley, R.H. J. Am. Chem. Soc. 1982, 104, 1235.
52.   Uppal, J.S.; Staley, R.H. J. Am. Chem. Soc. 1982, 104, 1238.
53.   Jones, R.W.; Staley, R.H. J. Am. Chem. Soc. 1982, 104, 2296.
54.   Corderman, R.R.; Beauchamp, J.L. J. Am. Chem. Soc. 1976, 98, 3998.
55.   Operti, L.; Tews, E.C.; Freiser, B.S. J. Am. Chem. Soc. 1988, 110, 3847.
56.   McLuckey, S.A.; Cameron, D.; Cooks, R.G. J. Am. Chem. Soc. 1981, 103, 1313.
57.   McLuckey, S.A.; Shoen, A.E.; Cooks, R.G. J. Am. Chem. Soc. 1982, 104, 848.
58.   Hettich, R.L.; Jackson, T.C.; Stanko, E.M.; Freiser, B.S. J. Am. Chem. Soc., 1986, 108, 5086.
59.   Lech, L.M., Ph.D. Thesis, Purdue University, 1988.
60.   Hettich, R.L.; Freiser, B.S. J. Am. Chem. Soc., 1987, 109, 3537.
61.   Forbes, R.A.; Lech, L.M.; Freiser, B.S. Int. J. Mass Spectrom. Ion Processes, 1987, 77, 107.
62.   Lehman, T.A.; Bursey, M.M. Ion Cyclotron Resonance Spectrometry; Wiley-Interscience, New York, 1976.
63.   Kang, H.; Beauchamp, J.L. J. Phys. Chem., 1985, 89, 3364.
64.   Dunbar, R.C. Gas Phase Ion Chemistry; Bowers, M.T., Ed.; Academic Press, New York, Vol. 2, 1979, Ch.14.
65.   Freiser, B.S.; Beauchamp, J.L. J. Am. Chem. Soc., 1976, 98, 3136.

RECEIVED March 22, 1990

Chapter 5

Gas-Phase Thermochemistry of Organometallic and Coordination Complex Ions

David E. Richardson, Charles S. Christ, Jr., Paul Sharpe, Matthew F. Ryan, and John R. Eyler

Department of Chemistry, University of Florida, Gainesville, FL 32611

Recent thermochemical results from gas-phase ion cyclotron resonance studies of organometallic compounds and coordination complexes are summarized. Reactions of dihydrogen, alkenes, alkynes, nitriles, and other substrates with d^0 $Cp_2ZrCH_3^+$(g) and d^1 Cp_2Zr^+(g) lead to estimates for minimum differences in dissociation enthalpies for various Zr-R bonds. Appearance potential studies have been used to provide a preliminary estimate of $D(Cp_2Zr^+$-$CH_3)$. Gas-phase charge-transfer bracketing and equilibrium methods have been used to determine the adiabatic free energies of electron attachment to species such as metallocenium ions (Cp_2M^+ (g)), Cp_2Ni(g), M(acetylacetonate)$_3$(g), and M(hexafluoroacetylacetonate)$_3$(g). When combined with other data in thermochemical cycles, these results can be used to obtain values for average gas-phase metal-ligand bond dissociation enthalpies and solvation free energies for organometallic ions and coordination complex ions.

Relatively few studies of the thermochemistry of gas-phase metal-containing ions with condensed-phase counterparts have been reported. Such data reveal aspects of intrinsic molecular properties in the absence of solvation, thus eliminating a very significant energetic component in the thermodynamics of dissolved ions. A perusal of the important compilation of gas-phase ion thermochemical data by Lias et al. (*1*) reveals only a few entries for ionic metal species that are also stable in solution or the solid phase. In our research we have focused on the kinetics and thermodynamics of gas-phase ion/molecule reactions that often feature metal-containing reactants that can be studied both in gas- and condensed-phases. Thus, the opportunity arises to investigate both intrinsic chemistry and thermodynamics of such

0097–6156/90/0428–0070$06.00/0
© 1990 American Chemical Society

species as well as the effect of solvation on the kinetics and energetics of fundamental chemical processes. In this article, we will review some recent results that pertain to the thermodynamics of metal-ligand bonds and charge-transfer processes involving the loss or gain of an electron at a metal center (i.e., oxidation-reduction). We will focus exclusively on reactants with direct condensed-phase counterparts, such as metallocenes and metal tris-chelate complexes.

Experimental

We have applied the Fourier transform ion cyclotron resonance mass spectrometry (FTICR-MS) technique in most of our studies. The basics of the technique have been reviewed (*2*), and applications of the method to gas-phase metal ion chemistry (*3*) and coordination compounds have been reviewed recently (*4*). Ion cyclotron resonance has been used for many years to study ion/molecule reactions involving traditionally organic reactants (*5*). Indeed, that body of work provides an important survey of the types of information that might be gained through studies of inorganic species. Here we will give only a brief overview of the capabilities of the instrument and techniques used in the work summarized.

The heart of the instrument is a six-sided ion trap located in a high magnetic field. The trap is maintained in a high vacuum chamber at background pressures of $\sim 10^{-9}$ torr, and neutrals are introduced into the system via leak valves or heated solids probes to give total pressures up to $\sim 10^{-5}$ torr. We use two FTICR instruments based on the Nicolet FTMS 1000 at the University of Florida, one with a 2 tesla field (maximum mass ~ 2000 amu) and one with a 3 tesla field (maximum mass ~ 3000 amu). The former instrument has a custom made vacuum system that allows rather accurate determination of pressures of neutrals in the ion trap. The latter instrument has a high m/z resolution and is particularly well-suited to studies of high-mass ions often encountered in studies of metal compounds.

Ions produced in the center of the trap (usually by electron impact) are constrained by the magnetic field and applied potentials on the trapping plates to follow orbits within the cell and only slowly diffuse toward the walls. During the time the ions are trapped, they may have many collisions with neutral molecules in the trap, and these collisions can lead to thermalization of the ions and may result in chemical reactions. At a neutral pressure of 10^{-6} torr, the approximate pseudo first-order rate constant for ion-molecule collisions is 30 s^{-1}. The ionic products of ionization and subsequent reactions in the cell can be detected at any time mass spectrometrically by application of a broad range of radio frequencies and detection of resonance at cyclotron frequencies corresponding to m/z values of the ions present. Techniques are available that allow isolation of one or more m/z value in the ion population by ejection of all other ions, and this allows the reactions of a particular ion to be followed directly. Fourier transform methods, introduced by Marshall and

Comisarow (6), allow the rapid detection and signal averaging for a complete mass spectrum of the ion population of the trap.

We have also applied more traditional mass spectrometric methods in the study of organometallic ions. For preliminary appearance potential determinations, compounds were admitted into a nearly field-free source and ionized by variable energy electron impact. Detection of ions is by acceleration into a quadrupole mass spectrometer equipped with a sensitive detector. Ramping of the electron ionization energy rapidly through the range of the appearance potential (AP) allows determination of the AP for constant source conditions. AP values determined on the current instrument are in good agreement ($\pm$ 0.1 - 0.2 eV) with literature values when comparisons are available (Berberich, D. W.; Hail, M. E.; Johnson, J. V.; Yost, R. A. *Int. J. Mass Spectrom. Ion Proc.*, in press).

Chemistry of $Cp_2ZrCH_3^+$ (g) and Zr-R Bond Dissociation Enthalpies

$Cp_2ZrCH_3^+$ (g) (**I**, Cp = cyclopentadienyl) can be produced by electron impact ionization of $Cp_2Zr(CH_3)_2$, and we have characterized the gas-phase reactions of this highly reactive ion with a number of substrates, including dihydrogen, alkenes, alkynes, and nitriles (7; C. Christ, J. R. Eyler, and D. E. Richardson, *J. Am. Chem. Soc.*, in press; C. Christ, J. R. Eyler, and D. E. Richardson, submitted). Interest in this particular ion derives from a number of factors, some of which are summarized here. First, the study of the solution chemistry of highly electrophilic d^0 and d^0f^n organometallic complexes has been rapidly expanding over the last several years (for leading references, see Ref. 8-12). In particular, these types of complexes, containing Ti(IV), Sc(III), or lanthanide(III) for example, are often efficient catalysts for C-H activation and the coordination polymerization of ethylene, and they are believed to be excellent homogeneous models for Ziegler-Natta catalysts (8). Second, $Cp_2ZrCH_3^+$ is thought to be the reactive complex in the solution chemistry of $[Cp_2ZrCH_3(S)_n]^+$ (S = solvent, n = 1,2), which has been extensively characterized by Jordan and coworkers (12). Third, the mechanistic chemistry of a d^0 ion is limited in the sense that oxidative addition is not generally an energetically viable option; thus, its reactivity is not expected to include the oxidative addition/reductive elimination pathways commonly encountered in the chemistry of bare metal ions with hydrocarbons (13). The overall reactivity of **I** in the gas-phase is very similar to the solution chemistry of related d^0 early transition metal compounds.

Relative Cp_2Zr^+-R Bond Energies. Thermochemical information from the study of ion/molecule reactions of **I** can be obtained through identification of thermoneutral or exothermic reactions with partially known thermochemistry. Although not all exothermic reactions are observed in the FTICR due to large kinetic barriers or competing pathways, all reactions that are observed must be exothermic or near thermoneutral (assuming that the reactant ions are thermalized). Endoergicity of more than a few kcal mol^{-1} will usually lead to reactions with efficiencies too low for detection in the ion trap prior to ion loss ($k_{forward}/k_{collision} < 10^{-3} - 10^{-4}$). Although this method is only semi-quantitative in nature, it has provided some insight into Zr-ligand bond dissociation enthalpies for products of reactions of **I** with various substrates.

As a simple example, consider the reaction of **I** with dihydrogen (Equation 1) (*7*).

$$Cp_2ZrCH_3^+ + H_2 \rightarrow Cp_2ZrH^+ + CH_4 \qquad (1)$$

The hydrogenolysis reaction proceeds at a near collisional rate, but the activation of methane, the reverse reaction, does not proceed at a measurable rate in the FTICR. Thus the equilibrium constant ($=k_f/k_r$) for Equation 1 is > 1, and the product Zr-H homolytic bond dissociation enthalpy, $D(Cp_2Zr^+-H)$, is therefore greater than $D(Cp_2Zr^+-CH_3)$ since $D(H-H) = D(H_3C-H) = 104$ kcal mol^{-1}. (Note that entropy changes in Equation 1 and following transformations are assumed to be negligible.) The difference in the Zr bond dissociation enthalpies cannot be derived from the data, but the order is typical for known condensed-phase organometallics and is opposite that usually observed for M-CH$_3^+$ and M-H$^+$ ions in bare metal ion chemistry (*13*). In particular, for the closely related zirconium(IV) complex $Cp^*_2ZrR_2$ (Cp^*=pentamethylcyclopentadienyl), $D(Zr-H) - D(Zr-CH_3) \approx 10$ kcal mol^{-1} (*14*).

Another example involves the reaction of $Cp_2ZrCH_3^+$ with benzonitrile to produce the phenyl complex (Equation 2). The organic part of this

$$Cp_2ZrCH_3^+ + C_6H_5CN \rightarrow Cp_2ZrPh^+ + CH_3CN \qquad (2)$$

transformation (Equation 3) is endothermic by 8 ± 2 kcal mol^{-1}. Thus, for

$$\cdot CH_3 + C_6H_5CN \rightarrow Ph \cdot + CH_3CN \qquad (3)$$

the homolytic Zr-R bond enthalpies, $D(Cp_2Zr^+-Ph) - D(Cp_2Zr^+-CH_3) \geq$ ca. 8 kcal mol^{-1}. This result is consistent with other thermochemical differences between M-Ph and M-CH$_3$ bond dissociation enthalpies in d^{0}f^n complexes. For example, Marks and co-workers found $D(Th-Ph) - D(Th-CH_3) = 13 \pm 3$ kcal mol^{-1} in $Cp^*_2ThR_2$ (*15*). From other observed reactions and thermochemical data, $D(Cp_2Zr^+-R) > D(Cp_2Zr^+-CH_3)$ for the following R groups in Cp_2ZrR^+ (minimum derived difference in kcal mol^{-1} in parentheses): -OH ($\sim$14), -C$\equiv$CH ($\sim$20), η^3-allyl ($\sim$2), -CH=C=CH$_2$ ($\sim$4). The results are generally in accord with experience for other organometallic thermochemistry, but the method is clearly incapable of giving absolute values of dissociation enthalpies for the various metal-ligand bonds.

Absolute Zr-R Bond Dissociation Enthalpies. It is desirable to determine the absolute Zr-R bond dissociation enthalpies in gas-phase Cp_2ZrR^+ ions. We have used two methods to estimate these quantities.

The first approach is to investigate the reactivity of the d^1 radical ion $Cp_2Zr^+(g)$ (**II**), which is also formed in the electron impact ionization of $Cp_2Zr(CH_3)_2(g)$. The general reaction of interest is given in Equation 4.

$$Cp_2Zr^+ + RX \rightarrow Cp_2ZrX^+ + R \cdot \qquad (4)$$

The abstraction of a radical X from the substrate will only occur efficiently if the reaction is exothermic or near thermoneutral (in the absence of significant entropy change). **II** was found to abstract Cl from CCl_4, CCl_2F_2, and $HCCl_3$, thereby placing a lower limit on the homolytic $D(Cp_2Zr^+ \text{-Cl})$ of ~ 81 kcal mol^{-1}. By observation of other atom and group abstraction reactions analogous to Equation 4, the following lower limits were established (kcal mol^{-1}): $D(Cp_2Zr^+ \text{-Br}) \geq 83$; $D(Cp_2Zr^+ \text{-OCH}_3) \geq 57$; $D(Cp_2Zr^+ \text{-SCH}_3) \geq 67$; $D(Cp_2Zr^+ \text{-NO}_2) \geq 70$. Many of these bond enthalpies are likely to be much larger than the lower limits, but an absence of atom and group transfer substrates with higher bond dissociation enthalpies and multiple reaction pathways found for many substrates limits the applicability of the method.

We have recently initiated appearance potential (AP) studies in an attempt to determine $D(Cp_2Zr^+ \text{-CH}_3)$. AP measurements have many well known limitations, but they are nevertheless a common source of thermochemical data (*1*). To obtain the desired Zr-C bond enthalpy, the AP values for eqs 5-7 must be considered.

$$Cp_2Zr(CH_3)_2 + AP_1 = Cp_2ZrCH_3^+ + CH_3^- \qquad (5)$$
$$Cp_2Zr(CH_3)_2 + AP_2 = Cp_2ZrCH_3^+ + CH_3 + e^- \qquad (6)$$
$$Cp_2Zr(CH_3)_2 + AP_3 = Cp_2Zr^+ + C_2H_6 + e^- \qquad (7)$$

AP_1 and AP_3 are expected to be the lowest energies required to produce $Cp_2ZrCH_3^+$ and Cp_2Zr^+, respectively. $D(Cp_2Zr^+ \text{-CH}_3)$ is then given by $AP_3 - AP_1 + D(H_3C\text{-CH}_3) + \Delta H_a(CH_3)$, where the last quantity is the enthalpy of electron attachment to methyl radical. Experimentally, $AP_1 \approx AP_3 = 7.7 \pm 0.2$ eV, and therefore $D(Zr\text{-CH}_3) = 86 \pm 7$ kcal mol^{-1}. The reaction for AP_1 probably has a lower cross section than that of AP_2, and if the observed threshold is in fact AP_2, $D(Cp_2Zr^+ \text{-CH}_3)$ would be 88 ± 7 kcal mol^{-1}. This value can be compared to the analogous bond dissociation enthalpy values given by Schock and Marks (*14*) for the neutral complex $Cp*_2Zr(CH_3)_2$. Experimentally, they determined $D(Zr\text{-CH}_3) = D(Zr\text{-Cl}) - 49$ kcal mol^{-1}, and, assuming $D(Zr\text{-Cl}) = 115.7$ kcal mol^{-1}, they estimate $D(Zr\text{-CH}_3) = 67 \pm 1$ kcal mol^{-1}. The AP-derived $D(M\text{-CH}_3)$ for **I** is much higher than usual values found for M-CH$_3$ bonds in both coordinatively saturated and unsaturated complexes, and application of alternate approaches to determination of the bond enthalpy in **I** and related ions clearly would be of interest.

Electron Attachment to Gas-phase Metal Complexes

For a particular redox couple, thermodynamic information on the intrinsic molecular properties of the oxidized and reduced forms of the complex can be obtained directly by determination of the thermodynamics for the gas-phase reaction given in Equation 8, where n is the overall charge on the oxidized form.

$$ML_y^n(g) + e^- \xrightleftharpoons{\quad \Delta X_a \quad} ML_y^{(n-1)}(g) \ (X = G, H, \text{or } S) \qquad (8)$$

The negative value of the enthalpy change for Equation 8 when n = 0 is defined (*1*) as the electron affinity (EA) of the oxidized species when the oxidized and reduced species are in their ground rotational, vibrational and electronic states (0 K). At any temperature for any value of n (0, positive, or negative), the thermodynamic state functions for Equation 8 are given by ΔX_a (X = G, H, or S), and the thermochemistry of electron attachment can be defined in the ion convention ("stationary electron convention") (*1*). The relationship between EA and ΔG_a is given by Equation 9. A similar relationship applies for adiabatic ionization energies.

$$EA = -\Delta G_a - T\Delta S_a + \int_0^T Cp \ (A^-) \ dT - \int_0^T Cp \ (A) \ dT \qquad (9)$$

Methods for obtaining values of ΔX_a for gas-phase neutrals and cations have been described in the literature (*1*). Estimates of adiabatic electron affinities ($M^{0/-}$) and adiabatic ionization potentials (aIP, $M^{0/+}$) of neutral molecules have been obtained in many cases through investigations of charge-transfer equilibria between pairs of neutral molecules and their parent ions. EA values for over two hundred compounds have been determined that span a range of ~0.8 to ~3 eV (*16*), and a large number of aIP values have also been reported (*17*). Charge-transfer equilibria (Equation 10) for cations or

$$A^{+/-}(g) + B(g) \xrightleftharpoons{\quad} A(g) + B^{+/-}(g) \qquad (10)$$

anions have generally been studied by using ion cyclotron resonance mass spectrometry or pulsed high pressure mass spectrometry (PHPMS). Values for ΔX_a (X = G, H, and S) for a new molecule (say, B) can be deduced from the temperature dependence of the equilibrium constant of Equation 10 if the corresponding values of ΔX_a for A are known. If no reference compounds are available that come to equilibrium with B, charge-transfer bracketing between two reference compounds can be used to estimate $\Delta G_a(B)$.

Experimental Results. The solution redox potentials (*18*) and gas-phase photoelectron spectroscopy of metallocenes (*19*) have received much attention in the literature. We have recently used charge-transfer equilibria in the FTICR-MS to determine the values of ΔG_a for several metallocenium ions (Figure 1). Reference compounds are not widely available for aIP values < 6.8 V, so the absolute values for many metallocenes of lower IP have not yet been determined. However, aIP values in the desired range are known (*17*), and we are continuing these studies.

We have also determined ΔG_a values for electron attachment to a number of neutral metal complexes by charge-transfer bracketing and equilibrium experiments with organic acceptors (Figure 2). One metallocene, Cp_2Ni, was known to form a stable negative ion from early studies by Beauchamp and coworkers (*20*). Many ΔG_a values have also been determined for electron attachment to the first row transition metal tris(acetylacetonate) ($M(acac)_3$) and tris(hexafluoracetylacetonate) ($M(hfac)_3$) complexes (P. Sharpe and D. E. Richardson, submitted). This process involves the formal reduction of the metals from the 3+ to 2+ oxidation states, and the variation of the $M(acac)_3^{0/-}$ values with atomic number is qualitatively similar to that observed for the $[M(H_2O)_6]^{3+/2+}$ couples.

Thermochemical Applications of Electron Attachment Energies. Although ΔX_a values for Equation 8 are of intrinsic interest, they also provide an important component of thermochemical cycles involving oxidation-reduction of metal complexes (Figure 3). The upper and lower cycles of Figure 3 can be used to obtain values of average heterolytic gas-phase M-L bond energies and the solvation thermochemistry of the species involved, respectively. Gas-phase homolytic M-L energies can also be derived from related cycles. Values of ΔX_a are then crucial to a more complete understanding of the various thermochemical contributions to electrode potentials for metal compounds since they allow a breakdown of the observed potentials into bonding and solvation components. Note that the solution redox couple in Figure 3 is defined by the single electrode potential (E_s^o) rather than the standard electrode potential (values of E_s^o can be estimated by adding 4.44 V to the value of E^o vs. NHE) (*21*). In view of the accepted convention for writing electrode potentials as electron attachments, we have adapted a convention in which all gas-phase processes involving a free electron are also written as electron attachments with thermochemistry defined by the notation ΔX_a.

Derived bond dissociation enthalpies for metallocene ions are given in Table I. These values assume that ΔS_a is negligible and, for the heterolytic D(M-Cp) values, that ΔH_a for $Cp^{0/-}$ = -40 ± 5 kcal mol^{-1} (*1*). Corrections for junction potentials in electrochemical data were not applied.

An interesting point concerning results in Table I is that for M = V, Ru, and Ni both oxidized and reduced forms of a metallocene have the same values of homolytic bond dissociation enthalpies within the estimated error

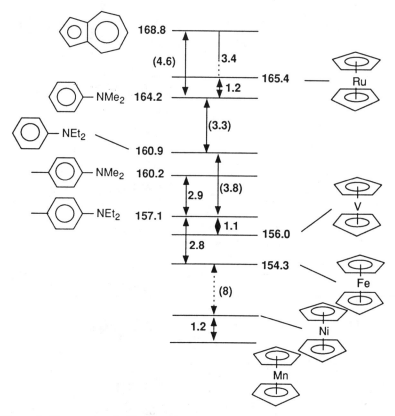

Figure 1. Free energy ladder for electron attachment to metallocenium ions, Cp_2M^+, as deduced by charge-transfer equilibria. Reference compounds (Ref. 17) are shown on the left with $-\Delta G_a$ values. Values of ΔG_{rxn} (Equation 8) are shown on the vertical lines for various equilibria, and derived values of $-\Delta G_a$ for the metallocenes are shown on the right. Values in parentheses are estimates. Temperature ~ 350 K.

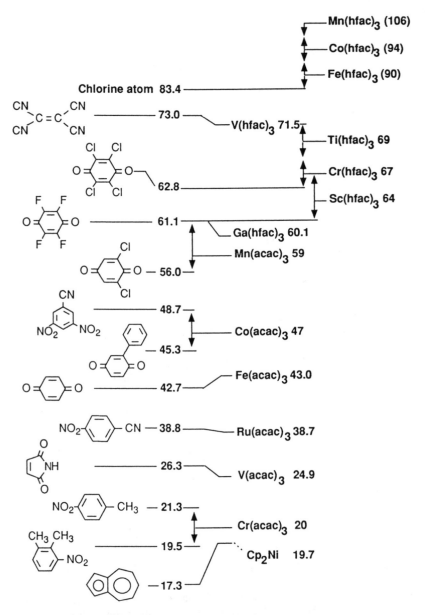

Figure 2. Free energy ladder for electron attachment to metal complexes deduced from charge-transfer equilibria and charge-transfer bracketing. Reference compounds are on the left, and derived values of $-\Delta G_a$ are on the right. Values of ΔG_{rxn} are not shown to simplify the graphic, but charge-transfer equilibria were determined for those complexes connected to reference complexes by a solid line. Temperature ~350 K.

limits. With respect to the thermochemical cycles used to derive the tabulated values, this arises from the nearly equal values for $-\Delta H_a$ for Cp_2M(oxidized) and the relevant ionization energies for the free metal atom (V^+/V, Ru^+/Ru, and Ni/Ni^-). In the case of ferrocene, ΔH_a (Fe^+ (g) + e = Fe(g)) is more exothermic than ΔH_a (Cp_2Fe^+) by ~ 27 kcal mol^{-1}, and this results in the difference of ~ 14 kcal mol^{-1} for the homolytic D values for Cp_2Fe and Cp_2Fe^+. Note that the larger homolytic bond enthalpy for Cp_2Fe^+ relative to Cp_2Fe would not be predicted from the increased M-C bond lengths in the cation relative to the neutral (22), thus illustrating the absence of a direct connection between bond energies and the ground state bonding in a molecule.

Table I. Mean Bond Dissociation Enthalpies and Differential Solvation Free Energies for Metallocenes

Cp_2M	M^{2+}-Cp^-	M^{3+}-Cp^-	M-$Cp^{\cdot}$	M^+-$Cp^{\cdot}$	$\Delta\Delta G_{solv}$
$Cp_2Fe^{+/0}$	317	593	79	93	41 (CH_3CN)
$Cp_2V^{+/0}$	302	562	95	95	61 (THF)
$Cp_2Ru^{+/0}$	331	577	94	96	--

	M^+-Cp^-	M^{2+}-Cp^-	M^--$Cp^{\cdot}$	M-$Cp^{\cdot}$	$\Delta\Delta G_{solv}$
$Cp_2Ni^{0/-}$	124	324	64	67	-49 (CH_3CN)
est errors	± 10	± 10	± 7	± 7	± 5

Differential solvation free energies ($\Delta\Delta G_{solv} = \Delta G_{solv}(Cp_2M^{n-1}) - \Delta G_{solv}(Cp_2M^n)$) for the $Cp_2M^{+/0}$ (n = 1) and $Cp_2Ni^{0/-}$ (n = 0) couples are also summarized in Table I. According to the Born equation, electrostatic solvation free energies (Equation 11) are modelled by a conducting sphere of radius r and charge q in a dielectric continuum with dielectric constant ϵ.

$$\Delta G_{el} = -q^2/2r(1 - 1/\epsilon) \tag{11}$$

This equation sometimes has been misinterpreted in the literature as predicting ΔG_{solv}(ion) values, e.g., for the process A^+ (g) = A^+ (solv). The proper definition for ΔG_{el} is the electrostatic free energy change when a charge from a conducting sphere in a vacuum is transferred to a sphere of equal radius in the dielectric medium. Thus, for couples of the type $Cp_2M^{+/0}$, $\Delta\Delta G_{solv}$ as defined above is given theoretically by $-\Delta G_{el}$ since the effective radii of metallocene neutrals and ions are roughly equal (22). Assuming r = 3.8 Å, which is very close to the average radius of a ferrocenium ion, Equation 11 predicts $\Delta\Delta G_{solv}^{o}$ = 43 kcal mol^{-1} for $Cp_2Fe^{+/0}$ in acetonitrile. For $Cp_2Ni^{-/0}$, using the same radius, theoretical $\Delta\Delta G_{solv}^{o}$ = -43 kcal mol^{-1}. The case of $Cp_2V^{+/0}$ is known to be anomalous (18), and the oxidized form probably is a bent metallocene with at least one coordinated solvent. Therefore, for the simple cases of Cp_2Ni and Cp_2Fe, the dielectric continuum model of the Born equation apparently provides a good estimate of the differential solvation energies given in Table I. This observation is in retrospect not so surprising given the roughly spherical structure of the metallocenes and the absence of specific interactions, such as H-bonding, between the solvent and the molecules.

Again assuming that ΔG_a is a good estimate for ΔH_a, average heterolytic bond disruption enthalpies, $D(M^{2+}-(acac)^{-})$, can be derived for the $M(acac)_3^{-}$ ions (Figure 4). It is noteworthy that the minimum $D(M^{2+}-(acac)^{-})$ in the series is found for M = Mn, which would be predicted for the high-spin d^5 ion by ligand field theory. Differential solvation free energies can also be estimated for the $M(acac)_3^{-}$ complexes from the difference ($\Delta G_a - \Delta G_s^{o}$), and an estimated value for $\Delta\Delta G_{solv}$ for $Ru(acac)_3^{-}$ in aqueous solution is found to be -60 ± 5 kcal mol^{-1}. This value is about twice that predicted by the Born equation (Equation 11) with r = 5.8 Å (the distance to the ligand periphery) and strongly suggests significant solvent interpenetration between the essentially planar acac ligands and, perhaps, specific interactions with solvent.

Entropy Effects

It should be noted that statistical mechanical considerations suggest that values of $T\Delta S_a$ for some transition metal complex couples are significant due to large changes in frequencies of M-L vibrational modes (e.g., low-spin Co(III) + e = high-spin Co(II); P. Sharpe and D. E. Richardson, in preparation). The contributions of $T\Delta S_a$ to ΔG_a may be large enough to invalidate the assumption used above that $\Delta G_a = \Delta H_a$, but the general points made here concerning D(M-L) values derived from the present work are expected to hold. We are currently investigating methods for studying the temperature dependence of charge-transfer equilibria in the FTICR-MS in the hope that we can further characterize the entropic contributions in these processes.

Conclusions

Thermochemical data for gas-phase neutrals and ions are fundamental to the understanding of the thermodynamics of molecular transformations. Furthermore, such data is valuable for comparisons to the results of quantum mechanical calculations of electronic energies since the theoretical models do

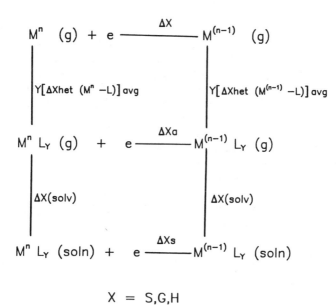

$$X = S,G,H$$

Figure 3. Thermochemical cycles for electron attachment to free metal ions, gas-phase metal complexes, and solvated metal complexes. ΔX_{het} refers to heterolytic bond disruption (M-L → M + :L).

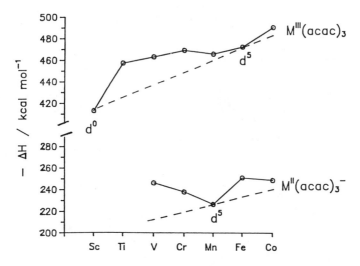

Figure 4. Average heterolytic bond enthalpies for $M(acac)_3$ and $M(acac)_3^-$ complexes. Dashed lines connect those complexes with no net ligand field stabilization (dashed line on anion data drawn arbitrarily for purposes of illustration). Values for $M(acac)_3$ taken from Ref. 27.

not normally include solvation. Comparisons of gas-phase and solution reactivity for metal complex ions can also be instructive about the role of solvent in the rates of reaction pathways (23-26). The techniques described here illustrate several approaches (and their limitations) for determination of relative and absolute bond energies in gas-phase organometallic ions. Other methods, such as beam techniques and photoionization, will also be important in defining these quantities experimentally. Appearance potential measurements have not been widely applied in organometallic thermochemistry, but the preliminary results reported here suggest that further efforts with sophisticated instrumentation may be profitable. The central role of gas-phase electron attachment thermodynamics in separating bonding and solvation contributions to solution electrode potentials can be clearly seen in this work. These various gas-phase ion techniques would appear to find virtually limitless applications to the thermochemistry of organometallic compounds and coordination complexes.

Acknowledgments

Support for inorganic gas-phase ion research at the University of Florida is provided by the National Science Foundation (CHE8700765) and is gratefully acknowledged. DER is an A. P. Sloan Foundation Research Fellow, 1988-1990. The authors thank R. Yost and J. Johnson for assistance with appearance potential measurements.

Literature Cited

1.   Lias, S. G.; Bartmess, J. E.; Liebman, J. F.; Holmes, J. L.; Levin, R. D.; Mallard, W. G., Eds. "Gas-Phase Ion and Neutral Thermochemistry"; American Institute of Physics: New York, 1988.
2.   Marshall, A. G. *Acc. Chem. Res.* **1985**, *18*, 316.
3.   Buckner, S. W.; Freiser, B. S. *Polyhedron* **1988**, *7*, 1583.
4.   Sharpe, P.; Richardson, D. E. *Coord. Chem. Rev.* **1989**, *93*, 59.
5.   See chapters in *Gas-Phase Ion Chemistry*; Bowers, M. T., Ed.; Academic Press: Orlando, 1979; Vol. 2.
6.   Comisarow, M. B. *Adv. Mass Spectrom.* **1980**, *8*, 1698.
7.   Christ, C. S.; Eyler, J. R.; Richardson, D. E. *J. Am. Chem. Soc.* **1988**, *110*, 4038.
8.   Borr, J. *Ziegler-Natta Catalysts and Polymerizations*; Academic Press: New York, 1979.
9.   Watson, P. L. *Acc. Chem. Res.* **1985**, *18*, 51.
10.   Bulls, A. R.; Bercaw, J. E.; Manriquez, J. M.; Thompson, M. E. *Polyhedron* **1988**, *7*, 1409 and references therein.
11.   Heddon, D.; Marks, T. J. *J. Am. Chem. Soc.* **1988**, *110*, 1647 and references therein.
12.   Jordan, R. F. *J. Chem. Ed.* **1988**, *65*, 285.
13.   Armentrout, P. B.; Beauchamp, J. L. *Acc. Chem. Res.* **1989**, *22*, 315.
14.   Schock, L. E.; Marks, T. J. *J. Am. Chem. Soc.* **1988**, *110*, 7701.

15. Bruno, J. W.; Marks, T. J.; Morss, L. R. *J. Am. Chem. Soc.* **1983**, *105*, 6824.
16. Kebarle, P.; Chowdhury, S. *Chem. Rev.* **1987**, *87*, 513.
17. Meot-Ner (Mautner), M. *J. Am. Chem. Soc.* **1989**, *111*, 2830.
18. Holloway, J. D. L.; Geiger, W. E. *J. Am. Chem. Soc.* **1979**, *101*, 2038.
19. Green, J. C. *Struct. Bonding 43*, 37.
20. Corderman, R. R.; Beauchamp, J. L. *Inorg. Chem.* **1976**, *15*, 665.
21. Parsons, R. In *Standard Electrode Potentials in Aqueous Solution*; Bard, A. J.; Parsons, R.; Jordan, J., Eds.; Marcel Dekker: New York, 1985; Ch. 2.
22. Haaland, A. *Acc. Chem. Res.* **1979**, *12*, 415.
23. Squires, R. *Chem. Rev.* **1987**, *87*, 623.
24. Richardson, D. E.; Christ, C. S.; Sharpe, P.; Eyler, J. R. *J. Am. Chem. Soc.* **1987**, *109*, 3894.
25. Richardson, D. E.; Christ, C. S.; Sharpe, P.; Eyler, J. R. *Organometallics* **1987**, *6*, 1819.
26. Sharpe, P.; Christ, C. S.; Eyler, J. R.; Richardson, D. E. *Int. J. Quantum Chem., Quantum Chem. Symp.* **1988**, *22*, 601.
27. Wood, J. L.; Jones, M. M. *Inorg. Chem.* **1964**, *3*, 1553.

RECEIVED November 15, 1989

Chapter 6

Ionization Energy–Bond Energy Relationships in Organometallic Chemistry

Dennis L. Lichtenberger and Ann S. Copenhaver

Laboratory for Electron Spectroscopy and Surface Analysis, Department of Chemistry, University of Arizona, Tucson, AZ 85721

The relationships between ionization energies and bond energies are implicated in numerous photoelectron investigations. The foundations of these relationships are developed and presented here. The quantitative accuracy and limitations of using ionization energies to obtain bond dissociation energies are illustrated with results on single- and multiple-bonded diatomic molecules. The principles that emerge from this analysis are then used to evaluate the bond energies in $Mn_2(CO)_{10}$, $Mn(CO)_5H$, $Mn(CO)_5CH_3$, $Cp_2Nb(CO)CH_3$ and $Cp_2Nb(CO)H$. The ionization energies are particularly useful because of their direct relationships to other thermodynamic quantities through energy cycles, and can be used as a check of other thermodynamic information. In some cases, bond energy information may be obtained from the ionization energies which is difficult to measure by other chemical or thermodynamic methods. The technique has the further advantage that it allows separation of individual symmetry orbital or electron distribution contributions to the total bond.

Knowledge of individual thermodynamic bond energies in molecules and fragments is essential for understanding and systematizing chemical behavior. Unfortunately, it has generally proven difficult to obtain quality thermodynamic information for wide ranges of molecules, especially organometallic complexes.(1) Alternative methods for obtaining thermodynamic information, particularly methods that can be independently related to other chemical or kinetic measurements, are necessary for developing consistent knowledge of bond energies. The ionization energy of an atom or molecule, at the most basic level, is a well defined thermodynamic quantity that can be precisely measured. Ionization energies are often used in thermodynamic cycles. Common examples are the use of ionization energies in Born-Haber cycles of ionic compounds(2) and the use of ionization energies with appearance potentials for determining proton affinities.(3) Detailed understanding of the chemically important valence ionizations of a species is closely related to its thermodynamic stability or chemical reactivity. Since any physical or chemical transformation will alter the valence

0097–6156/90/0428–0084$06.00/0
© 1990 American Chemical Society

electronic structure as bonds are compressed, elongated, broken, or formed, the amount of bonding lost in a valence ionization event is one factor that determines the ionization energy. Walsh pointed out over four decades ago that a molecular ionization energy characterizes both the orbital which is ionized as well as the contribution which that orbital makes to the bond strength.(4)

Often times the ionization potentials of a molecule are only loosely and qualitatively related to the strengths of the bonds. The fundamental question addressed here concerns the precise relationships between measured ionization energies and bond energies in molecules. The purpose here is to clarify the exact equations that relate ionization energies and bond energies and to illustrate some useful principles which emerge. The correlations are discussed in terms of simple bonding models.

Covalent Bonds and Ionization Energies.
The Hydrogen Molecule. The first step toward investigating the relationships between the bond dissociation energy of a molecule and the molecular ionization energy from photoelectron spectroscopy is to examine the simplest molecular case. The energy cycle which includes the ionization of the hydrogen molecule(5) and the bond dissociation energy is shown in Figure 1.(6) The vertical ionization energy is designated here as $IP(H_2)$. The potential curve for the neutral hydrogen molecule traces at large internuclear separation to the dissociation of two neutral hydrogen atoms. Similarly, the potential curve for the hydrogen molecule positive ion traces to the dissociation of one hydrogen atom and a proton. The energy separation between the two dissociation limits is the ionization energy of a hydrogen atom, designated $IP(H\cdot)$.

These energies are related simply by the following exact equation:

$$IP(H_2) + D(H_2^+) = IP(H\cdot) + D(H_2) \qquad (1)$$

$D(H_2)$ is the bond dissociation energy of molecular hydrogen, and $D(H_2^+)$ is the bond dissociation energy of the H_2^+ molecular cation. To be consistent, $IP(H_2)$ and $D(H_2^+)$ are measured from the same vibrational level of the H_2^+ ion. Collecting the ionization energies on one side and the dissociation energies on the other side gives:

$$IP(H_2) - IP(H\cdot) = D(H_2) - D(H_2^+) \qquad (2)$$

Equation 2 shows that the difference between the ionization potential of the H_2 molecule and the ionization potential of the dissociated hydrogen atom(7) is the bond energy lost upon ionization. Knowledge of any three of the quantities in Equation 2 provides knowledge of the fourth quantity. This relationship is used extensively by Herzberg to determine the bond dissociation energies of molecular cationic species from knowledge of the ionization energies and the dissociation energy of the neutral molecule.(9) As another example, the dissociation energy of the bond in the molecular cation is often obtained from the appearance potential of the dissociated cation fragment (measured by mass spectrometry) minus the molecular ionization energy, and these are in turn used to obtain the dissociation energy of the bond in the neutral molecule.

If there is no independent information that gives the bond energy in the neutral molecule or in the positive ion state, an additional relationship must be introduced to obtain the bond energies from the ionization energies. A simple first approximation is that the ionization of an electron from the covalent electron pair bond destroys half the bond.($\underline{8}$)

$$D(H_2) \approx 2 \cdot [D(H_2^+)] \tag{3}$$

This will be called the bond order approximation. It might be noted that this approximation is at the heart of the Hückel MO method. Using this approximation:

$$
\begin{aligned}
IP(H_2) - IP(H\bullet) \quad &= D(H_2) - D(H_2^+) \\
&\approx D(H_2) - 1/2 \ D(H_2) \\
&\approx 1/2 \ D(H_2)
\end{aligned}
\tag{4}
$$

Note that this bond order approximation applies without changing or relaxing the distance between the two atoms. Thus the vertical ionization energy, rather than the adiabatic ionization energy, is the correct energy to use in the application of this bond order approximation to Equation 4. Using the vertical ionization potential of H_2 and the atomic ionization potential of H, a bond dissociation energy of 4.70 eV is obtained. Despite the extreme simplicity of this bond order approximation, this value is in good agreement with the known $D_0(H_2)$ of 4.478 eV.($\underline{9}$) This approximation provides a bond energy estimate that is correct to within about 5% of the true bond energy for H_2. We find later that this assumption generally applies well provided that the covalent bond is reasonably strong. In the absence of other approximations, this bond order approximation provides an upper bound to the true dissociation energy.

A better estimate of the bond energy from the ionization energies is obtained from a more complete understanding of the nature of the bond. Theoretical models represent the bond in the hydrogen molecule with a 94% contribution of a covalent term and a 6% contribution of ionic terms.($\underline{10}$) The bond order approximation applies only to the covalent portion. As we will show again later, the contribution of the ionic portion to the bond energy is essentially lost with ionization. Equation 2 can be generalized as follows:

$$
\begin{aligned}
IP(H_2) - IP(H\bullet) \quad &= D(H_2) - D(H_2^+) \\
&= D(H_2) - k \cdot D(H_2) \\
&= (1-k) \cdot D(H_2)
\end{aligned}
\tag{5}
$$

where k is the ratio of the bond energy in the positive ion to the bond energy in the neutral molecule, $D(H_2^+)/D(H_2)$. This is still an exact equation provided that the value of the proportional bond strength, k, is precisely known. The value of k depends on the degree of bond covalency. If the bond energy is decomposed into covalent energy ($D_{covalent}$) and ionic energy (D_{ionic}) portions, then:

$$k \approx \frac{1/2 \ D_{covalent}}{D(H_2)} = \frac{1/2 \ D_{covalent}}{D_{covalent} + D_{ionic}} \tag{6}$$

If an electron is removed from a completely covalent bond, half of the bond order is removed and the remaining bond is approximately half as strong as the original. In this case $k \approx 0.5$. If ionization occurs from a completely ionic bond, then the negative ion of the ionic bond becomes neutral, and the ionic bond is completely lost. In this case $k \approx 0$. Using the 6% ionic character for H_2 molecule, k is then found to be 0.47

and the estimated bond dissociation energy is 4.43 eV versus the experimentally determined 4.478 eV value. This is within 1% of the correct value. Thus, these very simple models and concepts of bonding lead to a close agreement with the experimentally determined bond energy of H_2.

Halogens. The same principles can be applied to the ionization energies and bond energies of the halogens, although there is a greater number of valence electrons to be considered and there is also the presence of σ and π symmetry interactions. The low energy photoelectron spectra of the halogens,(11-16) X_2 (X= F, Cl, Br, I), exhibit three different valence ionization events corresponding to different states of the positive molecular ions. The first and second lowest ionizations, the Π_g and Π_u respectively, correspond to the antibonding and bonding π combinations on the halogens. The ionization at highest ionization energy is assigned to the sigma bond, Σ_g^+ (see Table I).

The appropriate states of the dissociated atoms and ions for application of the bond order approximation that was illustrated for the H_2 molecule must be identified. Using the Cl_2 molecule as an example, the energies of the appropriate molecular and atomic electronic states are illustrated in Figure 2. The bond dissociation of the Cl_2 molecule results in two neutral Cl atoms in the ground state. In dissociating a Cl_2^+ positive ion, there are two choices for the electronic state of dissociated Cl^+ to be used with the bond order approximation. The electron could have originated from one of the two molecular π combinations. Using the first ionization potential of Cl_2, corresponding to the $^2\Pi_{g3/2}$, and Equation 1, the bond dissociation energy of Cl_2^+ is reported by Herzberg(9) to be 3.9 eV as compared with the neutral Cl_2 bond dissociation energy of only 2.5 eV. The bond's energy has increased by 1.4 eV upon removal of an electron from the ionization of an antibonding π orbital to form the $^2\Pi_{g3/2}$ state. This bond energy change is given directly by the difference in energy between the molecular ionization and the atomic ionization to the $^3P_2$ state. Thus the $^3P_2$ state is appropriate for correlation to the molecular π ionizations. Note in Figure 2 that the molecular π bonding and antibonding states essentially cancel each other with respect to the atomic $^3P_2$ state. Thus, the π bonding combination is stabilized the same amount as the π antibonding combination is destabilized from the atom's $^3P_2$ state. The average energy of the molecular Π_u and Π_g states are within 0.1 eV of the respective atomic $^3P_2$ states for all the halides.

If the hole that is produced on ionization corresponds to the molecular sigma bond, the final state of the dissociated Cl^+ cation should appropriately correlate with this molecular state for application of the bond order approximation. The atomic cation state which best reflects the loss of the sigma bound electron is the $^1D_2$ state. This state corresponds to the atomic electron paired $2s^22p^4$ valence occupation that results from removal of the unpaired electron that formed the bond. Using the ionization values for the $^2\Sigma_g^+$ molecular state and the final state $^1D_2$ for Cl^+ in Equation 2, a bond dissociation energy for Cl_2^+ from this state is found to be 0.8 eV. This is less than half the dissociation energy of the neutral molecule. As a consequence, application of Equation 4 to estimate the bond dissociation energy from the ionization energies results in a value which is too large (Table I).

The nature of the molecular ionization and the final state of the dissociated cation must be understood for quantitative applications of these principles. In Table I, the bond dissociation energies for several homonuclear diatomic molecules are calculated using Equation 4 and are compared to the dissociation energies found by spectroscopic means. The dissociation energy determined from ionization energies decreases from

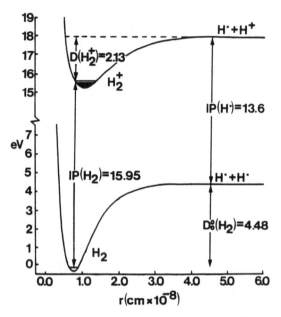

Figure 1. Molecular hydrogen bond dissociation and ionization diagram.

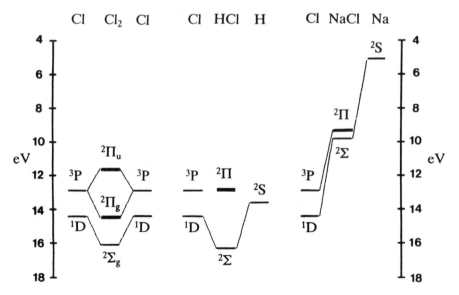

Figure 2. Experimental ionization correlation diagram for Cl_2, HCl and NaCl.

Table I Experimental and Calculated Bond Dissociation Energies (in eV) for Homonuclear Diatomics

	$D_{0(exp)}$[a]	$^2\Sigma_g^+$	$^2\Pi$	Atomic[e]		$D_{(calc)}$[‡]	% Err
H_2	4.478(1)	15.95(2)	-	$^2S_0$	13.60	4.70(3)	6
F_2	1.602(1)	20.8[†]	15.83(3)[b]	$^3P_2$	17.43	-	-
			15.87(3)	$^1D_2$	20.01		
			18.98(3)	$^1S_0$	22.99		
			19.02(3)				
Cl_2	2.479(1)	16.08(2)[b]	11.59(2)[b]	$^3P_2$	12.97	3.32(3)	34
			11.67(2)	$^1D_2$	14.42		
			14.42(2)	$^1S_0$	16.42		
			14.50(2)				
Br_2	1.971(1)	14.60(2)[c]	10.56(2)[c]	$^3P_2$	11.82	2.58(3)	31
			10.91(2)	$^1D_2$	13.31		
			13.08(2)	$^1S_0$	15.26		
			13.34(2)				
I_2	1.542(1)	12.95(2)[d]	9.36(2)[d]	$^3P_2$	10.45	1.60(3)	4
			9.98(2)	$^1D_2$	12.15		
			11.03(2)	$^1S_0$	14.50		
			11.82(2)				

[†] The photoelectron spectrum of F_2 does not show the σ ionization, although it is known to be slightly greater that 20 eV. This value is the predicted ionization potential found by using the known dissociation energy.
[‡] $D_{(calc)} = 2 \cdot [IP(X_2, {}^2\Sigma_g^+) - IP(X, {}^1D)]$
[a] Ref. 9. [b] Ref. 11. [c] Ref. 13. [d] Refs. 13 and 16. [e] Ref. 7.

Cl_2 to Br_2 to I_2, in agreement with the trend in experimentally measured bond dissociation energies. This treatment gives very good agreement with the experimental measured bond dissociation energies for I_2 and H_2 and suggests that the bonding is primarily covalent in character. The bond dissociation energies of Cl_2 and Br_2 are estimated too high by some $\approx 35\%$. These are the largest errors we have observed for reasonable covalent bonds. Ionic terms and electron correlation may play a role, as these contributions will adjust the estimated bond energy downward. It may also be that the valence s electrons and other atomic terms are contributing, as will be shown shortly in the case of N_2.

<u>Halogen Acids.</u> In the previous discussions, attention was given to homonuclear diatomic molecules. The bonds were largely covalent, although ionic contributions have a small influence on the energy relationships. For a bond formed between two differing atoms, one atom will inevitably possess a greater electronegativity and ionic terms will become more important. The bonding picture is now better viewed as some contribution of covalent bond energy and a remainder of ionic bond energy. As the difference in electronegativity increases, the bond is considered more ionic in character, as illustrated in Figure 2.

An example is the case of the halogen acids.(<u>14</u>) Consider the estimation of HF,(<u>17</u>) HCl,(<u>18</u>) HBr(<u>19</u>) and HI(<u>20</u>) bond dissociation energies from ionization

Table II Experimental and Calculated Dissociation Energies (in eV) for the Halogen Acids and Interhalogens

$D_{0(exp)}^a$		$^2\Sigma$	$^2\Pi$	Atomic[b]		Atomic[b]		$D_{(calc)}^{\ddagger}$	% Err
HF	5.869	20.00[†c]	16.05(3)[c]	H	$^2S_0$ 13.60	F	$^3P_2$ 17.43	6.39[†]	9
							$^1D_2$ 20.01		
							$^1S_0$ 22.99		
HCl	4.433	16.28[d]	12.75(3)[d]	H	$^2S_0$ 13.60	Cl	$^3P_2$ 12.97	4.54	3
			12.85(3)				$^1D_2$ 14.42		
							$^1S_0$ 16.42		
HBr	3.894	15.60[e]	11.71(3)[e]	H	$^2S_0$ 13.60	Br	$^3P_2$ 11.82	4.29	10
			12.03(3)				$^1D_2$ 13.31		
							$^1S_0$ 15.26		
HI	3.054	14.25[c]	10.38(3)[c]	H	$^2S_0$ 13.60	I	$^3P_2$ 10.45	2.75	10
			11.05(3)				$^1D_2$ 12.15		
							$^1S_0$ 14.50		
ICl	2.153	14.26[f]	10.10(3)[f]	I	$^3P_2$ 10.45	Cl	$^3P_2$ 12.97	1.95	9
			10.68(3)		$^1D_2$ 12.15		$^1D_2$ 14.42		
			12.88(3)		$^1S_0$ 14.50		$^1S_0$ 16.42		
IBr	1.817	13.70[f]	9.85(3)[f]	I	$^3P_2$ 10.45	Br	$^3P_2$ 11.82	1.94	7
			10.42(3)		$^1D_2$ 12.15		$^1D_2$ 13.31		
			11.99(3)		$^1S_0$ 14.50		$^1S_0$ 15.26		
			12.38(3)						

[†]Because the photoelectron bands are very broad, the value is uncertain.
[‡]$D_{calc} = [2 \cdot IP(HX, {}^2\Sigma_g^+) - IP(X, {}^1D) - IP(H, {}^0S)]$
[a]Ref. 9 [b]Ref. 7. [c]Ref. 20. [d]Ref. 18. [e]Ref. 19. [f]Ref. 11.

energies.   Again, the correct molecular and atomic positive ion states must be correlated for application of the bond order approximation.   The dissociative limit for $[HCl]^+$ may be either to H^+ and atomic chlorine or to Cl^+ and atomic hydrogen. Further, Cl^+ possesses several possible atomic states.[(7)]

The lowest energy dissociative state of $[HCl]^+$ is to the $^3P_2$ of Cl^+ and atomic H. However, this set of dissociation limits correlates most directly with the removal of a lone pair electron on the Cl atom in molecular HCl.   This is also the lowest IP of HCl.   As seen in Figure 2, the final dissociation state for $[HCl]^+$ which correlates with the σ bond requires either H^+ and neutral Cl or atomic hydrogen and $^1D_2$ Cl^+.   The bond order approximation applies most closely to the averaged final atomic states. The HCl bond energy is estimated by summing the energy stabilization of the σ bonded hydrogen atom and the σ bonded Cl atom.   This is calculated by the following equation:

$$D_{(calc)} = [IP(HCl^+, {}^2\Sigma^+) - IP(Cl^+, {}^1D_2)] + [IP(HCl^+, {}^2\Sigma^+) - IP(H^+, {}^2S_0)] \qquad (7)$$

where the energy stabilization from the Cl^+, $^1D_2$ state to the molecular HCl^+, $^2\Sigma^+$ state is summed with the stabilization from the H^+, $^2S_0$ state to the molecular HCl^+, $^2\Sigma^+$ state.   Table II lists the calculated dissociation energies for the halogen acids and two interhalogens, (ICl and IBr). In all cases the calculated dissociation energies are within 10% of literature values of dissociation energies.(9) Further, the trend from greatest bond strength to weakest bond strength is followed by the calculated bond dissociation energies. This suggests that for predominantly covalent interactions, the molecular bond energy may be reasonably obtained from ionization energies.

Salts. The bonding stabilization of salts is attributed to an ionic contribution and as such does not give a net stabilization of an electron pair shared between both atoms. Instead, the electron levels on the cationic atom are stabilized by its net positive atomic charge and the electron levels on the anionic atom are destabilized by its net negative atomic charge. Because of the net stabilization and destabilization of one atom versus the other, the "molecular" Σ ionization energy nearly averages between the atomic cation and atomic anion, as shown in Figure 2.

Other workers(21,22) have estimated the bond energies of salts by taking the difference between the first ionization potential, (the $^2\pi$, corresponding to a anion lone pair) and the ionization potential of the atom corresponding to the cation in the salt. They realized that most all the bond is destroyed with removal of an electron, as we mentioned earlier, and leaves only a small stabilization due to the polarizability of the halide. This stabilization, (increasing from F to I) is small relative to the other energy contributions. Thus, with the removal of a π electron from a salt, the bond dissociation energy of the molecular cation, $D(MX^+)$, is nearly zero. These bond dissociation energies(22) are in excellent agreement to the values reported by Herzberg.(9)

General Relationships between Ionization Energies and Dissociation Energies.
A general relationship which includes bonds ranging from homonuclear diatomic molecules to salts would be useful. The bond energies in covalent molecules are estimated as the sum of the stabilizations from the atomic states to the molecular state. The bond energies in ionic complexes are estimated by just the stabilization from the metal atomic state to the molecular state. The nature of a bond between these two extremes is dependent on the difference in electronegativity of the atoms. Using the difference in electronegativities, a single equation may be written which relates the ionization energies to the bond energies:

$$D(MX) = [IP(MX,\Sigma)-IP(M)] + [IP(MX,\Sigma)-IP(X)] \cdot [1-(\Delta\chi/\Delta\chi_{ionic})] \qquad (8)$$

The dissociation energy, $D(MX)$, is the sum of the "M" atomic stabilization and the "X" atomic stabilization, weighted by a measure of the bond's covalent character (see Table III). Here we define the "M" atom as the atom with the lowest atomic ionization potential (for the atomic state which correlates to the bond of interest). The "X" atom is the atom with the larger atomic ionization potential (for the atomic level involved with the bond of interest). The weighting factor of $[1-\Delta\chi/\Delta\chi_{ionic}]$ is an estimate of the bond's covalent character. The difference in electronegativity, $\Delta\chi$, is divided by a constant, $\Delta\chi_{ionic}$. $\Delta\chi_{ionic}$ is a parameter that defines the value of $\Delta\chi$ corresponding to a completely ionic bond. For this case, $\Delta\chi_{ionic}$ was chosen to equal 2.70. The ratio of $\Delta\chi$ to $\Delta\chi_{ionic}$ equals one if the diatomic is completely ionic and the weighting factor will be zero (no covalent character). If the difference in electronegativity is zero, the weighting factor is one.

Figure 3 illustrates the results for a wide range of diatomic molecules with widely different electronegativities. The numerical results are listed in Table III. The

Table III   Measured and Calculated Bond Dissociation Energies (in eV) for a Series
of Diatomics

MX	$D_{0(exp)}$	$D_{(calc)}$[†]	IP[$^2\Sigma$]	IP(M)	IP(X)	χ(M)	χ(X)	% err
H$_2$	4.480	4.5	15.95	13.60	13.60	2.20	2.20	-
Cl$_2$	2.479	3.1	16.10	14.46	14.46	3.16	3.16	25
Br$_2$	1.970	2.5	14.62	13.31	13.31	2.96	2.96	27
I$_2$	1.542	1.5	12.93	12.15	12.15	2.66	2.66	3
ICl	2.153	2.0	14.29	12.15	14.46	2.66	3.16	7
IBr	1.817	1.9	13.70	12.15	13.31	2.66	2.96	5
HF	5.856	5.8	19.50	13.60	20.01	2.20	3.98	1
HCl	4.433	4.2	16.61	13.60	14.46	2.20	3.16	5
BrH	3.755	3.6	15.60	13.31	13.60	2.20	2.96	4
IH	3.054	2.6	14.25	12.15	13.60	2.20	2.66	15
NaCl[a]	4.230	4.2	9.80	5.14	14.46	0.93	3.16	1
NaBr[a]	3.740	3.7	9.45	5.14	13.31	0.93	2.96	1
NaI[a]	3.000	3.2	9.21	5.14	12.15	0.93	2.66	7
KI[a]	3.310	3.5	8.66	4.34	12.15	0.82	2.66	6
RbI[a]	3.370	3.4	8.48	4.18	12.15	0.82	2.66	1

[†]Equation 8 was used to calculate D_{calc} (± 0.1 eV). All these values were shifted a constant -0.22 eV. This constant was found by evaluating D_{calc} for the H$_2$ molecule and calibrating this value to correspond to the experimentally measured D(H$_2$) value.
[a] Ref. 22.

electronegativities used in Table III are Pauling's electronegativities.(10) We have not made any attempts to improve the correlations with more appropriate electronegativity scales. It should also be stressed that Figure 3 represents a direct application of Equation 8. The are no parameters in Equation 8 to be optimized in order to improve the fit, except perhaps the choice of $\Delta\chi_{ionic}$. As can be seen in Figure 3, this one general equation gives good values for the bond energies in a wide range of diatomic molecules.

Multiple Bonds, N$_2$.
The examination of multiple bonded diatomics illustrates additional relationships between ionization energies and bond energies. One important distinction is the ability to separate the different symmetry σ and π bonding components of the total bond. This is not possible by other methods of thermodynamic measurements because these methods affect the entire bond and do not separate the individual symmetry components.

Again, the first step toward calculating bond energies with the bond order approximation is identifying the atomic states which correlate with the dissociation products for ionization of a specific molecular state. The molecular orbital view of N$_2$ consists of two degenerate π levels and a set of sigma levels. Figure 4 is a correlation diagram of the experimentally measured ionization energies of the appropriate molecular and atomic states. The final atomic nitrogen states in Figure 4 are simply labeled as 2s and 2p. As stated previously, the bond order approximation applies to

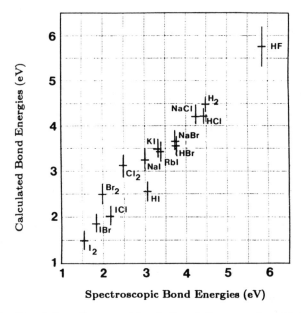

Figure 3. Correlation diagram of bond dissociation energies with Equation 8 versus measured bond dissociation energies.

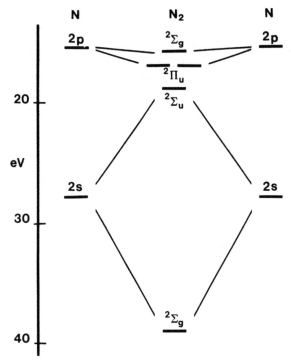

Figure 4. Experimental ionization energy correlation diagram of N_2.

the average configuration of the atomic states. All the atomic states which correspond to a $2s^2 2p^2$ electron filling of the nitrogen cation (the atomic $^3P_{0,1,2}$, $^1D_2$ and $^1S_0$ states) are statistically averaged to give a value of 15.46 eV as the ionization potential of an atomic nitrogen 2p electron for use with the bond order approximation. Similarly, the 2s ionization to the nitrogen cation produces the $2s^1 2p^3$ valence filling, and the corresponding states (the atomic $^5S_2$, $^3S_1$, $^3D_{3,2,1}$, $^3P_{2,1,0}$, $^1D_2$ and $^1P_1$ states) are statistically averaged to give a value of 27.84 eV.

The sigma bond is a sum of the $1\sigma_g$, $1\sigma_u$ and $2\sigma_g$ molecular valence levels. The σ contribution to the total bond is then given by the equation below:

$$\begin{aligned}
D(\sigma) \approx\; & 2 \cdot [IP(N_2, \tilde{D}^2\Sigma_g^+) - IP(N,2s)] \\
& + 2 \cdot [IP(N_2, \tilde{B}^2\Sigma_u^+) - IP(N,2s)] \\
& + 2 \cdot [IP(N_2, \tilde{X}^2\Sigma_g^+) - IP(N,2p)]
\end{aligned} \tag{9}$$

where $D(\sigma)$ is the sum of the stabilizing energies attributed to the σ_g interactions and the destabilizing energy attributed to the σ_u molecular level. Note that ionization from each σ bond results in the loss of "half the bond" which then requires the factors of two in the above equation. To determine the π contribution, the stabilization energy between atomic cation N^+ $2s^2 2p^2$ state (2p) and molecular $\tilde{A}^2\Pi_u$ state is found. Since this represents a fourth of the entire two π bonding contributions, the following equation gives $D(\pi)$:

$$2 \cdot D(\pi) = 4 \cdot [IP(N,2p) - IP(N_2, \tilde{A}^2\Pi_u^+)] \tag{10}$$

and

$$D(total) = 2 \cdot D(\pi) + D(\sigma) \tag{11}$$

Table IV list the atomic([7]) and molecular states([23,24]) upon ionization for N_2.

The calculated bond dissociation energy for N_2 of 10.28 eV is 5% greater than the experimentally determined value of 9.759 eV. This is the same error found in the analysis of the H_2 molecule. The sigma contribution to the overall bond strength is 4.20 ± 0.05 eV and is greater than the single π contribution of 3.04 ± 0.03 eV, as theory would predict.

Table IV Experimental and Calculated Bond Dissociation Energies (in eV) for N_2

$D_{0(exp)}$	Molecular		Atomic[†]		$D_{(calc)}$	
N_2 9.759						10.28(6)
	$\tilde{X}^2\Sigma_g^+$	15.57(3)	2p	15.46(1)	$2\sigma_g$	0.22(3)
	$\tilde{A}^2\Pi_u^+$	16.98(3)	2s	27.84(1)	$1\pi_u$	3.04(3)
	$\tilde{B}^2\Sigma_u^+$	18.75(3)			$1\sigma_u$	-18.18(3)
	$\tilde{D}^2\Sigma_g^+$	38.92(1)			$1\sigma_g$	22.16(3)

[†]See section on N_2 for discussion.

Organometallics.

The principles developed in the previous sections apply directly to the study of bond energies in organometallic molecules, and should help the reader in gaining useful information from publications of the photoelectron spectra of organometallic molecules. There is much less thermodynamic information available for organometallic molecules than for the diatomic molecules used to illustrate these principles. The thermodynamic information for organometallic molecules is also much less precise. Widely different results are often obtained by different techniques. In this section it will be shown that the ionization energies can contribute to the determination of more consistent collections of bond dissociation energies. Even when little additional information is available, it is possible to obtain simple determinations of relative bond energies.

One of the most extensively studied areas of organometallic bond energies involves the metal-ligand bonds in $(CO)_5Mn$-ligand complexes. These bond energies are determined through thermodynamic cycles which depend upon the accurate evaluation of the first ionization potential of $Mn(CO)_5\bullet$ and/or the metal-metal bond energy in $Mn_2(CO)_{10}$. There have been many estimates of the $Mn(CO)_5\bullet$ ionization energy from other techniques.([25]) In many cases, small differences in large numbers and additional assumptions have led to clearly erroneous values. A common example is mass spectrometric determinations of bond energies, particularly through appearance potentials, which are subject to errors of measurement and interpretation.([26]) An estimate of the ionization energy of the $Mn(CO)_5\bullet$ fragment reported from calorimetric data is 8.1 eV,([27]) and a mass spectrometry appearance potential measurement indicates a value of 8.4 eV.([28]) These values are greater than the initial ionization energy of $Mn_2(CO)_{10}$ (7.99 eV vertical and 7.69 eV adiabatic). The first ionization energy of molecular $Mn_2(CO)_{10}$ is from the metal-metal σ bond, which correlates with the bond formation from combination of the $Mn(CO)_5\bullet$ radicals. It is not reasonable that ionization of the Mn-Mn bond should be less stable than ionization of the $Mn(CO)_5\bullet$ radicals. The ionization energy of $Mn(CO)_5\bullet$ must be less than 7.69 eV.

We have previously estimated, by two different approaches, that the vertical ionization energy of $Mn(CO)_5\bullet$ is about 7.4±0.2 eV.([29]) Precise measurements of molecular ionization energies improve the confidence in the value of this ionization energy. For instance, a third approach to obtaining the ionization energy of $Mn(CO)_5\bullet$ depends on the appearance potential of $Mn(CO)_5^+$ from $Mn(CO)_5H$, which is a combination of the dissociation energy of $Mn(CO)_5H$ into $Mn(CO)_5\bullet$ and $H\bullet$ plus the ionization energy of $Mn(CO)_5\bullet$ to $Mn(CO)_5^+$. This appearance potential is reported to be 10.3 eV.([27]) However, we have often found that reported appearance potentials are too high by several tenths of an electron volt.([30]) For instance, the appearance potential of $Mn_2(CO)_{10}^+$ is reported to be 8.42 eV,([31]) rather than the 7.69 eV adiabatic ionization energy cited above from photoelectron spectroscopy. It is seen from the photoelectron spectrum of $Mn_2(CO)_{10}$ that the appearance potential value corresponds to the second ion state rather than the first ion state with a vertical IP of 8.05 eV.([27]) Assuming at least a 0.3 eV overestimate in the appearance potential of $Mn(CO)_5^+$ and knowing that the homolytic dissociation energy of the Mn-H bond in $Mn(CO)_5H$ is about 2.6 eV (60 kcal/mole ([27],[32])), the ionization energy of $Mn(CO)_5\bullet$ is again estimated to be 7.4 eV.

The most direct determination of the ionization energy of $Mn(CO)_5\bullet$ comes from Equation 1 applied to $Mn(CO)_5H$:

$$IP(Mn(CO)_5\bullet) = IP(Mn(CO)_5H) + D(Mn(CO)_5H^+) - D(Mn(CO)_5H) \quad (12)$$

The best current values (vertical $IP(Mn(CO)_5H) = 8.85$ eV ($\underline{33}$); $D(Mn(CO)_5H^+) = 1.0$ eV ($\underline{34}$); $D(Mn(CO)_5H) = 2.6$ eV (above)) yield a vertical ionization energy for $Mn(CO)_5 \bullet$ of 7.25 eV.

An estimate of the Mn-Mn bond strength in $Mn_2(CO)_{10}$ can now be made using the ionization energies and Equation 4. The assumption that leads to this equation requires the vertical ionization energy of the Mn-Mn bond of $Mn_2(CO)_{10}$ (8.05 eV) and the adiabatic ionization energy of the $Mn(CO)_5 \bullet$ fragment. If we select 7.2 eV as the adiabatic ionization energy of $Mn(CO)_5 \bullet$ (the lower end of the estimates discussed above), we obtain 1.7 eV for the bond dissociation energy. This agrees with the value of 1.7 ±0.4 eV reported by Beauchamp.($\underline{34}$) The uncertainty in our value is also several tenths of an electron volt, but it supports the larger values for the metal-metal dissociation energy of $Mn_2(CO)_{10}$.

In many cases there is not sufficient information to obtain estimates of absolute bond energies. However, by comparing the bonds of two different species with a common fragment, such as comparing the hydride and methyl bonds with the same metal fragment, relative bond energies can be obtained. In these cases, two equations of type 2 can be written, one for the M-H bond and one for the M-C bond. Taking the difference between these equations, the ionization energy of the common metal fragment cancels out, and one obtains:

$$IP(M\text{-}H) - IP(M\text{-}C) = \{D(M\text{-}H) - D(M\text{-}H^+)\} - \{D(M\text{-}C) - D(M\text{-}C^+)\} \quad (13)$$

where (M-L) represents the particular metal fragment-ligand bond for L = H or L = C(alkyl). This equation is further simplified if it is recognized that the dissociation energies of the ions, $D(M\text{-}C^+)$ and $D(M\text{-}H^+)$, are generally small in comparison to the dissociation energies of the neutral molecules. The general trend of weaker bonds with increasing metal oxidation state is a characteristic feature of organometallic chemistry, where an oxidation of the metal center is a common chemical approach to effect disproportionation of organic ligands from metal complexes. As shown above, the dissociation of a hydrogen atom from the $Mn(CO)_5H^+$ ion is 1.0 eV in comparison to the 2.6 eV value for dissociation from the neutral molecule. It should further be noted that the ground ion state of $Mn(CO)_5H$ for this dissociation energy refers to removal of a metal-based electron from $Mn(CO)_5H$ that is nonbonding with respect to hydrogen. The ionization in Equation 13 above refers to removal of an electron from the predominantly Mn-H bond, and the dissociation energy from this ion state will be considerably less than 1.0 eV. Thus the difference between $D(M\text{-}H^+)$ and $D(M\text{-}C^+)$ is generally the difference between two relatively small numbers, and Equation 13 can be reduced to:

$$IP(M\text{-}H) - IP(M\text{-}C) \approx D(M\text{-}H) - D(M\text{-}C) \quad (14)$$

An example of relationships between ionization energies and bond energies is illustrated with the Mn-H and Mn-CH$_3$ bond energies in $Mn(CO)_5H$ and $Mn(CO)_5CH_3$. The adiabatic ionization energies associated with the Mn-H and Mn-CH$_3$ bonds of these complexes are approximately 10.0 eV(33) and 9.3 eV($\underline{35}$) respectively. These bond energies have been reported to be 2.6 eV and 2.0 eV respectively on the basis of microcalorimetric measurements.(27) The relative stabilities of the bonds and the ionizations are the same within the uncertainties of the experiments.

We have recently reported similar comparisons from the photoelectron data of niobium and tantalum complexes.($\underline{36}$) From comparison of the spectra of $Cp_2Nb(CO)H$ and $Cp_2Nb(CO)CH_3$ as well as $Cp_2Ta(C_2H_4)H$ and $Cp_2Ta(C_2H_4)(C_2H_5)$, the relative ionization energies of early transition metal M-H and M-C(alkyl) bonds is observed to be M-H > M-C. It can in turn be concluded from this trend that the M-H bond energy is greater than the M-C bond energy. In the case of the Nb(CO)H and Nb(CO)CH$_3$ complexes the energy of the Nb-H ionization is about 0.5 eV greater than that of the corresponding Nb-CH$_3$ ionization. This indicates that the Nb-H bond is about 12 to 14 kcal mol^{-1} more stable than a Nb-C(methyl) bond in these molecules. This result is in good agreement with other thermochemical results involving group IV and actinide bent metallocenes, where M-H > M-C by 11-15 kcal mol^{-1}.($\underline{37}$)

Conclusions.
It has been shown that bond dissociation energies are closely related to photoelectron ionization energies. There are several advantages as well as limitations to using ionization energies for evaluating bond dissociation energies. Among the present limitations are: (1) If photoelectron ionization information is the only information available, the ionization energies of radical species are necessary to obtain absolute bond dissociation energies. It is often difficult to obtain photoelectron information on radical species. However, between complexes with similar fragment radicals, trends in bond strengths between the complexes may be evaluated without the ionization energies of the radicals. (2) Spectroscopic methods are inherently precise in determining bond dissociation energies. However, uncertainty in determination of the ionization energies leads directly to uncertainty in the determination of the bond energies. An uncertainty of a few tenths of an eV in an ionization energy leads to an uncertainty of several kcal/mole in the bond energy. Nonetheless, the uncertainties of this technique are comparable to those of other techniques. (3) The ionizations must be assigned correctly. For polyatomic molecules, ionizations often correspond to delocalized states that do not correlate to a single localized bond. (4) Key ionizations may be difficult to observe.

Among the advantages of relating ionization energies and bond energies are: (1) If other thermodynamic information, such as appearance potentials or other calorimetric data, is available, then additional bond dissociation energies may be determined directly by Equation 2. (2) The ionization information can be used to check the consistency and validity of results from other methods. (3) The ionization information is unique in being able to separate different symmetry and orbital contributions to the total bond. The nitrogen molecule is one example. We are presently studying similar relationships in multiple metal-metal bonds. (4) The technique applies to gas phase molecules without complications from solvents and other side effects or chemical precesses. (5) The technique provides additional information on the nature of the bond (ionic, covalent, etc.) and the factors that contribute to the strength of bonding in organometallic molecules.

Acknowledgments.
D.L.L. acknowledges support by the U.S. Department of Energy (Division of Chemical Sciences, Office of Basic Energy Sciences, Office of Energy Research, DE-FG02-86ER13501), the National Science Foundation (CHE8519560), and the Materials Characterization Program, Department of Chemistry, University of Arizona. We would like to thank Dr. Mark Jatcko for supplying accurate ionization values for the low lying N_2, $\tilde{D}^2\Sigma_g^+$ state.

Literature Cited.
1. Bruno, J. W.; Marks, T. J.; Morss, L. R. J. Am. Chem. Soc. 1981, 103, 190-2.
2. Purcell, K. F.; Kotz, J. C. Inorganic Chemistry; W. B. Saunders Company: Philadelphia, Pennsylvania, 1977, Chapter 6.
3. Douglas, B. E.; McDaniel, D. H.; Alexander, J. J. Concepts and Models of Inorganic Chemistry; 2nd. ed.; Wiley: New York, New York, 1983, p. 514.
4. Walsh, A. D. Trans. Faraday Soc. 1946, 42, 779-89.
5. Al-Joboury, M. I.; May, D. P.; Turner, D. W. J. Chem. Soc. 1965, 616-22.
6. Herzberg, G. Molecular Spectra and Molecular Structure. I. Spectra of Diatomic Molecules; Van Nostrand Reinhold: New York, New York, 1950.
7. Moore, C. Atomic Energy Levels; Vol I, II and III, Nat. Stand. Ref. Data Ser., Nat. Bur. Stand.: Washington, D.C., (U.S.), 1971.
8. Brundle, C. R.; Baker, A. D. Electron Spectroscopy: Theory, Techniques and Applications; Vol 2, Academic Press: New York, New York, 1978.
9. Huber, K. P.; Herzberg, G. Molecular Spectra and Molecular Structure. IV. Constants of Diatomic Molecules; Van Nostrand Reinhold: New York, New York, 1979.
10. Huheey, J. E. Inorganic Chemistry; 3rd Ed., Harper and Row: New York, New York, 1983.
11. Potts, A. W.; Price, W. C. Trans. Faraday Soc. 1971, 67, 1242-52.
12. Evans, S.; Orchard, A. F. Inorg. Chim. Acta 1971, 5, 81-5.
13. Cornford, A. B.; Frost, D. C.; McDowell, C. A.; Ragle, J. L.; Stenhouse, I. A. J. Chem. Phys. 1971, 54, 2651-7.
14. Frost, D. C.; McDowell, C. A.; Vroom, D. A. J. Chem. Phys. 1967, 46, 4255-9.
15. Imre, D; Koenig, T. Chem. Phys. Lett. 1980, 73, 62-6.
16. Higginson, B. R.; Lloyd, D. R.; Roberts, P. J. Chem. Phys. Lett. 1973, 19, 480-2.
17. Walker, T. E. H.; Dehmer, P. M.; Berkowitz, J. J. Chem. Phys. 1973, 59, 4292-8.
18. Weis, M. J.; Lawrence, G. M.; Young, R. A. J. Chem. Phys. 1970, 52, 2867-70.
19. Kimura, K.; Katsumata, S.; Achiba, Y.; Yamazaki, T.; Iwata, S. Handbook of He I Photoelectron Spectra of Fundamental Organic Molecules; Halsted Press: New York, New York, 1981.
20. Lempka, H. J.; Passmore, T. R.; Price, W. C. Proc. Roy. Soc. Lond. A 1968, 304, 53-64.
21. Berkowitz, J.; Dehmer, J. L.; Walder, T. E. H. J. Chem. Phys. 1973, 59, 3645-53.
22. Potts, A. W.; Williams, T. A.; Price, F. R. S. Proc. Roy. Soc. Lond. A 1974, 341, 147-61.
23. Turner, D. W.; Baker, C.; Baker, A. D.; Brundle, C. R. Molecular Photoelectron Spectroscopy; Wiley-Interscience: London, England, 1970.
24. Gardner, J. L.; Samson, J. A. R. J. Electr. Spectrosc. Relat. Phenom. 1973, 2, 259-66.
25. Simoes, J. A. M.; Schultz, J. C.; Beauchamp, J. L. Organometallics 1985, 4, 1238-42.
26. Pilcher, G.; Skinner, H. A. The Chemistry of the Metal-Carbon Bond; Hartley, F.R.; Patai, S. Eds.; Wiley: New York, New York, 1982.
27. Connor, J. A.; Zafarani-Moattar, M. T.; Bickerton, J.; El Saied, N. I.; Suradi, S.; Carson, R.; Takhin, G. A.; Skinner, H. A. Organometallics 1982, 1, 1166-74.
28. Bidinosti, D. R.; McIntyre, N. S. J. Chem. Soc., Chem. Commun. 1966, 555.
29. Russell, D. H. Gas Phase Inorganic Chemistry; Plenum Press: New York, New York, 1989, Chapter 8, 245.
30. Lichtenberger, D. L.; Kellogg, G. E. Acc. Chem. Res. 1987, 20, 379.

31. Bidinosti, D. R.; McIntyre, N. S. Can. J. Chem. 1970, 48, 593-7.
32. Halpern, J., Amer. Chem. Soc. Abstr. 1989, 198, INOR, 8.
33. Hall, M. B. J. Am. Chem. Soc. 1975, 97, 2057-65.
34. Stevens, A. E.; Beauchamp, J. L. J. Am. Chem. Soc. 1981, 103, 190-2.
35. Lichtenberger, D. L.; Fenske, R. F. Inorg. Chem. 1974, 13, 486-8.
36. Lichtenberger, D. L.; Darsey, G. P.; Kellogg, G. E.; Sanner, R. D.; Young, V. G., Jr.; Clark, J. R. J. Am. Chem. Soc. 1989, 111, 5019-28.
37. Bruno, J. W.; Marks, T. J.; Morss, L. R. J. Am. Chem. Soc. 1983, 105, 6824-32.

RECEIVED December 15, 1989

Chapter 7

Thermodynamics and Kinetics of Transition Metal–Alkyl Homolytic Bond Dissociation Processes

Jack Halpern

Department of Chemistry, The University of Chicago, Chicago, IL 60637

It is only during the past ten years that reliable and widely applicable methods for determining homolytic metal-alkyl bond dissociation energies of stable organometallic compounds in solution have been developed and that information about such bond dissociation energies has become available. Today about one hundred transition metal-alkyl bond dissociation energies have been determined, the majority for cobalt-alkyl complexes. Most of these have been from kinetic measurements. The scope, limitations and results of such determinations are discussed.

This paper is concerned with certain aspects of the thermodynamics and kinetics of transition metal-alkyl homolytic bond dissociation processes, notably of stable, ligated complexes in solution (L_nM-R, where L is a ligand and R = alkyl, benzyl, etc.)(1). The metal-alkyl bond dissociation energy of such a complex (BDE, strictly bond dissociation enthalpy) is defined as the enthalpy of the process represented by Equation 1.

$$L_nM\text{-}R \underset{k_{-1}}{\overset{k_1}{\rightleftharpoons}} L_nM\cdot \ + \ R\cdot \qquad (1)$$

A knowledge of such BDE's is important for an understanding of the kinetic and thermodynamic stabilities of organometallic compounds and of the thermodynamics of the many organometallic reactions that involve the formation or dissociation of transition metal-alkyl bonds, for example, insertion or oxidative addition-reductive elimination (Equations 2 and 3, respectively)(1-3).

$$L_nM\!\!\begin{array}{c} {}^{CO} \\ {}_{R} \end{array} \ + \ L \ \rightleftharpoons \ L_{n+1}M\text{-}\overset{\overset{O}{\|}}{C}\text{-}R \qquad (2)$$

$$L_nM \ + \ RH \ \rightleftharpoons \ L_nM\!\!\begin{array}{c} {}^{R} \\ {}_{H} \end{array} \qquad (3)$$

0097–6156/90/0428–0100$06.00/0

Certain biological processes, notably coenzyme B_{12}-dependent rearrangements (4), also are triggered by enzyme-induced homolytic Co-C bond dissociation of the coenzyme, Equation 4 (coenzyme B_{12} = 5'-deoxyadenosyl cobalamin, abbreviated Ado-B_{12}; B_{12r} = cob(II)alamin).

$$Ado-B_{12} \longrightarrow Ado. + cob(II)alamin \ (B_{12r}) \tag{4}$$

Notwithstanding their importance, it is only within the past ten years or so that reliable and widely applicable methods for determining such BDE's have been developed and applied. To date approximately sixty absolute BDE values have been determined, the majority for Co complexes (5), but also for a few complexes of Mn (6), Fe (7), Ru (8), Mo (7,9), W (7), U (10), Sm (10) and Ir (11). In addition, a number (ca 40) of relative BDE's. I.e., $D(L_nM-R)-D(L_nM-X)$ where X = R', I, OR', H, etc., have been determined for complexes of Mn (12), Sc (13,14), Hf (14), Th (15,16), Zr (16,27), Rh (18-20) and Ir (21).

Of related interest are extensive measurements in the gas phase, by a variety of techniques such as ion cyclotron resonance and guided ion beam experiments, of the M-C bond dissociation energies of unligated ("bare") ionic and neutral metal alkyls, e.g., M^+-CH_3 and M-CH_3, where M = Sc, Ti, V, Cr, Mn, Fe, Co, Ni, Cu, Zn, Ru, Rh, etc.(22).

Determination of Bond Dissociation Energies from Kinetic Measurements: Macroscopic Description

In 1982 (23) we described a method of determining transition metal-alkyl bond dissociation energies, based on measuring the kinetics of reaction 1 and using the relation,

$$D_{L_nM-R} \ (\approx \Delta H_1^{\ddagger}) = \Delta H_1^{\ddagger} - \Delta H_{-1}^{\ddagger} \tag{5}$$

where $\Delta H_1^{\ddagger}$ and $\Delta H_{-1}^{\ddagger}$ are the activation enthalpies of the forward and reverse reactions of Equation 1, derived from the temperature dependencies of the macroscopic rate constants, k_1 and k_{-1}. Since then, this has been widely applied by ourselves and many others (8) and continues to be the most widely used method for determining such BDE's. In particular, the great majority of known absolute transition metal-alkyl BDE's (including virtually all those for cobalt-alkyl complexes) have been determined by this method (8). At this stage it probably constitutes also the most accurate ($\pm \sim$ 1-2 kcal/mol) method of determining such BDE's.

In favorable cases the determinations of k_1 and $\Delta H_1^{\ddagger}$ are straightforward. An obvious requirement is that the compound decompose (not necessarily exclusively) by metal-alkyl bond homolysis. This may be established by adding an appropriate scavenger (TRAP, for example, n-C_8H_{17}SH (23) or 2,2,6,6-tetramethylpiperidineoxy, TEMPO (24) to trap the initially-formed free radical in accord with the scheme of Equations 1, 6 and 7, where L_nM. is, for example, cobalt(II).

$$L_nM\text{-}R \underset{k_{-1}}{\overset{k_1}{\rightleftharpoons}} L_nM. \ + \ R. \tag{1}$$

$$R. \ + \ TRAP \xrightarrow{k_2} R\text{-}TRAP \tag{6}$$

OVERALL: $L_nM\text{-}R \ + \ TRAP \longrightarrow L_nM. \ + \ R\text{-}TRAP \tag{7}$

Identification of the product R-TRAP and fitting of the kinetic data to the steady-state rate law of Equation 8 serve to confirm the mechanism and to evaluate k_1 and k_{-1}/k_2. $\Delta H_1^{\ddagger}$ may be deduced from the temperature-dependence of k_1.

$$\frac{-d\ln[L_nM\text{-}R]}{dt} = k_{obsd} = \frac{k_1 k_2 \ [TRAP]}{k_{-1}[L_nM.] + k_1[TRAP]} \tag{8}$$

In favorable cases k_{-1} may be evaluated either directly (e.g., from the recombination of flash photolytically-generated $L_nM.$ and R.) ($\underline{25,26}$) or from the value of k_{-1}/k_2 where k_2 can be independently evaluated (e.g., R. + n-$C_8H_{17}S\bar{H} \xrightarrow{} RH + n\text{-}C_8H_{17}S.$ ($\longrightarrow \frac{1}{2}$(n-$C_8H_{17}S)_2$) ($\underline{23}$) or R. + Tempo $\longrightarrow$ R-Tempo ($\underline{27}$)). As elaborated below, this usually is not necessary. Additional decomposition pathways give rise to additional terms in the rate-law which are not inhibited by $L_nM.$ and which typically result in the formation of distinctive products. As elaborated below, a common such pathway, when the alkyl group contains β-hydrogen atoms, is olefin elimination according to Equation 9 ($\underline{23,28\text{-}30}$).

$$L_nM\text{-}CH_2CH_2R' \underset{}{\overset{k_9}{\rightleftharpoons}} L_nM. \ + \ CH_2\text{=}CHR' \ + \frac{1}{2} H_2 \tag{9}$$

The majority of BDE determinations of the type described above have been made on cobalt-alkyl complexes. Among the reasons for this emphasis are: (a) the relevance of such complexes as models for coenzyme B_{12} whose action is triggered by enzyme-induced cobalt-carbon bond homolysis (Equation 4) ($\underline{4}$), (b) the typically low range ($\underline{ca}$ 20-30 kcal/mol) in which most cobalt-alkyl BDE's lie, permitting homolysis to be studied at relatively low temperatures, often without intervention of other reactions ($\underline{5}$), and (c) the monomeric nature of most cobalt(II) complexes. This permits variation of the concentration of the latter (i.e., of $L_nM.$) and hence determination of the full rate law of Equation 8 ($\underline{5}$). The monomeric nature of cobalt(II) permits the determination of absolute BDE's of $L_nCo\text{-}R$ compounds also by other methods (e.g., calorimetric or from determination of the equalibrium constants of Equation 9)($\underline{28,29,31}$) in contrast, say to $(CO)_5Mn\text{-}R$ where such measurements yield only relative values whose magnitudes depend on a knowledge of the Mn-Mn BDE of $Mn_2(CO)_{10}$ ($\underline{12}$).

Microscopic Description

A more detailed description of the homolytic bond dissociation process in solution, the simplest case depicted by Equation 10, takes account of the fact that the initial products of dissociation are a geminate (caged) radical pair, $L_nM.$, $R.$, which is subject to competing cage reactions. (The subscripts, DISS, RECOMB, SEP, DIF refer to dissociation, recombination, separation and diffusional encounter, respectively.)(5,32).

$$L_nM-R \; \underset{k_{RECOMB}}{\overset{k_{DISS}}{\rightleftharpoons}} \; \overline{L_nM., R.} \; \underset{k_{DIF}}{\overset{k_{SEP}}{\rightleftharpoons}} \; L_nM. + R. \; \xrightarrow{k_{TRAP}} \; R-TRAP \qquad (10)$$

In the limit of high TRAP concentration, i.e., $k_{TRAP}[TRAP] \gg k_{DIF}[L_nM.]$,

$$\frac{-d\ln [L_nM-R]}{dt} = k_1 = \frac{k_{DISS} \, k_{SEP}}{k_{RECOMB} + k_{SEP}} \qquad (11)$$

Two limiting cases may be identified:

(a) $k_{SEP} \gg K_{RECOMB}$; $k_1 = k_{DISS}$

$$\Delta H_1^{\ddagger} = \Delta H_{DISS}^{\ddagger} = D_{L_nM-R} + \Delta H_{RECOMB}^{\ddagger} \qquad (12)$$

(b) $k_{RECOMB} \gg k_{SEP}$; $k_1 = \dfrac{k_{DISS} \, k_{SEP}}{k_{RECOMB}}$

$$\Delta H_1^{\ddagger} = D_{L_nM-R} + \Delta H_{SEP}^{\ddagger} - \Delta H_{RECOMB}^{\ddagger} \sim D_{L_nM-R} + \Delta H_{SEP}^{\ddagger} \qquad (13)$$

Case (b) corresponds to the diffusion-controlled limit.

In determining BDE's from kinetic measurements, instead of actually evaluating k_{-1} and $\Delta H_{-1}^{\ddagger}$ as described above, the following assumptions and approximations commonly are made (5):
(i) The recombination reaction (k_{-1}) is diffusion-controlled; therefore, from Equation 13,

$$D_{L_nM-R} = \Delta H_1^{\ddagger} - \Delta H_{SEP}^{\ddagger} \qquad (14)$$

(ii) $\Delta H_{SEP}^{\ddagger}$ ($\sim \Delta H_{DIF}^{\ddagger}$) $\sim \Delta H^{\ddagger}$(viscosity) $\qquad (15)$

Assumption (i) has been verified directly in a few cases. Indeed for virtually every case where the rate constant of combination of a metal radical ($L_nM.$) and an alkyl radical has been estimated the value ($\sim 10^8$-10^9 M^{-1} sec^{-1}, Table I) has been found to be close to the diffusion controlled limit. Moreover, the rate constants for self-combination of metal radicals, such as $Mn(CO)_5.$ or $Mo(CO)_3(C_5H_5).$, which might be expected to be reduced by steric barriers, also typically are found to be close to diffusion-controlled ($k \sim 10^8 - 10^9$ M^{-1} sec^{-1}, Table I). Thus, the assumption of diffusion-controlled recombination of $L_nM.$ and $R.$ would appear to be widely justified.

Assumption (ii) above, namely that $\Delta H^{\ddagger}_{DIF} \sim \Delta H^{\ddagger}$(viscosity), also appears to be consistent with available data, e.g., those in Table II for the diffusion-controlled self-reaction of $(CH_3)_2COH$ radicals. (More rigorously, $\Delta H^{\ddagger}_{DIF}$ is expected to correlate with the diffusion coefficient ($\underline{39}$) which, in turn, is related to viscosity ($\underline{40}$)).

For many common organic solvents of low viscosity, e.g., benzene, toluene, acetone, and for water in the range $50°\text{-}100°C$, $\Delta H^{\ddagger}$(viscosity) $\sim 2 \pm 1$ kcal/mol. These are the media in which the majority of determinations of BDE's from kinetic measurements have been made ($\underline{5}$). The above considerations would appear to justify the relation that has generally been used to deduce BDE's from such measurements, namely Equation 16, and suggest that such BDE values are reliable to within ± 2 kcal/mol (except in solvents of high viscosity such as ethylene glycol for which $\Delta H^{\ddagger}_{DIF}$ is expected to be larger--see below).

$$D_{L_n M-R} \sim \Delta H^{\ddagger}_1 - 2\text{kcal/mol} \qquad (16)$$

It has been pointed out ($\underline{32}$) that, because of contributions from S G and H may have maxima at different points along the reaction coordinate. In such cases, $\Delta H^{\ddagger}$ (i.e., the value of H at

TABLE I. Rate Constants for Metal-Alkyl Radical and Metal-Metal
          Radical Combination Reactions

Reaction[a]	k_{-1} ($M^{-1}sec^{-1}$)	Method	Ref.
$Co(Saloph)(Py) + CH_3CH_2CH_2\cdot$	10^8	b	($\underline{23}$)
$Co(Saloph)(Py) + (CH_3)_2CH\cdot$	10^9	b	($\underline{23}$)
$Co(Saloph)(Py) + (CH_3)_3CCH_2\cdot$	10^8	b	($\underline{23}$)
$Co(Saloph)(Py) + C_6H_5CH_2\cdot$	10^9	b	($\underline{23}$)
$Co(TPP)(THF) + C_6H_5CH_2\cdot$	3×10^9	c	($\underline{33}$)
$Co(OEP)(THF) + C_6H_5CH_2\cdot$	7×10^9	c	($\underline{33}$)
$Co(OEP)(PMe_2Ph) + C_6H_5CH_2\cdot$	5×10^9	c	($\underline{33}$)
$Co(DH)_2(PPh_3) + C_6H_5CH_2\cdot$	1×10^8	c	($\underline{33}$)
$Cob(II)alamin (B_{12r}) + R.$[d]	$(4\text{-}20) \times 10^9$	e	($\underline{25,26}$)
$Cob(II)alamin (B_{12r}) + Ado.$	$10^8 - 10^9$	e	($\underline{34}$)
$Mo(Cp)(CO)_3 + C_6H_5CH(CH_3)$	9×10^9	f	($\underline{7}$)
$Fe(Cp)(CO)_2 + C_6H_5CH(CH_3)$	1×10^9	f	($\underline{7}$)
$2Mn(CO)_5\cdot$	2×10^9	e	($\underline{35}$)
$2Re(CO)_5\cdot$	2×10^9	e	($\underline{35}$)
$2Co(CO)_4\cdot$	4×10^8	e	($\underline{36}$)
$2Co(CO)_4(PBu_3^n)\cdot$	9×10^7	e	($\underline{36}$)
$Fe(Cp)(CO)_2\cdot$	3×10^9	e	($\underline{35}$)
$2Mo(Cp)(CO)_3\cdot$	3×10^9	e	($\underline{35}$)
$2W(Cp)(CO)_3\cdot$	2×10^9	e	($\underline{35}$)

[a]TPP = tetraphenylporphyrin; OEP = octaethylporphyrin; DH_2 = dimethylglyoxime; [b]Competition with trapping by $n\text{-}C_8H_{17}SH$ ($\underline{37}$). [c]Competition with trapping by TEMPO ($\underline{27}$). [d]R = CH_3, C_2H_5, C_3H_7. [e]Direct. [f] Competition with self trapping.

Table II. Comparison of $\Delta H^{\ddagger}$(viscosity) and $\Delta H^{\ddagger}$ for the
Diffusion-Controlled Self-Reaction of $(CH_3)_2\dot{C}OH$ Radicals

	$\Delta H^{\ddagger}$(Kcal/mol)	
Solvent	Viscosity[a]	$2(CH_3)_2COH$[b]
Acetonitrile	1.2	0.9
Heptane	1.3	2.1
Benzene	1.9	0.6
Tetradecane	3.1	3.1
Water	3.6	3.7
2-Propanol	4.6	2.6
2-Butanol	4.3	4.0

[a]0-50°C, from Ref. (38). Other values (0-50°C): acetone, 0.9;
toluene, 1.5; ethylene glycol, $\sim$7; ethanol, 2.7. [b]Ref. (39).

the point where G is a maximum) may not correspond to the maximum
value of H. This may give rise to a correction of BDE's deduced by
the above method but the discrepancy is expected to be small,
usually within the experimental uncertainty of the method (typically
± 2 kcal/mol).

Solvent Effects

Solvent effects may influence the magnitudes of BDE's and their
determination in several ways.
        First of all the BDE itself may depend on the solvent as a re-
sult of contributions of solvation to the enthalpy of the bond dissociation process, i.e., differences between the solvation enthalpies of the reactant (L$_n$M-R) and products (L$_n$M· + R·) of Equation 1.
There is not much direct information about this. Most of the applications of the kinetic method to date have involved low spin
octahedral d^6 complexes (notably of Co^{III}) where the product of
homolysis is a 17 electron low spin five-coordinate d^7 complex.
Such complexes generally exhibit little tendency to add a sixth
ligand so that solvent influences due to coordination are likely to
be small.   Support for this is provided by invariance of the Co-C
BDE of PhCH(CH$_3$)-Co(DH)$_2$Py in solvents of considerably different
polarity and coordinating ability, e.g., toluene and acetone. Bond
dissociation processes that are accompanied by substantial changes
in charge distribution (unexpected for the homolytic dissociation of
neutral radicals) are, of course, expected to exhibit significant
medium effects.
        Even when the BDE itself is unaffected by solvent variation,
determination of the BDE from kinetic measurements through Equation
14 must take account of the solvent dependence of $\Delta H^{\ddagger}_{SEP}$ (approximated by $\Delta H^{\ddagger}_1$ (viscosity).   From data such as those in Table II, it

is expected that ΔH_{SEP}^{+} (hence, $\Delta H_1^{\ddagger}$) will increase in going to solvents of higher viscosity. This is supported by Table III which compares $\Delta H^{\ddagger}$ for the homolytic Co-C bond dissociation of several cobalt alkyl complexes in ethylene glycol with the corresponding values in solvents of considerably lower viscosity, notably toluene and water. Clearly the widely assumed value of ca 2 kcal/mol (8,23,24,43) for ΔH_{SEP}^{+} (Equation 16) must be modified for measurements in solvents of higher viscosity such as ethylene glycol. Comparison with $(CH_3)_3CCH_2-B_{12}$ (Table III) suggests that the ca 5 kcal/mol difference between the reported values of $\Delta H^{\ddagger}$ for the thermolysis of Ado-B_{12} in ethylene glycol and aqueous solutions (42,43) is due to such a solvent effect.

Comparison of BDE's Determined by Different Methods

In addition to the procedure described above for determining transition metal alkyl BDE's in solution from kinetic measurements, several other methods now are available for determining such BDE's. The most important of these are:

Equilibrium Measurements. Measurements of the temperature dependence of equilibrium constants of reactions involving transition metal-alkyl bond disruption yield values of $\Delta H\pm$ from which BDE's can be deduced through appropriate thermodynamic cycles. The first example of this application involved the determination of the Co-C BDE of $Py(DH)_2Co-CH(CH_3)Ph$ from measurements of the equilibrium constant of reaction 17, according to Equations 17-19 (28,29).

Table III.   Solvent Effects on Activation Enthalpies
of Co-C Bond Homolysis

	$\Delta H^{\ddagger}$ (kcal/mol)			
Compound	Ethylene Glycol	Toluene	Difference	Ref.
$PhCH_2-Co(OEP)(PMe_2Ph)$	33.3	29.1	4.2	(41)
$PhCH_2-Co(DH)_2(PPh_3)$	34.1	27.8	6.3	(41)
$PhCH_2-Co(DH)_2(PPh_3)$	29.1	24.8	4.3	(41)
$PhCH_2-Co(DH)_2(PEtPh_2)$	34.8	28.8	6.0	(41)
	Ethylene Glycol	Water		
Neopentyl-B_{12} Uncorr.	26.8	23.1	3.7	(30)
Neopentyl-B_{12} Corr.[a]	33	27	6	(30)
Ado-B_{12} Uncorr.	30.6	36.3	4.3	(42,43)
Ado-B_{12} Corr.[a]	34.5	28.6	5.9	(42,43)

[a]Corrected for equilibrium between reactive "base-on" form and unreactive protonated "base-off" form.

$$Py(DH)_2Co\text{-}CH(CH_3)Ph \xrightleftharpoons{K_{eq}} PhCH=CH_2 + Py(DH)_2Co^{II} + \tfrac{1}{2} H_2 \qquad (17)$$

$$PhCH=CH_2 + \tfrac{1}{2} H_2 \longrightarrow PhCHCH_3 \qquad (18)$$

OVERALL: $\quad PhCH(CH_3)\text{-}Co(DH)_2Py \;\text{---}\; PhCH(CH_3) + Co^{II}(DH)_2Py \qquad (19)$

$$D_{Co\text{-}C} = \Delta H_{19} = \Delta H_{17} + \Delta H_{18} = 22 - 2 = 20 \text{ kcal/mol} \qquad (20)$$

Other applications include determination of <u>relative</u> values of the $Cp_2^*HfH\text{-}H$ and $Cp_2^*HfH\text{-}C_6H_5$ BDE's from measurements of the equilibrium constants of reaction 21 (<u>14</u>).

$$Cp_2^*HfH_2 + C_6H_6 \rightleftharpoons Cp_2^*HfH\text{-}C_6H_5 + H_2 \qquad (21)$$

$$\Delta H_{20} = D(Cp_2^*HfH\text{-}H) - D(Cp_2^*HfH\text{-}C_6H_5) + D(C_6H_5\text{-}H)$$

$$-D(H\text{-}H) = 6 \text{ kcal/mol} \qquad (22)$$

$$D(Hf\text{-}H) - D(Hf\text{-}C_6H_5) = 1 \text{ kcal/mol} \qquad (23)$$

<u>Calorimetric Measurements</u>. In a few favorable cases, <u>absolute</u> BDE's can be determined from calorimetric measurements of appropriately coupled bond disruption processes. An example is the determination of the Co-C BDE of $Py(DH)_2Co\text{-}CH_3$ from thermochemical measurements of ΔH_{24} and ΔH_{25} (<u>31</u>).

$$Py(DH)_2Co\text{-}CH_3 + I_2 \longrightarrow Py(DH)_2Co\text{-}I + CH_3I; \; \Delta H_{24} = 20 \text{ kcal/mol} \qquad (24)$$

$$Py(DH)_2Co^{II} + \tfrac{1}{2} I_2 \longrightarrow Py(DH)_2Co\text{-}I; \; \Delta H_{25} = -3 \text{ kcal/mol} \qquad (25)$$

$$D(Py(DH)_2Co\text{-}CH_3) = \Delta H_{24} - \Delta H_{25} + D(CH_3\text{-}I) - \tfrac{1}{2}D(I\text{-}I)$$

$$= 33 \text{ kcal/mol} \qquad (26)$$

Similarly, calorimetric measurements (<u>10 a</u>) of ΔH of,

$$Cp_3'' U + \tfrac{1}{2} I_2 \longrightarrow Cp_3'' U\text{-}I \qquad (27)$$

and $\qquad Cp_3'' U\text{-}R + I_2 \longrightarrow Cp'' U\text{-}I + RI \qquad (28)$

(where $Cp'' = Me_3SiC_5H_4$) yield the following values (kcal/mol) for $D(Cp_3''U\text{-}R)$: R = CH_3 (45), $CH_2C_6H_5$ (26), $n\text{-}C_4H_9$ (29). A similar approach has yielded absolute BDE's for $Cp_2^*Sm\text{-}H$ and $Cp_2^*Sm\text{-}CH(SiMe_3)_2$ (52-57 and 46-48 kcal/mol, respectively) (<u>10b</u>).

In other cases <u>relative</u> BDE's can be determined from calorimetric measurements; for example, $D(Ir\text{-}R) - D(Ir\text{-}R')$ from measurements of ΔH of the oxidative additions of RI and R'I to $IrI(CO)(PMe_3)_2$ (<u>20</u>); $D(Th\text{-}R) - D(Th\text{-}OR')$ from measurements of ΔH for the alcoholysis of $Cp_2ThR(OR')$ (Equation 29) (<u>15,16</u>), etc.

$$Cp_2^*ThR(OR') + HOR' \longrightarrow Cp_2^*Th(OR')_2 + RH \qquad (29)$$

Independent estimation of D(Th-OR') premits derivation of absolute values of D(Th-R) (15,16).

Combination of kinetic and calorimetric measurements with photoaccoustic calorimetry has been used to determine the D(Ir-H) and D(Ir-cyclo-C_6H_{11}) BDE's of Cp*(PMe$_3$)IrH(cyclo-C_6H_{11}) (11). Photoaccoustic calorimetry also has been employed to determine the Mn-Mn BDE of Mn$_2$(CO)$_{10}$ (45).

Comparison of BDE's Determined by Different Methods

Only a few transition metal-alkyl BDE's, namely those listed in Table IV, have been determined by more than one of these methods. In each of these cases other determinations are in excellent agreement with those deduced from kinetic measurements supporting the validity of the latter method.

Olefin Formation Accompanying Homolytic Transition Metal-Alkyl Bond Dissociation

For metal complexes of alkyls containing β-hydrogen atoms, metal-alkyl homolytic bond dissociation often is accompanied by olefin formation. An example (23) is provided by Equation 30 (Saloph = N,N'-bid(salicylidene)-o-phenylenediamine).

$$\longrightarrow Py(Saloph)Co^{II} + CH_3CH_2CH_2 \cdot$$

$$(\xrightarrow{TRAP} CH_3CH_2CH_2\text{-}TRAP) \qquad (30a)$$

Py(Saloph)Co-CH$_2$CH$_2$CH$_3$

$$\longrightarrow Py(Saloph)Co^{II} + CH_3CH=CH_2$$

$$+ \tfrac{1}{2} H_2 \qquad (30b)$$

Table IV. Comparisons of Bond Dissociation Energies of
Alkyl-Cobalt Complexes Determined by Different Methods

Complex	Bond Dissociation Energy (Kcal/mol)		
	Kinetic	Calorimetric	Equilibrium
$C_6H_5CH_2$-Co(DH)$_2$(Py)	29[a]	31[b]	
$C_6H_5CH(CH_3)$-Co(DH)$_2$(Py)	20[c]		20[c]
$C_6H_5CH_2$-Mo(Cp)(CO)$_3$	32[d]	32[e]	

[a]Ref. (33).   [b]Ref. (31).   [c]Ref. (28,29).   [d]Ref. (7).   [e]Ref. (9).

Similar behavior has been observed for $Py(Saloph)Co-CH(CH_3)_2$ (23), $Py(DH)_2Co-CH(CH_3)C_6H_5$ (28,29) and $cyclo-C_5H_9CH_2-B_{12}$ (30). Table V reveals that in every such case ΔH for homolysis and olefin formation are similar for the two processes, suggesting that they proceed through a common rate-determining step. The most likely such step is metal-alkyl bond homolysis followed by competition of cage escape and separation of the radical pair with β-H transfer between the geminate alkyl and metal radicals (Equation 31). Concerted olefin elimination is a less likely processes. Such concerted elimination is expected to require a vacant coordination site which is not present in these cases. Additional evidence favoring this interpretation has been advanced (46).

$$L_nM-CH_2CH_2R \longrightarrow \overline{L_nM.,\ RCH_2CH_2}. \begin{cases} \longrightarrow L_nM. + RCH_2CH_2.(\xrightarrow{TRAP} \\ \qquad\qquad RCH_2CH_2-TRAP) \qquad (31a) \\ \\ \longrightarrow RCH=CH_2 + L_nM-H\ (\longrightarrow \\ \qquad\qquad L_nM. + \tfrac{1}{2}H_2) \qquad (31b) \end{cases}$$

Table V. Simultaneous Radical Formation and Olefin Elimination During Thermolysis of Cobalt Alkyl Complexes

Compound	$\Delta H^{\ddagger}$ (kcal/mol)		Ref.
	Homolysis	Olefin Formation	
$(CH_3)_2CH-Co(Saloph)(Py)$	19.8	21.8	(23)
$CH_3CH_2CH_2-Co(Saloph)(Py)$	23.4	27.1	(23)
$PhCH(CH_3)-Co(DH)_2(Py)$	21.3	22.4	(28,29)
$Cyclo-C_5H_9CH_2-B_{12}$	25.4	25.2	(30)

Available data (Table VI) now permit comparison of the relative rates of competitive cage reactions, depicted by Equation 32, for several systems.

$$L_nM.,\ R. \begin{cases} \xrightarrow{k_{RECOMB}} L_nM-R & (32a) \\ \xrightarrow{k_{OLEFIN}} Olefin + L_nM-H\ (\longrightarrow L_nM. + \tfrac{1}{2}H_2) & (32b) \\ \xrightarrow{k_{SEP}} L_nM^{\cdot} + R^{\cdot}\ (\xrightarrow{TRAP} R-TRAP) & (32c) \end{cases}$$

The fact that the relative rates for these cage processes all are close to unity suggests that all, including k_{SEP}, have very low activation barriers and is consistent with the assumption of diffusion-controlled recombination.

Table VI.   Relative Rates of Competitive Cage Reactions

$L_nM.$	R•	k_{SEP}/k_{OLEFIN}	k_{SEP}/k_{RECOMB}	Method	Ref.
$Mn(CO)_5$	$C_6H_5\overset{\bullet}{C}(CH_3)_2$	~ 1 (70°)		a	(47)
Cob(II)alamin	Ado.		0.3 (25°)	b	$\overline{(48)}$
Cob(II)alamin	$c\text{-}C_5H_9CH_2.$	1.3 (58-78°)		c	$\overline{(30)}$
$Co(DH)_2(Py)$	$C_6H_5\overset{\bullet}{C}H(CH_3)$	4(8°)-6(26°)		c	$(28,\overline{29})$
$Co(Saloph)(Py)$	$(\overset{\bullet}{C}H_3)_2CH.$	3(70°)		c	$\overline{(23)}$
$Co(Saloph)(Py)$	$CH_3\overset{\bullet}{C}H_2CH_2.$	2(70°)		c	$\overline{(23)}$

^{a}CIDNP/Isotopic tracer. ^{b}Picosecond spectroscopy.
^{c}Kinetics/Product distribution.

Acknowledgments

Support of this research through grants from the National Science Foundatiion, the National Institutes of Health and the Petroleum Research Fund administered by the American Chemical Society is gratefully acknowledged.

Literature Cited

1. Halpern, J. Acc. Chem. Res. 1982, 15, 238-244, and references cited therein.
2. Stoutland, P. O.; Bergman, R. T.: Nolan, S. P.; Hoff, C. D. Polyhedron 1988, 7, 1429-1440.
3. Marks, T. J. Pure Appl. Chem. 1989, 61, 1665-1672.
4. Halpern, J. Science 1985, 227, 869-875.
5. Halpern, J. Polyhedron 1988, 7, 1483-1490.
6. Nappa, M.; Santi, R.; Halpern, J. Organometallics 1985, 4, 34-41.
7. (a) Mancuso, C.; Halpern, J. Unpublished results. (b) Mancuso, C. Ph.D. Dissertation, The University of Chicago, 1990.
8. Collman, J. P.; McElwee-White, L.; Brothers, P. J.; Rose, E. J. Am. Chem. Soc. 1986, 108, 1332-1333.
9. Nolan, S. P.; Lopez de la Vega, R.; Mukerjee, S. L.; Gonzalez, A. A.; Zhang. K.; Hoff, C. D. Polyhedron 1988, 7, 1491-1498.
10. (a) Schock, L. E.; Seyam, A. M.; Sabat, M.; Marks, T. J. Polyhedron 1988, 7, 1517-1529. (b) Nolan, S. P.; Stern, D.; Marks, T. J. J. Am. Chem. Soc. 1989, 111, 7844-7853.
11. Nolan, S. P.; Hoff, C. D.; Stoutland, P. O.; Newman, L. J.; Buchanan, J. M.; Bergman, R.; Yang, G. K.; Peters, K. S. J. Am. Chem. Soc. 1987, 109, 3143-3145.
12. Connor, J. A.; Zafarani-Moattar, M. T.; Bickerton, J.; El Saied, N. I.; Suradi, S.; Carson, R.; Al Takhin, G.; Skinner, H. A. Organometallics 1982, 1, 1166-1174.
13. Thompson, M. E.; Baxter, S. M.; Bulls, A. R.; Berger, B. J.; Nolan, M. C.; Santarsiero, B. D.; Shefer, W. P.; Bercaw, J. E. J. Am. Chem. Soc. 1987, 109, 203-219.
14. Bulls, A. R.; Bercaw, J. E.; Manriquez, J. M.; Thompson, M. E. Polyhedron 1988, 7, 1409-1428.

15. Bruno, J. W.; Stecher, H. A.; Morss, L. R.; Sannenberger, D. C.; Marks, T. J. J. Am. Chem. Soc. 1986, 108, 7275-7280.
16. Sonnenberger, D. C.; Morss, L. R.; Marks, T. J. Organometallics 1985,4, 352-355.
17. Bruno, J. W.; Marks, T. J.; Morss, L. R. J. Am. Chem. Soc. 1983, 105, 6824-6832.
18. Wax, M. J.; Stryker, J. M.; Buchanan, J. M.; Kovac, C. A.; Bergman, R. G. J. Am. Chem. Soc. 1984, 106, 1121-1122.
19. Buchanan, J. M.; Stryker, J. M.; Bergman, R. G. J. Am. Chem. Soc. 1986, 108, 1537-1550.
20. Yoneda, G.; Blake, D. M. Inorg. Chem. 1981, 20, 67-71.
21. Jones, W. D.; Feher, F. J. J. Am. Chem. Soc., 1984, 106, 1650-1663.
22. Armentrout, P. B.; Beauchamp, J. L. Acc. Chem. Res. 1989, 22, 315-321 and references cited therein.
23. Tsou, T. T.; Loots, M.; Halpern, J. J. Am. Chem. Soc. 1982, 104, 623-624.
24. Finke, R. G.; Smith, B. L.; Mayer, B. J.; Molinero, A. A. Inorg. Chem. 1983, 22, 3677-3679.
25. Ferraudi, G. J., Endicott, J. F. J. Am. Chem. Soc. 1977, 99, 243-245.
26. Bakac, A.; Espenson, J. H. Inorg. Chem. 1989, 28, 4319-4322.
27. Chateauneuf, J.; Lusztyk, J.; Ingold, K. U. J. Org. Chem. 1988, 53, 1629-1632.
28. Halpern, J.; Ng, F. T. T.; Rempel, G. L. J. Am. Chem. Soc. 1979, 101, 7124-7126.
29. Ng, F. T. T.; Rempel, G. L. Mancuso, C.; Halpern, J. Organometallics, submitted.
30. Kim, S. H.; Chen, H. L.; Feilchenfeld, N.; Halpern, J. J. Am. Chem. Soc. 1988, 110, 3120-3126.
31. Toscano, P. J.; Seligson, A. L.; Curran, M. T.; Skrobutt, A. T.; Sonnenberger, D. C. Inorg. Chem. 1989, 28, 166-168.
32. Koenig, T. W.; Hay, B. P.; Finke, R. G. Polyhedron 1988, 7, 1499-1516.
33. Geno, M. K.; Halpern, J. Unpublished results.
34. Chen, E.; Chance, M. R. J. J. Inorg. Biochem. 1989, 36, 264. J. Biol. Chem. submitted.
35. Meyer, T. J.; Caspar, J. V. Chem. Rev. 1985, 85, 187-218 and references cited therein.
36. Wegman, R. W.; Brown, T. L. Inorg. Chem. 1983, 22, 183-186.
37. Encinad, M. V.; Wagner, P. J.; Scaiano, J. C. J. Am. Chem. Soc. 1980, 102, 1357-1360.
38. Visanath, D. S.; Natarajan, G. Data Book on the Viscosity of Liquids; Hemisphere Publishing Corp. New York, 1989.
39. Lehni, M.; Fischer, H. Int. J. Chem. Kinetics 1983, 15, 733-757.
40. Hildebrand, J. H. Viscosity and Diffusivity; Wiley, New York, 1977.
41. Geno, M. K.; Halpern, J. J. Chem. Soc. Chem. Comm. 1987, 1052-1053.
42. Hay, B. P.; Finke, R. G. Polyhedron 1988, 7, 1469-1481.
43. Halpern, J.; Kim, S. H., Leung, T. W. J. Am. Chem. Soc. 1984, 106, 8317-8319.

44. Ohgo,   Y.; Orisaku, K.; Hosegawa, E.; Takeuchi, S.   Chem. Lett.
     1986, 27-30.
45. Goodman,   J.   L.;   Peters,   K.   S.;   Vaida, V.   Organometallics
     1986, 5, 815-816.
46. Derenne,   S.; Gaudemer, A.; Johnson, M. D.   J. Organomet. Chem.
     1987, 322, 229-238.
47. Sweany,   R. L.; Halpern, J.   J. Am. Chem. Soc.   1977, 99, 8335-
     8337.
48. Endicott, J.   F.; Netzel, T. L.   J. Am. Chem. Soc.   1979, 101,
     4000-4002.

RECEIVED March 16, 1990

Chapter 8

Cage Pair Intermediates and Activation Parameters

T. Koenig[1], T. W. Scott[2], and James A. Franz[3]

[1]Department of Chemistry, University of Oregon, Eugene, OR 97403
[2]Corporate Research Laboratories, Exxon Research and Engineering Company, Clinton Township, Route 22 East, Annandale, NJ 08801
[3]Pacific Northwest Laboratories, P.O. Box 999, Richland, WA 99352

Equations that describe the effect of cage pair intermediates on apparent activation parameters for bond homolysis and recombination are presented. These equations demonstrate the curvature that is present in ln(k/T) versus 1/T relationships which depends on the activation parameters for the recombination and diffusive separation reactions of the cage pair intermediates. Recent results on laser photolysis of diphenyl disulfide in decalin solution are analyzed in terms of a simple chemical model and the radiation boundary hydrodynamic model for the cage effect. The resulting activation parameters suggest that the effective activation enthalpy for diffusive separation of the thermal cage pair is equal to that for diffusion of the free radical intermediates. This result is used to illustrate the predicted curvature for self-termination of the phenylthiyl free radicals and for the reactions of reactive free radicals with trapping agents. Curvature effects for reversible bond homolyses, as determined by observed rate constants for disappearance of organometallic precursors in the presence of the appropriate excess of free radical trapping agent, are also discussed.

A description of a bond thermolysis in solution requires at least two elementary steps. This is due to the fact that the immediate product of the bond cleavage step is a pair of radicals which are initially restricted to remain as neighbors by the surrounding solvent molecules. We follow Rabinowitch and Wood and term this reactive species the cage pair intermediate[1,2]. Scheme 1 depicts a simple chemical model for this kinetic sequence. Subsequent to the

0097–6156/90/0428–0113$06.00/0
© 1990 American Chemical Society

$$R\text{-}R \underset{k_c}{\overset{k_h}{\rightleftharpoons}} [R\cdot \ \cdot R]_c \underset{2k_D}{\overset{k_d}{\rightleftharpoons}} 2\ [R\cdot]$$

Precursor   Cage Pair   Free Radicals
                            Collisional

SCHEME 1   Chemical Model Collisional Cage Pair

homolysis step (k_h, Scheme 1), the collisional cage pair of Scheme
1 ($[R\cdot \ \cdot R]_c$) can be partitioned by the competition between the
reformation of the covalent precursor (k_c, Scheme 1) and diffusive
separation (k_d, Scheme 1) which leads to free radical
intermediates. In our[3] version of Scheme 1, both the cage
recombination (k_c) and diffusive separation (k_d) processes are
kinetically first order in cage pair. The process of reformation of
cage pairs from the free radical intermediates (k_D, Scheme 1) is
kinetically second order in free radicals. The latter rate
constant (k_D) must be kept distinct from that for diffusive
separation of the cage pair (k_d).
       The center of our focus is the effective energy surface
leading to and from cage pair intermediates. Figure 1 shows a
schematic comparison of the Gibbs free energy and enthalpy profiles
for a Scheme 1 reaction in the case where the barrier to
recombination of the cage pair is entirely entropic ($\Delta S^*_c < 0$)[4].
The transition state structure for the covalent precursor/cage pair
inter-conversion ($\ddagger_c$, Figure 1) is thus determined by the
corresponding maximum in the free energy profile. On the other
hand, the transition structure ($\ddagger_d$) for the diffusive separation
of the cage pair is determined by the position of the top of the
enthalpy barrier for relative movement of one of the cage pair
radicals past the solvent and away from its cage partner. This part
of the energy profile is uniquely a property of the liquid phase.
If the reaction coordinate is reduced to a simple radial separation
of the cage pair, then the transition structure for diffusive
separation ($\ddagger_d$) can be identified with the cage radius (R_c, Figure
1).
       The determination of differences in the activation
enthalpies[5] for the separation/combination steps ($\Delta H^*_d - \Delta H^*_c$) of
Figure 1 is quite simple but, to our knowledge, no determination of
the individual values of the activation parameters for the k_c and
k_d processes have previously been available. One of the primary
purposes of the present work is to discuss an analysis that yields
such values. These activation parameters, in turn, are used to
illustrate the curvatures that exist in Eyring or Arrhenius
treatments of the temperature dependences of the observed rate
constants for free radical recombination, trapping and formation by
thermolysis of a covalent precursor in solution.
       In order to establish a base for our analysis, the equations
that relate the activation parameters for the elementary steps of
Scheme 1 (Figure 1) to the observables for free radical self-
termination are presented first. Next, the distinctions between
collisional cage pairs and photochemical cage pairs are

established. The results section shows the correspondence that can exist between hydrodynamic and chemical models for the cage effect. The implications of an approximate equivalence between the activation enthalpy for cage pair separation (k_d, ΔH^*_d) and that for free radical diffusion (k_D, ΔH^*_D) are then developed.

Free Radical Recombination--A Scheme 1 system can be studied experimentally[6] by measuring the second order rate constants for self-termination of the free radical species. In other words, the reverse of Scheme 1 system can be studied starting with the free radicals which form ($2k_D$, Scheme 1) the collisional cage pair through diffusion. The observed second order rate constants for free radical self-termination ($2k_t$obs) depend on the value of k_D and the efficiency of the cage combination reaction (k_c) compared to the competing diffusive (re)separation (k_d) at the collisional cage pair stage. In quantitative terms, the experimental rate constant ($2k_t$obs) depends on the fractional cage efficiency [$F_c(T)$] which, for Scheme 1, is defined by equation (1). Equations (2)-(4) give the relationships of the observable self-termination rate constants ($2k_t$obs) and their activation parameters (ΔH^*_tobs, ΔS^*_tobs) with those for the elementary processes of Scheme 1. The fractional cage efficiency factor [$F_c(T)$] is temperature dependent to the extent that k_c/k_d is temperature dependent[7].

$$F_c(T) = k_c/(k_c + k_d); \quad [1/F_c(T)]-1 = k_d/k_c \quad (1)$$

$$2k_t\text{obs} = F_c(T) \cdot (2k_D) \quad (2)$$

$$\Delta H^*_t\text{obs} = \Delta H^*_D + [1-F_c(T)] \cdot (\Delta H^*_c - \Delta H^*_d) \quad (3)$$

$$\Delta S^*_t\text{obs} = \Delta S^*_D + R\ln[F_c(T)] + [1-F_c(T)][(\Delta H^*_c - \Delta H^*_d)/T] \quad (4)$$

These equations [(3)-(5)] are simply implied by composite mechanisms, such as Scheme 1, but have not received much previous attention. They express the curvature in Eyring ($\ln(k/T)$ vs $1/T$) treatments of observed rate constants for two step mechanisms where neither step is clearly rate determining. For the cage pair case, this includes the domain where the cage combination efficiency is from 10 to 90 percent ($0.1 < F_c(T) < 0.9$). A notational clarification, which is required by this curvature, is to introduce the apparent activation parameters (ΔH^*_tapp, ΔS^*_tapp) that would be obtained from the $\ln(2k_t\text{obs}/T)$ versus $1/T$ linearized fit. As we will show, the apparent activation parameters must be kept distinct from the ΔH^*_tobs and ΔS^*_tobs of equations (3) and (4). The latter values are labelled "obs" because they give the observed rate constant ($2k_t$obs) at any temperature in the range of observation. The apparent activation parameters give approximate values for $2k_t$obs that differ from the actual values by the extent of the curvature.

The values for ΔH^*_tapp and ΔS^*_tapp are not, by themselves, physically significant and could be quite misleading. In spite of this feature equation (5) shows a useful approximation for

$$F_c(T) \simeq 1 - [(\Delta H^*_t\text{app} - \Delta H^*_D)/(\Delta H^*_c - \Delta H^*_d)] \quad (5)$$

estimating the cage efficiency factor at the mean temperature
$[F_c(T)]$ in the range used to determine ΔH^*_tapp, presuming that the
values for ΔH^*_D and the $(\Delta H^*_c - \Delta H^*_d)$ differences are known.
The activation parameters for the purely diffusion controlled
formation of the collisional cage pair from the free radicals
(Scheme 1, k_D , ΔH^*_D, ΔS^*_D) are relatively well established[6b,9,10].
For nonpolar radicals in non-associating alkane solvents, the value
of ΔH^*_D should be very close to the Andrade energy [E_n, equation
(6)] for viscous flow of the solvent as predicted by the

$$n = A_n \exp(E_n/RT) \tag{6}$$

Stokes-Einstein-Schmoluchowski[11] (SSE) equation. This result, cast
in the present notation, is given in equations (7) where the
constant C incorporates the collision diameter of the collisional
cage pair and the numerical constants.

$$k_D = (kT/h)(C)(1/n) \; ; \tag{7}$$
$$\Delta H^*_D = E_n \; ; \quad \Delta S^*_D = R \ln[C/(A_n)]$$

The critical analysis by Shuh and Fischer[6b] on corrections to
equation (7) showed that, for alkyl radicals in alkane solvents,
the prescription of Spernol and Wirtz[12] (SW) gives adequate[13]
modifications to the simple SSE prediction. The SW recipe can
generally be re-expressed as in equation (8) where

$$k_D = (kT/h) \cdot mf \cdot (C)(1/n)^\alpha \; ; \tag{8}$$
$$\Delta H^*_D = \alpha E_n \; ; \quad \Delta S^*_D = R \ln[mf \cdot C/(A_n^\alpha)]$$

the α values are constants that depend on the solute/solvent system
and vary between 0.9 and 1. The major effect of the SW correction
comes from the temperature independent multiplier [mf, equation
(8)] which, in transition state models, affects the activation
entropy (ΔS^*_D). The activation parameters for the diffusion of free
radicals (k_D) are thus not too much of a problem. The main
difficulty is with effective activation parameters for reactions
leading out of the cage pair intermediates and our main goal is to
elucidate these values.

Photochemical Cage Pairs--Laser methods have progressed to the
point where direct observations of cage pair intermediates are
feasible. With pulse widths in the picosecond (psec) regime, the
time course of the formation and decay of cage pair intermediates
can be recorded as first demonstrated by Endicott and Netzel[14] for
aqueous solutions of coenzyme B$_{12}$. A recent report[15] on the 354.7
nm (35psec) pulsed photolysis of cis-decalin solutions of diphenyl
disulfide showed that the observed kinetic behavior for photolysis
of this precursor/solvent system was due to formation and reaction
of the cage pair species (Scheme 2). The present work provides a
further analysis of the temperature dependence exhibited by this
system which extracts the effective activation parameters for

$$C_6H_5S\text{-}SC_6H_5 \underset{k_{cp}}{\overset{h\nu}{\rightleftharpoons}} [C_6H_5S\cdot\cdot SC_6H_5]_p \xrightarrow{k_{dp}} 2\ [\ C_6H_5S\cdot\ ]$$

Precursor   Cage Pair   Free Radicals
                     Photochemical

Scheme 2   Chemical Model   Photochemical Cage Pairs

the reactions of the collisional cage pair from observations of the photochemical cage pair. It is important to remember that the photochemical cage pairs, like that of Scheme 2, are not identical to their collisional (Scheme 1) counterparts and the notation of Scheme 2 has added the subscript p to differentiate the photochemical cage pair reactions. Hydrodynamic models, like the radiation boundary model (RBM)[15] that was used in our preliminary account[16] of the behavior of the Scheme 2 system, can provide a basis for accounting for the differences in the two related cage pair types and for predicting the behavior of the collisional cage pair of Scheme 1 ($R = C_6H_5S$) from the observations on the photochemical cage pair. The following sections demonstrate results that can be derived from this type of analysis.

Results

The Radiation Boundary Model (RBM)-- The optical density of the Scheme 2 system at 435nm, produced by 35-40 psec (gaussian fwhm) laser pulses (0.5mJ per pulse at 354.7 nm), is due to the sum of the free and cage pair phenylthiyl radical concentrations. The expression for the time dependence of this sum [$SR\cdot(t)$], according to the RBM, is equation (9). This equation would be directly applicable if the temporal widths of the exciting and probing

$$SR\cdot(t) = 1-[L/K(1+L)\{erfc_1 - (exp_1 * exp_2 * erfc_2)\};$$

$$erfc_1 = erfc[(K-1)/2\tau^{1/2}]\ ,\ \ exp_1 = exp[(1+L)(K-1)],$$

$$erfc_2 = erfc\{[(1+L)/\tau^{1/2}]+[(K-1)/2\tau^{1/2}]\},$$

$$exp_2 = exp[(1+L)2\tau] \tag{9}$$

pulses were small compared to the time constant for the cage pair decay. This condition does not pertain to the phenylthiyl system in decalin with 35-40 psec pump and probe widths. Convolution of the decay function [equation (9)] and the pulse/probe temporal shapes is therefore necessary and Scott and Liu have established[16] that the numerically convoluted time dependence, based on equation (9), gives good agreement with experimental observations[17].
    The variables of equation (9) are the ratio (K) of the initial separation distance (d_0) to the sum of the hard sphere collision radii (s), the ratio (L) of the proportionality constant for the combination reaction (k_R) to the second order rate constant for pair diffusion (k_{d2})   and the ratio (τ) of the time elapsed since pair formation to the diffusive time constant (s^2/D). The fractional cage efficiency is the complement to the pair

survival [equation (9)] at infinite time. Equation (10) gives the
RBM expression for cage combination efficiency and simply involves
K and L.

$$F_c(T) = (1/K)*L/(1+L) \quad ; \quad \text{Photochemical: } K > 1$$
$$\text{Collisional: } \quad K = 1 \tag{10}$$

The distinction between the photochemical cage of Scheme 2
and the collisional cage of Scheme 1 is the additional
momentum[1,2,18] available for separating the incipient photochemical
pair by virtue of the (26 kcal/mole)[19] energy in the photon that is
in excess of that required for bond cleavage. According to the RBM,
the difference between the photochemical and collisional cases is
in the value of K. For the collisional case, K is unity while in
the photochemical case it is greater than 1. The greater than one
condition in the photochemical case results from the extra
separation momentum which is not available in the collisional case.
The adsorption of a 354.7 nm photon yields an excited diphenyl
disulfide molecule which is 80.6 kcal/mole above the ground state
and 26 kcal/mole above the dissociation limit[19] for formation of
two separated phenylthiyl radicals if both are in their electronic
ground state. The excess energy in the excitation event is
dissipated in the fsec regime, partly by (translational) separation
of the cage pair.

Our present focus is on the values of $F_c(T)$ that are contained
in the convoluted fits to equation (9) obtained in cis-decalin over
the temperature range from 237 K to 429 K. These results give cage
efficiencies for the photochemical case directly and predict those
for the corresponding collisional case where K may be set equal to
one. The viscosity of this solvent has been characterized (Table I)
and fits the Andrade equation (6) with an E_n of 3685 cal/mole and a
pre-exponential constant (A_n) of 6.45×10^{-3} cP. These results are
summarized in Table I and Figure 2.

The fitting procedure[16] was remarkably simple for the
photolysis in cis-decalin. A single value of K (1.3) and a single
proportionality constant (k_R) were sufficient throughout the 192
degree temperature range. Furthermore, the variations of the
measured diffusional time constant (s^2/D) were directly
proportional to the changes in solvent viscosity. A hard sphere
collision diameter of 360pm, estimated from the x-ray structure of
diphenyl disulfide, gives absolute diffusion coefficients for
phenylthiyl radicals that are uniformly 7.3 times greater than the
Stokes-Einstein value.

The combination efficiencies, predicted for the collisional
case by the RBM, can be converted to the simple chemical models
(Scheme 1) through the effective local concentration of the initial
pair and the observed diffusive time constants. The k_{dp} of Scheme 2
is thus given as $3D/d_o^2$. The $F_c(T)$ and $F_{cp}(T)$ values of Table I
yield the corresponding k_{cp} and k_c values at each temperature. The
corresponding activation parameters for diffusive separation and
combination of the collisional cage pairs are summarized in Table
II.

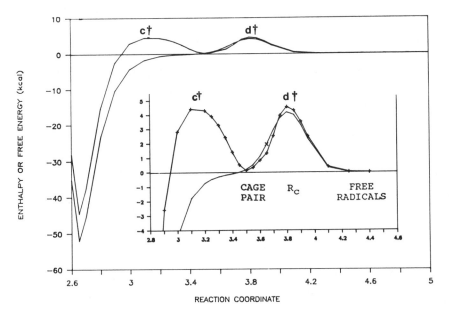

Figure 1. Free energy and enthalpy for bond dissociation in solution. (+)Gibbs Free Energy, (-) Enthalpy.

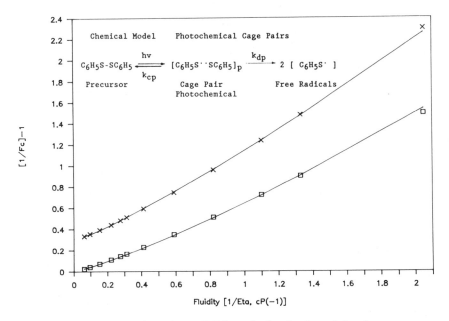

Figure 2. Radiation boundary (RBM) and chemical models for phenylthiyl cage pairs. (x)RBM (photochemical), (□)Chemical model, (—)RBM (collisional)

Table I
Fractional Cage Efficiencies in cis-Decalin

T [$^{\circ}$K]	Viscosity[a] n [cP]	Fluidity 1/n [cP^{-1}]	Photochemical Efficiency Fcp(T)[b]	Collisional Efficiency Fc(T)
237	16.21	0.584	0.75	0.98
251	10.47	0.524	0.74	0.96
268	6.55	0.457	0.72	0.93
283	4.54	0.404	0.69	0.90
293	3.63	0.371	0.67	0.88
299	3.20	0.352	0.66	0.86
313	2.42	0.312	0.63	0.82
333	1.70	0.260	0.57	0.74
354	1.22	0.231	0.51	0.66
375	0.910	0.173	0.45	0.58
390	0.751	0.148	0.40	0.52
429	0.488	0.097	0.30	0.39

a) Landolt-Bornstein. Zahlewerte Und Functionen; Springer: Berlin, 1969: Vol.II, Part 5a, p.169.   b) Ref. 16

Table II
Activation Parameter Summary
Phenylthiyl radical

Collisional Cage Pair Recombination and Diffusion			
Process	Rate Constant	ΔH^*	ΔS^*
Collisional Cage Combination	(k_c)	-600 cal/mole	-11.4 cal/mole-K
Collisional Cage Separation	(k_d)	3705 cal/mole	- 0.5 cal/mole-K
Free Radical Diffusion	($2k_D$)	3447 cal/mole	- 4.4 cal/mole-K

Figure 2 compares the ratios of the rate constants for diffusive separation to cage combination (kd/kc as 1/F_c(T) - 1) for both the photochemical (Scheme 2) and collisional cage pairs (Scheme 1). The points on the collisional curve show the agreement between the activation parameter derived chemical (Scheme 1) model with the hydrodynamic RBM. This agreement suggests that the simple chemical model gives an adequate account of the collisional cage pair, at least in this case.

Free Radical Self-Termination. The cage efficiencies and activation parameters for the phenylthiyl collisional cage pair provide the basis for illustrating some of the important features of equations (3)-(5) and for predicting the observed rates of self-termination of phenylthiyl free radicals[19]. Application of the SW procedure[19] to the completely diffusion controlled step of Scheme 1 (k_D) for phenylthiyl free radicals in cis-decalin can be expressed by the transition state equation with a ΔH^{*}_D of 3448 cal/mole and a ΔS^{*}_D of -4.3 cal/mole-K. The corresponding activation enthalpy (ΔH^{*}_D) from the Stokes-Einstein-Schmoluchowski relationship is 3685 cal/mole (E_n for cis-decalin) so that the α of equation (8) is 0.94. The micro-frictional multiplier (mf, equation 8 above), which is incorporated into the SW activation entropy (ΔS^{*}_D), is 2.4. The SW activation entropy for a truly diffusion controlled self-termination of phenylthiyl free radicals ($2k_t$obs $= 2k_D$, $F_c = 1$ at all temperatures) is -5.72 cal/mole-K after correction for the 1/4 spin statistical factor[10]. These values are also included in Table II. Figure 3 (uppermost curve) shows this SW/($F_c=1$) prediction of the self-termination rate constants in this case. This part of Figure 3 provides a reference that is helpful in illustrating the effects of less than infinitely effective cage combination reactions ($F_c<1$) derived above.

As indicated by equation (2) above, the observed values for the rate constants of the second order self-termination of Scheme 1 free radicals ($2k_t$obs) are the product $2k_D \cdot F_c(T)$. The RBM has provided the required values for $F_c(T)$ and SW has given $2k_D$ so that the values of $2k_t$obs can be predicted. The effective activation parameters for the k_d/k_c competition and equations (3) and (4) offer the chemical model alternative for calculating $2k_t$obs. The lower two curves of Figure 3 are the results of the two model calculations and simply reconfirm the equivalence of the RBM and chemical models shown in Figure 2. The important feature of Figure 3 is the pronounced curvature in the Eyring treatment of the predicted rate constants ($2k_t$obs). As mentioned at the outset, this is a result of the composite mechanism where neither diffusive separation nor cage combination is clearly rate dominant.

If the self-termination rates were measured over a more limited temperature range, then apparently or approximately linear activation parameter plots would be observed. For example, treatment of the calculated $2k_t$obs values from 237 °K to 268 °K by linear regression yields a ΔH^{*}_tapp of 3265 cal/mole with a correlation coefficient (r^2) of 0.999. This value of ΔH^{*}_tapp and equation (5) gives $F_c(T)$ as 0.96 at 252 °K (the mean of the temperature interval), the same as calculated by the RBM. Values for $2k_t$obs from 375 °K to 429 °K also give acceptable linear regression statistics ($r^2 = .995$) but a ΔH^{*}_tapp of 1235 cal/mole and an $F_c(T)$ at the mean temperature (402 °K) equal to 0.44. These results highlight the importance of the very wide temperature range over which laser methods can operate. The degree of curvature predicted here (Figure 3) appears to be sufficient to be detected with modern nanosecond pulse methods.[13a,22,22a]

These results (Table II, Figure 3) reveal the caution that is required in interpreting the temperature dependence of free radical

self-termination. Curvature in the ln(k/T) versus 1/T is certainly
expected but might not be visible, especially over a limited
temperature range. The apparent activation parameters that would be
obtained in such a case could be misleading. The major general
conclusion, suggested by from these results, is that simple
chemical models can describe cage pair phenomena in a manner that
is equivalent to more sophisticated hydrodynamic formalisms such as
the RBM. The activation entropy for the (collisional) cage
recombination process (ΔS^{*}_{c}) is negative and, as indicated in
Figure 1, is the source of the barrier for recombination[23]. The
activation enthalpies for diffusive separation of cage pairs (ΔH^{*}_{d})
and diffusive collision of free radicals (ΔH^{*}_{D}) are essentially the
same as the Andrade energy for viscous flow of the solvent. This
approximate equality is used in the succeeding sections that treat
trapping reactions and bond homolysis.

Free Radical Trapping Reactions--Scheme 3 represents a reversible
bond thermolysis starting from an organometallic precursor (R-M)
reacting by trapping of the radical intermediates (R·, M·) with a
trapping agent (T). This scheme is formally identical to Scheme 1
with the addition of the trapping processes ($k_{Tpair}·[T]$, $k_{Tf}·[T]$).
The free radical trapping process ($k_{Tf}·[T]$) is of fundamental
importance in allowing the observation of the kinetics for
formation of free radicals for such systems.

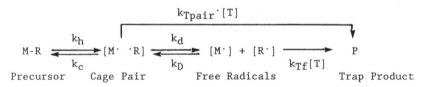

$$k_{Tpair}·[T]$$

M-R $\underset{k_c}{\overset{k_h}{\rightleftharpoons}}$ [M· ·R] $\underset{k_D}{\overset{k_d}{\rightleftharpoons}}$ [M·] + [R·] $\xrightarrow{k_{Tf}[T]}$ P

Precursor   Cage Pair   Free Radicals   Trap Product

Scheme 3   Reversible Bond Homolysis

The free radical trapping reaction (k_{Tf}) of Scheme 3
involves a collisionally formed cage pair (where the trapping agent
(T) and the alkyl (R·) radical are the chemically reactive
components) which is formally identical to that for free radical
self-termination discussed above. Scheme 4 provides this

$$k_{TD}$$

[R·] + [T] $\underset{k_{Td}}{\overset{k_{TD}}{\rightleftharpoons}}$ [R· T] $\xrightarrow{k_{Tc}}$ P

Free Radicals   Cage Pair   Trap Product
          (Reactive radical/Trap)

Scheme 4 - k_{Tf} Step of Scheme 3

elaboration of the free radical trapping (k_{Tf}) process of Scheme 3.
The overall k_{Tf} reaction of Scheme 3 is again a composite of the
elementary steps of Scheme 4. The chemical reaction of the
radical/trap cage pair is denoted as k_{Tc}. The diffusive steps for

formation and (re)separation of the radical/trap pair are denoted as k_{TD} and k_{Td} respectively.

The relationships corresponding to equations (1)-(5) above are as follows. The observed rate constants (k_{Tf}obs) and activation parameters (ΔH^*_{Tf}obs, ΔS^*_{Tf}obs) for disappearance of R· by reaction with T depend on the second order rate constant for diffusive collision to form the cage pair (k_{TD}, equation 12) and the fractional cage efficiency (F_{Tc}, equation 11). The latter is determined by the first order rate constants for the chemical reaction (k_{Tc}) and diffusive (re)separation (k_{Td}) of the T/R· collisional cage pair. The associated activation parameters are given by equations (13) and (14).

$$F_{Tc} = k_{Tc}/(k_{Tc} + k_{Td}) \tag{11}$$

$$k_{Tf}\text{obs} = F_{Tc}\, k_{TD} \tag{12}$$

$$\Delta H^*_{Tf}\text{obs} = \Delta H^*_{TD} + [1-F_{Tc}(T)]\,(\Delta H^*_{Tc} - \Delta H^*_{Td}) \tag{13}$$

$$\Delta S^*_{Tf}\text{obs} = \Delta S^*_{TD}+[1-F_{Tc}(T)][(\Delta H^*_{Tc}-\Delta H^*_{Td})/(T)]-R\ln[F_{Tc}(T)] \tag{14}$$

These equations are quite useful in the analysis of recent observations on free radical trapping. The diffusive activation enthalpies in equation (13) (ΔH^*_{TD}, ΔH^*_{Td}) certainly tend to cancel and if they do so exactly, then the value of ΔH^*_{TD} can be used with equation (13) to obtain ΔH^*_{Tc} which is intrinsic to the radical/trap cage pair. The recent work[26] on reaction of butyl radicals with thiophenol provides a useful example. In this case, enough information[26] is available to evaluate k_{TD} and hence F_{Tc} (F_{Tc} = k_{Tf}obs/k_{TD} = 0.02) for the n-butyl (NBU·)/thiophenol reaction in n-nonane. This low value, the $\Delta H^*_{Td} = \Delta H^*_{TD}$ equality (suggested by the psec laser/RBM results above) and equation (13) immediately explain the observed[26] activation parameters. The 1-F_{Tc} factor, in this case, is essentially 1 throughout the temperature range of observation and ΔH^*_{Tf}obs becomes equal to ΔH^*_{Tc} (when ΔH^*_{TD} is presumed to be equal to ΔH^*_{Td}).

In cases where the fractional cage efficiency is this low (< 0.1), equation (14) can be replaced with the more intuitive approximate form shown as equation (14a). This equation together with the value of ΔS^*_{TD} can be used to obtain the value of ΔS^*_{Tc}

$$\Delta S^*_{Tf}\text{obs} = \Delta S^*_{TD} + [1-F_{Tc}(T)][(\Delta S^*_{Tc}-\Delta S^*_{Td}) \tag{14a}$$

relative to ΔS^*_{Td}. The results obtained above (Table II) suggest that ΔS^*_{Td} should be near 0. We wish to emphasize that the values for ΔH^*_{Tc} and ΔS^*_{Tc} are those which should be transferred for a given R·/trap system to a different solvent rather than ΔH^*_{Tf}obs and ΔS^*_{Tf}obs. Changing solvent will involve a new ΔH^*_{TD} and ΔS^*_{TD} and possibly a new ΔS^*_{Td}. It is ΔH^*_{Tc} and ΔS^*_{Tc} that are properties of the R·/T pair with the solvent effects stripped (or assumed) away.

If the $F_{Tc}(T)$ values are not so near 0 (or 1), then the composite mechanism curvature problem arises. As in the free radical self-termination case discussed above, it is necessary to introduce the apparent activation parameters that are obtained from the standard $\ln(k/T)$ versus $1/T$ treatment of the observed rate constants (k_{Tf}obs). These activation parameters must be kept distinct from those of equations (13) and (14) above. The apparent activation enthalpy (ΔH^{*}_{Tf}app) gives an approximation to k_{Tf}obs which neglects the curvature while ΔH^{*}_{Tf}obs and ΔS^{*}_{Tf}obs give the correct values of the rate constants throughout. The apparent activation enthalpy is useful in estimating the value of the fractional cage efficiency for the reactive radical/trap cage pair at the mean of the temperature range of observation ($F_{Tc}(T)$). Equation (15), which is analogous to equation (5) above, expresses this possibility.

$$F_{Tc}(T) \simeq 1 - [(\Delta H^{*}_{Tf}app - \Delta H^{*}_{TD})/(\Delta H^{*}_{Tc} - \Delta H^{*}_{Td})] \qquad (15)$$

This use of equation (15) can be illustrated using recently published[27] results on the trapping of the 5-hexenyl radical (5-HE·) with the Beckwith nitroxide (BEN)[27] in cyclohexane. The reported[27] observations correspond to an apparent activation enthalpy (ΔH^{*}_{Tf}app) of .2 kcal/mole. The $\Delta H^{*}_{Td} = \Delta H^{*}_{TD} = E_n$ equivalence, suggested by the laser photolytic results on diphenyl disulfide discussed above, gives $\Delta H^{*}_{TD} = 2.9$ kcal/mole. Assuming a temperature independent k_{Tc} gives a ΔH^{*}_{Tc} of ca. -.5 kcal/mole. These values and equation (15) give $F_{Tc}(T) = 0.2$ at 360 K, the mean temperature of the observation range. This value, as $F_{Tc}(T)/[1-F_{Tc}(T)]$, fixes the $\Delta S^{*}_{Tc} - \Delta S^{*}_{Td}$ difference (12.1 cal/mole-K) and an observed rate constant allows the determination of ΔS^{*}_{TD} (-5.8 cal/mole-K).

Figure 4 shows the comparison of the F_{Tc}=1 line with that was generated by the present formalism for the BEN/5-hexenyl pair in cyclohexane. The temperature range of this figure is much wider that in the experiment and linear regression over the experimental temperature interval (333 to 397 K) reproduces the apparent activation parameters observed by the Australian group in that interval. The curvature imparted by the varying values of $F_{Tc}(T)$ gives a ΔH^{*}_{Tf}app which is much lower than the E_n of the solvent.

The transference property of the radical/trap cage pair activation parameters can be illustrated by recent data on trapping cyclopropylmethyl (CPM) radical[28] with BEN. Transferring the $\Delta H^{*}_{Tc}/\Delta S^{*}_{Tc}$ values for BEN/5-HE· to C_6F_6 and using the E_n of C_6F_6[29] for ΔH^{*}_{TD} together with Warkentin's[29] nicely conceived BEN/ CPM· product ratios gives rate constants for the CPM· isomerization (k_i Figure 5) in close agreement with the Franz[26]-Newcomb[28]-Beckwith[27] results. Figure 5 shows this comparison after very slight adjustment (-0.83 cal mole^{-1}K^{-1}) of ΔS^{*}_{TD} in the C_6F_6 solvent. The slightly more negative value for ΔS^{*}_{TD} in C_6F_6 compared to cyclohexane is expected from the relative Andrade pre-exponential factors [A_n, equation (7)] of the two solvents. The activation parameters for these trapping reactions are summarized in Table III.

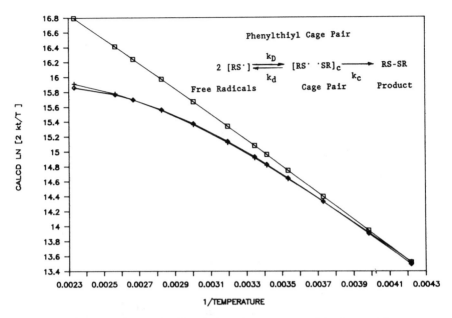

Figure 3. Calculated ln(2k_tobs) values versus 1/T for self termination of phenylthiyl free radicals. (□) Spernol-Wirtz (SW, $F_C = 1$), (◇)radiation boundary (RBM), (+) Chemical model.

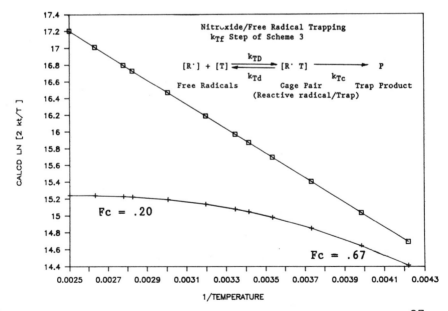

Figure 4. Trapping of 5-hexenyl radical with Beckwith nitroxide[27] in cyclohexane. (□) $F_C = 1$, (+) Scheme 4.

Table III[a]

R·/Trap	ΔH^{*}_{Tfapp}	ΔS^{*}_{Tfapp}	ΔH^{*}_{TD}	ΔS^{*}_{TD}	ΔH^{*}_{Tc}	ΔS^{*}_{Tc}[b]	F_{Tc}[c]
NBU·/C$_6$H$_5$SH[26]	1.04	-18.32	2.00[d]	-5.23[e]	1.04	-13.1	.02
5-HE·/BEN[27]	0.26	-16.29	2.90[f]	-5.77	-.50	-12.1	.42
CPM·/BEN[29]			2.52[g]	-6.60	-.50	-12.1	.30

a) Enthalpies in kcal mole-1, Entropies in cal/ mole-K.
b) $\Delta H^{*}_{Td} = \Delta H^{*}_{TD}$, ΔS^{*}_{Td} ca. 0. Addition of the same constant to both ΔS^{*}_{Tc} and ΔS^{*}_{Td} does not affect F_{Tc} or k_{Tc}^{obs}. c) 25 C. d) Nonane solvent. e) No cage singlet/triplet spin selection. f) Cyclohexane solvent. g) Hexafluorobenzene solvent.

Finally, we should note that the rate constants for the *cage pair* trapping reaction (k_{Tpair}) in Scheme 3 are also determined by values for ΔH^{*}_{Tc} and ΔS^{*}_{Tc} such as those of Table III. The form of the k_{Tpair} rate constant differs from that of k_{Tf} in that it applies to relatively high trap concentrations where the trapping agent can be one of the 4-8 molecules that make up the cage wall. A diffusive step need not be involved in the k_{Tpair} process in this concentration range. The cage pair trapping reaction rate constant could be approximated as the sum of a diffusional path and a non-diffusional cage wall path. The activation parameters for the chemical cage reaction for reactive radical/trap cage pairs (ΔH^{*}_{Tc}, ΔS^{*}_{Tc}, Table III) are the fundamental values that govern both of these cage pair trapping processes. The cage pair trapping reaction is experimentally[3] accessible through careful measurements of the rate constants of Scheme 3 homolyses in the 0.1M - 1M trap concentration region. The temperature dependance of those results can yield the activation enthalpy differences for the cage recombination (ΔH^{*}_{c}, k_{c}, Scheme 3) and cage pair trapping (k_{Tpair}, Scheme 3) processes. Values for ΔH^{*}_{c} are the most important unknown in the solution phase equations[8] for bond dissociation energies of organometallic systems.

Bond Homolysis reactions.--We have previously reported[8] an approximate version of the present formalism as it pertains to the activation parameters for disappearance of an organometallic precursor through reaction with a suitable free radical trapping agent (Scheme 3). That preliminary account was incomplete and did not make the distinction between apparent activation parameters and what we term "observed" activation parameters. This distinction is necessary, in principle, for the rate constants observed for homolytic decay of R-M of Scheme 3 ($k_{h}obs$) at scavenger concentrations such that all free radicals and no cage pairs are being trapped. As discussed above for free radical self-termination, we use the subscript "obs" to denote the activation parameters that give the observed rate constants at any temperature in the interval of observation. These values should be distinguished from apparent activation parameters ($\Delta H_{h}^{*}app$, $\Delta S_{h}^{*}app$) derived from linear regression of $\ln(k_{h}obs/T)$ against $1/T$. The latter values will give approximate values for $k_{h}obs$ which differ from the true ones by the extent of the curvature.

Equations (17) and (18) below give the activation parameters which correspond to the observed rate constants [$k_h$obs, equation (16)] for a bond homolysis leading to reaction of reactive radicals with a trapping agent at the *free radical stage* (Scheme 4).

$$k_h obs = [1 - F_c(T)] \cdot k_h \tag{16}$$

$$\Delta H^*_h(T)obs = \Delta H^*_h + F_c(T)[\Delta H^*_d - \Delta H^*_c] \tag{17}$$

$$\Delta S^*_h(T)obs = \Delta S^*_h + F_c(T)[(\Delta H^*_d - \Delta H^*_c)/(T)] + R\ln[1 - F_c(T)] \tag{18}$$

Figure 6 illustrates the distinction between ΔH^*_h, $\Delta H^*_h(T)_{obs}$ and ΔH_h^*app for a hypothetical M-R and solvent with the activation parameters shown. The line corresponding to ΔH^*_happ was generated by linear regression of $\ln(k_h obs/T)$ against $1/T$ at the three highest temperatures and extended to the (experimentally inaccessible) low temperature region. The curved line is the ratio of the rate constants calculated from the apparent activation parameters ($k_h$app) to those calculated from equations (17) and (18) with an added graphical multiple (-15) to allow the display of the $k_h$app/$k_h$obs measure of the curvature on the same scale as the Eyring plots. The curvature effect is present but of much less importance than that shown in Figure 3 above for free radical self-termination. The difference between ΔH^*_happ and ΔH^*_hobs is not significant for rate constants of ordinary accuracy over ordinary temperature ranges.

If differential solvation effects are ignored[8], then the bond dissociation energy for the R-M bond can be approximated as in equation (19). The example shown in Figure 6 corresponds to a case where $F_c(T)$ approaches unity. Equation (19) shows that,

$$BDE \approx \Delta H_h^*app + [1 - F_c(T)] [\Delta H^*_d - \Delta H^*_c] - \Delta H^*_D \tag{19}$$

in the $F_c(T) = 1$ limit, the BDE is simply the difference between ΔH_h^*app (34.80 kcal/mole in Figure 6) and ΔH^*_D (5 kcal/mole in Figure 6 if $\Delta H^*_D = \Delta H^*_d$). For the example of Figure 6, this difference is 29.80 kcal/mole which compares favorably with the actual (30 kcal/mole) value. This procedure is equivalent to the one put forth by Halpern[31] and the present results show that it can give quite good results if $F_c(T)$ is close to one.

The Halpern procedure becomes less accurate when the fractional cage efficiency is not so close to one. A system with the activation parameters of Figure 6, except for a reduction of ΔH^*_d, models a more fluid solvent. Diffusive escape becomes faster, relative to cage combination, and this leads to reduced values of $F_c(T)$. A reduction of the value of ΔH^*_d to 2.6 kcal/mole gives $F_c(T)$ as 0.2 at the mean of the temperature range that would apply to a ΔH_h^*app value of 30.6 kcal/mole. The Halpern procedure leads to an estimated 28 kcal/mole for the BDE in this case, somewhat less than the actual 30 kcal/mole value. The value of $F_c(T)$, the

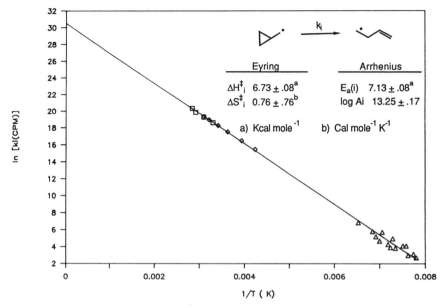

Figure 5. Eyring plot for the ring opening isomerization (k_i) of cyclopropylmethyl free radical.
(□) Ref. 29, (◇) Ref. 28, (△) Ref. 30.

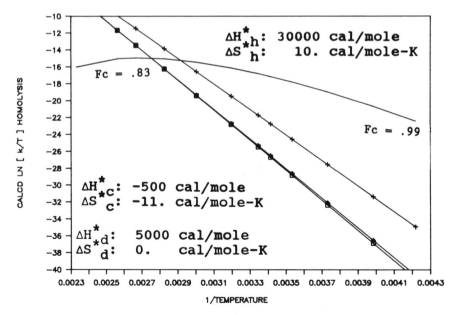

Figure 6. Temperature dependence (1/T) of apparent ($k_h$app) and observed ($k_h$obs) rate constants for a bond homolysis (k_h) in solution. (+)k_h, (□)k_h(obs), (◇)k_h(app), (—) -15*k_h(app)/k_h(obs)

activation parameters for the cage pair reactions and equation (19) give the correct 30 kcal/mole value for the BDE. The solution phase method of estimating BDE values requires some attention to the value of the fractional cage efficiency and the activation parameters for the reactions of the cage pair if precise values are the objective. It should also be noted that the apparent activation entropies show a relatively strong response to differences in $F_c(T)$. This response needs to be considered in interpretations of variations of experimental ΔS_h^* app values for reactions of organometallic systems of varying structure.

Discussion

The major objective of the present work is to provide a form for treating the effects of cage pair intermediates on observed activation parameters which brings consistency to observations on composite   free radical formation and termination processes. The free radical trapping process is completely analogous to self-termination in that significant curvature in Eyring or Arrhenius treatments of observed rate constants is expected if measurements are made over a wide enough temperature range. One of the most important results, suggested by the *preliminary* laser photolytic results on phenylthiyl cage pairs[16], is that cage combination rate constants that are essentially temperature independent. The corresponding activation entropies for the cage combination of both phenylthiyl cage pairs and primary alkyl/nitroxide (BEN[27]) cage pairs support the idea depicted in Figure 1; the barrier to cage combination may be entirely entropic.

The present results suggest a correspondence between transition state and hydrodynamic models for cage effects. They also suggest an equivalence between ΔH^*_d, ΔH^*_D and E_n. The major difference between the rate constants for diffusive separation (k_d) and diffusive collision (k_D) is in the activation entropies. However, we emphasize that more experimental data and examination of a wider range of models[32] is necessary before this suggestion is accepted. Our emphasis is on demonstrating that cage reactions are directly approachable over a wide range of temperatures and solvents via psec laser methods. We do not imply that the present results establish the generality of either the temperature independence of rate constants for cage combination reactions or the $\Delta H^*_d/\Delta H^*_D/E_n$ equivalence. Instead, we hope to have provided a framework that will encourage and improve future direct determinations of the nature of the reactions of cage pair radicals and organometallic thermolyses. We expect that such studies will soon provide a fairly precise picture of temperature effects of cage pair reactions of varying structure.

Acknowledgment The support of J. A. Franz by the Office of Basic Energy Sciences, Chemical Sciences Division, Office of Energy Research, U. S. Department of Energy (Contract DE- ACO6-76RLO 1830) is gratefully acknowledged. We also thank the National Science Foundation (CHE-8815745) for partial support of this work.

Literature Cited

1. Frank, J. and Rabinowitch, E. *Trans. Faraday Soc.*
   1934, *30*, 120; Rabinowitch, E. and Wood, W. C.
   Trans. Faraday Soc. **1936**, *32*, 1381; Rabinowitch, E.
   Trans. Faraday Soc. **1937**, *33*, 1225.
2. Koenig, T. and Fischer, H. "'Cage' Effects", in
   "Free Radicals", Vol. 1, J. K. Kochi (Ed.), J. Wiley
   and Sons, New York, 1973. Chapter 4, pp. 157-189.
3. Koenig, T. "Cage Effects in Peresters and
   Hyponitrites" in "Organic Free Radicals", ACS
   Symposium Ser No. 69, W. A. Pryor (Ed.), Ch 3,
   pp.134-160; Koenig, T. and Owens, J. *J. Am. Chem
   Soc.* **1974**, *96*, 4052; *J. Am. Chem Soc.* **1973**, *95*,
   8484.
4. For a very recent example of the activation free
   energy/enthalpy structural effects see Sciano, J.
   C.; Wintgens, V.; Haider, K. and Berson, J. A. *J.
   Am. Chem Soc.***1989**, *111*, 8732 and references therein.
5. Herkes, F., Friedman, J. and Bartlett, P. D. *Int.
   J. Chem. Kinetics*, **1969**, *1*, 193; Chakravorty, K.,
   Pearson, J. and Szwarc, M. *J. Am. Chem. Soc.* **1968**,
   90, 283.
6. (a) Weiner, S. A. and Hammond, G. S. *J. Am. Chem.
   Soc.* **1969**, *91*, 986.
   (b) Shuh, H-H. and Fischer, H. *Helv. Chim. Acta*
   1978, *61*, 2130.
7. Koenig, T. and Finke, R. G. *J. Am. Chem. Soc.* **1988**,
   110, 2657.
8. Koenig, T.; Finke, R. G. and Hay, B. P. *Polyhedron*
   1988, *7*, 1499-1516.
9. Burshtein, A. I.; Khudyakov, I. V. and Yakobson, B.
   I. *Prog. Reaction Kinetics*, **1984**, *13*, 221-305.
10. Saltiel, Jack and Atwater, B. W. *Adv. Photochem.*
    1989, *14*, 1-90. The spin statistical effect may not
    be operative with thiyl radical pairs (McPhee, D.
    J.; Campredon, M.; Lesage, M. and Griller, D. *J. Am.
    Chem. Soc.* **1989**, *111*, 7563).
11. Einstein, A. "Investigations on the Theory of the
    Brownian Movement", R. Furth (Ed.), A. D. Cowper
    (Trans.), E. P. Dutton, New York, 1956, p. 38. It
    might be worth noting that, as shown by equation
    (7), the Stokes-Einstein relationship implies that
    the activation *enthalpy* for the purely diffusion
    controlled formation of the cage pair is directly
    equal to E_n. This point is not clear in the footnote
    (on p.1510) of our recent discussion [8] of cage
    effects in organometallic chemistry. For non-
    associating solvents, the ΔH^*_n of equation (11a) in
    that paper is equal to E_n. For associating solvents,
    the simple Andrade relationship for viscosity
    doesn't work well and the situation is more
    complicated.

12. Spernol, A. and Wirtz, K. Z. *Naturforsch.* 1953, *8a*, 522.

12a. The SW method requires the ratio of the solute and solvent radii, which come from molar volumes, and reduced temperatures that involve the melting points and boiling points of the solute and solvent. Reference 6b gives a clear comparison of this and other corrections to the Stokes-Einstein-Schmoluchowski equation.

13. The Battelle group (JAF) has recently measured the diffusion coefficient of benzyl phenyl sulfide in nonane and compared it with the value calculated by the Spernol-Wirtz (SW) method. The agreement between experiment an SW is within 3.8 % in this case. The calculated (SW) termination rate constants for phenylthiyl radicals in heptane are 30% higher than observed[13a]. The direction of this discrepancy could be explained by a 70% efficiency of the collisional cage in heptane.

13a. Burkey, T. J. and Griller, D. *J. Am. Chem. Soc.* 1985, *107*, 246-249

14. Endicott, J. F. and Netzel, T. L. *J. Am. Chem. Soc.* 1979, *101*, 4000. The cage efficiency for the photochemical cage pair of the B_{12} system has recently be reconfirmed by Chen and Chance (Chen, E. and Chance, M. R. *J. Inorg. Biochem.* 1989, *36*, 264.)

15. Shin, K. J. and Kapral, R. *J. Chem. Physics* 1979, *69*, 3685.

16. Scott, T. W. and Liu, S. N. *J. Phys. Chem.* 1989, *93*, 1393-1396; Scott, T. W. and Liu, S. N. in "Ultrafast Phenomena V"; Fleming, G. R. and Stiegman, A. E., Eds.; Springer-Verlag: Berlin, 1986; pp. 338-340.

17. Professor C. B. Harris has informed us that he has found quite similar cage efficiencies for diphenyl disulfide photolysis using his ultrashort pulse system in which convolution is less important.

18. Noyes, R.M. *Prog. Reaction Kinetics* 1961, *1*, 129.

19. The gas phase heat of formation of diphenyl disulfide is 58 kcal/mole[20]. The heat of formation of the phenylthiyl radical is 56.8 kcal/mole[21]. The bond dissociation energy for diphenyl disulfide is thus 55 kcal/mole.

20. Thermochemical Data of Organic Compounds, Pedley, J. B.; Naylor, R. D. and Kirby, S. P.; Eds. Chapman and Hall, New York, 1986 p.205.

21. Colussi, A. J. and Benson, S. W. *Int. J. Chem. Kinetics* 1977, *9*, 295.

22. Lipscher, J. and Fischer H. *J. Phys. Chem.* 1984, *88*, 2555.

22a. We hope to provide an experimental test of this prediction in the near future.

23. Olson and Koch[24] have previously pointed out this type of feature in their studies of the (Scheme 1)

formation of 3,5,5-trimethyl-2-oxomorpholin-3-yl
radicals although the complications of the cage pair
were neglected there. Negative temperature
dependences in the solution phase decay of triplet
diradicals has also been recently discussed[25].

24. Olson, J. B. and Koch, T. H. *J. Am. Chem. Soc.*
    1986, *108*, 756-761)
25. Wang, J.; Doubleday, C. and Turro, N. J. *J. Am.*
    Chem. Soc. **1989**, *111*, 3962-3965)
26. Franz, J. A.; Bushaw, B. A.; Alnajjar, M. S. *J. Am.*
    Chem. Soc. **1989**,   *111*, 268.
27. Beckwith, A. L. J.; Bowry, V. W.; Moad, G. *J. Org.*
    Chem. **1988**, *53*, 1632; BEN: 1,1,1,3,3-
    tetramethylisoindolin-2-yloxyl.
28. Newcomb, M.; Glenn, A. G. *J. Am. Chem. Soc.* **1989**,
    111, 275.
29. Mathew, L.; Warkentin, J. *J. Am. Chem. Soc.* **1986**,
    108, 7981.
30. Maillard, B.; Forrest, D. and Ingold, K. U. *J. Am.*
    Chem. Soc. **1976**, *98*, 7024.
31. Ng, F. T. T.; Rempel, G. L. and Halpern, J. *J. Am.*
    Chem. Soc. **1982**, *104*, 623.
32. It might be noted that convoluted fits to the decay
    of laser generated phenylthiyl radicals by simple first
    order reactions gives activation entropies for the
    reations of the cage pair that are about four
    cal/mole-K more negative than those discussed here.
    These values give unchanged fractional cage
    efficiencies. The exact form of the temporal decay
    and the high and low viscosity limits of this form
    should be studied with a wider range of models.

RECEIVED January 4, 1990

Chapter 9

Thermodynamic and Kinetic Studies of Binding Nitrogen and Hydrogen to Complexes of Chromium, Molybdenum, and Tungsten

Alberto A. Gonzalez[1], Kai Zhang[1], Shakti L. Mukerjee[1], Carl D. Hoff[1], G. Rattan K. Khalsa[2], and Gregory J. Kubas[2]

[1]Department of Chemistry, University of Miami, Coral Gables, FL 33124
[2]Los Alamos National Laboratory, Los Alamos, NM 87545

The enthalpies and entropies of binding of molecular nitrogen and hydrogen to the complexes $(PCy_3)_2M(CO)_3$, M = Cr, Mo, W, have been measured by a combination of solution calorimetry and high pressure FTIR studies. For all three metals nitrogen is the preferred ligand with regard to enthalpy of binding, however hydrogen has a less negative entropy of binding and is the preferred ligand at higher temperatures. Stopped-flow kinetic studies have been performed for ligand displacement reactions of $(PCy_3)_2W(CO)_3(L)$, L = H_2, D_2, and N_2. Rate constants and activation energies have been measured for these complexes and can be used to construct complete reaction profiles for binding of nitrogen and binding and oxidative addition of hydrogen. Thermodynamic and kinetic data have been measured by NMR techniques for the equilibrium between the dihydride and molecular hydrogen tautomers $H_2W(CO)_3(PR_3)_2$ and $W(CO)_3(PR_3)_2(H_2)$ for R = iso-propyl and the newly synthesized cyclopentyl complex. These data are in reasonable agreement with data for the analogous cyclohexyl complex which was determined independently by the solution calorimetric and stopped-flow kinetic methods as described above.

It has only recently been discovered that H_2 can form stable complexes with transition metals in which the H-H bond is retained (1). The general series of reactions shown in eqn.(1) probably reflects the course of many reactions in which a dihydride complex is formed as part of a catalytic cycle:

$$L_nM + H_2 \underset{k_{-1}}{\overset{k_1}{\rightleftharpoons}} L_nM(H_2) \underset{k_{-2}}{\overset{k_2}{\rightleftharpoons}} L_nM(H)_2 \qquad (1)$$

The first step involves binding of molecular hydrogen to a "coordinatively unsaturated" or weakly ligated metal center. The second step involves intramolecular cleavage of the H-H bond to form a dihydride complex. At the

0097–6156/90/0428–0133$06.00/0
© 1990 American Chemical Society

present time it seems most likely that catalytic reactions such as olefin hydrogenation probably proceed through sequential addition of hydride ligands from the final dihydride product of reaction (l). No convincing evidence has been presented to date that hydrogen transfer can occur via concerted addition of intact H-H in a molecular hydrogen complex, but this remains a possibility. While molecular hydrogen complexes may be shown in the future to have a unique chemistry of their own, there can be no doubt that they are key intermediates in oxidative addition and reductive elimination reactions. Detailed understanding of the rates and equilibria in reaction (l) as a function of the metal and its ligands is important to understanding catalyzed reactions of hydrogen.

The individual rate constants shown in eqn.(l) play an important role in determining the nature of a metal's interaction with hydrogen. A number of coordinatively unsaturated complexes don't react with hydrogen at all, implying that both k_1/k_{-1} and $k_1k_2/k_{-1}k_{-2} \ll 1$. The majority of complexes that do react with hydrogen proceed quantitatively to the dihydride form, implying that both $k_1k_2/k_{-1}k_{-2}$ and $k_2/k_{-2} \gg 1$. In the complexes $(PR_3)_2W(CO)_3$, R = cyclohexyl, isopropyl, cyclopentyl, the individual rate constants are balanced so that all species can be observed as a function of temperature and hydrogen pressure. This allows a unique opportunity to study the factors involved in binding and oxidative addition of hydrogen. The analogous molybdenum (2) and chromium (3) complexes also bind H_2, but in solution they do so quantitatively only under pressure. In this paper we report studies of all the rate and equilibrium constants shown in eqn.(l) for molecular hydrogen complexes of tungsten, and of the k_1/k_{-1} equilibrium for analogous chromium and molybdenum complexes.

Key molecules in nitrogen fixation are complexes of N_2, H_2, N_2H_2, N_2H_4, and NH_3. As part of a systematic study of the thermodynamic factors determining stability of these complexes (4) we also report results for enthalpies and entropies of binding of nitrogen to the series of complexes described above. For each of the metals nitrogen is the preferred ligand with respect to enthalpy of binding. However, entropic factors favor binding of hydrogen and at high temperatures H_2 binds more strongly than N_2.

Experimental

All operations were performed under an argon atmosphere using standard techniques. Tetrahydrofuran and toluene were distilled from sodium benzophenone ketyl under an argon atmosphere into flame dried glassware. The complexes $(PR_3)_2M(CO)_3$ were prepared as described in the literature: R = cyclohexyl, M = Mo, W (2),and Cr (3), M = W, R = isopropyl (2) and cyclopentyl (5). Due to the highly air sensitive nature of the solutions, great care had to be taken to exclude even traces of oxygen from all operations. Solution calorimetric measurements were made using either a Guild isoperibol or a Setaram Calvet calorimeter as described previously (6). Infrared spectra were measured on a Perkin-Elmer 1850 FTIR spectrometer. High pressure FTIR spectra were obtained using a special thermostated cell purchased from Harrick Scientific. The cell was loaded in the glove box and then transferred to a special optical bench/glove box described in detail elsewhere (7,8). NMR measurements were obtained on an IBM AF-250 spectrometer equipped with a Bruker VT-1000 temperature controller. A Bruker BSV3 decoupler unit was used for single-frequency ^{31}P saturation. Absolute intensity values were determined by use of an Aspect 3000 computer which monitored intensities relative to the noninteracting tetracarbonyl complexes $(PR_3)_2W(CO)_4$. Full details of these measurements will be published (5). Stopped-flow kinetic measurements were performed on a Hi-Tech Scientific SF-51 apparatus equipped with an SU-40 spectrophotometer unit as described previously (9).

Results

Enthalpies of Binding of Ligands to $(P(C_6H_{11})_3)_2W(CO)_3$

The dark blue complex $(P(C_6H_{11})_3)_2W(CO)_3$ binds a selective "menu" of ligands and calorimetric data has been reported for a range of donors (6). Typical reactions studied by solution calorimetry are the reactions with pyridine shown in eqns.(2) and (3) below:

$$(P(C_6H_{11})_3)_2W(CO)_3 + py \text{ ---> } (P(C_6H_{11})_3)_2W(CO)_3(py) \qquad (2)$$

$$(P(C_6H_{11})_3)_2W(CO)_3(H_2) + py \text{ --> } (P(C_6H_{11})_3)_2W(CO)_3(py) + H_2 \qquad (3)$$

The enthalpy of reaction (2) was determined to be -18.9 ± 0.4 kcal/mole with all species in toluene solution. The enthalpy of reaction (3), measured under an atmosphere of H_2, was found to be -9.5 ± 0.5 kcal/mole. The difference between these two reactions, shown in eqn.(4), corresponds to binding of hydrogen and is calculated to be exothermic by 9.4 ± 0.9 kcal/mole:

$$(P(C_6H_{11})_3)_2W(CO)_3 + H_2 \text{ ---> } (P(C_6H_{11})_3)_2W(CO)_3(H_2) \qquad (4)$$

Utilizing a series of measurments of this type in both THF and toluene, average heats of binding were measured (6) for: $H_2 = -10$, $N_2 = -15$, $NCCH_3 = -15$, $py = -19$, $P(OMe)_3 = -26$, and $CO = -30$ kcal/mole. These data are shown schematically in Figure 1.

Equilibrium Measurments for Binding of Nitrogen and Hydrogen to $(P(C_6H_{11})_3)_2M(CO)_3$ Complexes

The calorimetric methods outlined above could not be used to determine heats of binding of H_2 and N_2 to the chromium and molybdenum complexes. In solution, even under an atmosphere of H_2 and N_2, binding is not quantitative for these metals. In order to obtain thermodynamic parameters for binding of these ligands the equilibrium reactions (5) and (6) were studied as a function of temperature and pressure (7).

$$(P(C_6H_{11})_3)_2Cr(CO)_3 + H_2 \rightleftharpoons (P(C_6H_{11})_3)_2Cr(CO)_3(H_2) \qquad (5)$$

$$(P(C_6H_{11})_3)_2Cr(CO)_3 + N_2 \rightleftharpoons (P(C_6H_{11})_3)_2Cr(CO)_3(N_2) \qquad (6)$$

Typical variable temperature spectral data for eqns.(5) and (6) are shown in Figure 2. This data can be used to calculate equilibrium constants and plots of ln K_{eq} versus $1/T$ are shown in Figure 3. Enthalpies and entropies of binding calculated from these plots are collected in Table I.

As discussed above, enthalpies of binding of H_2 and N_2 to $(P(C_6H_{11})_3)_2W(CO)_3$ were determined by solution calorimetry. It was not practical to measure equilibrium constants for the tungsten complex since binding of both H_2 and N_2 is near quantitive under normal conditions. Using N_2/H_2 mixtures we have measured (7) the thermodynamic parameters for the ligand exchange reaction shown below:

$$(P(C_6H_{11})_3)_2W(CO)_3(H_2) + N_2 \rightleftharpoons (P(C_6H_{11})_3)_2W(CO)_3(N_2) + H_2 \qquad (7)$$

The enthalpy of reaction (7) calculated from this equilibrium data, -4.4 ± 0.4 kcal/mole, agrees within experimental error with that calculated from the calorimetric data described above, -3.5 ± 2.0 kcal/mole. The entropy of reaction (7), -13.8 ± 3.5 cal/mole deg serves to offset the favorable enthalpy of reaction and leads to a small favorable free energy for reaction (7) as summarized in Table I and discussed in more detail later.

Stopped-Flow Kinetic Study of Displacment of H_2, D_2, and N_2 from $(P(C_6H_{11})_3)_2W(CO)_3(L)$

Displacement of coordinated dinitrogen by pyridine as shown in reaction (8) was too rapid to measure by conventional techniques:

$$(P(C_6H_{11})_3)_2W(CO)_3(N_2) + py \longrightarrow (P(C_6H_{11})_3)_2W(CO)_3(py) + N_2 \qquad (8)$$

Under pseudo first order conditions this reaction could be studied by stopped-flow methods and plots of ln(abs) versus time were linear over four to five half lives as shown in Figure 4. This data was analyzed in terms of the following mechanism ($W_k = (P(C_6H_{11})_3)_2W(CO)_3$):

$$W_k\text{-}N_2 \underset{k_{-1}}{\overset{k_1}{\rightleftharpoons}} N_2 + W_k + py \overset{k_2}{\longrightarrow} W_k\text{-}py \qquad (9)$$

Using standard procedures the following rate constants were determined at 25 ^oC, $k_1 = 75$ s^{-1}, and $k_2/k_{-1} = 1.6$. The value of k_2 was independently determined as 8.0×10^5 $M^{-1}s^{-1}$ (9) allowing calculation of $k_{-1} = 5.0 \pm 1.0 \times 10^{-5}$ $M^{-1}s^{-1}$. The enthalpy of activation for dissociation of N_2 was determined as 17.8 ± 0.7 kcal/mole. Since the enthalpy of binding of nitrogen, as measured by solution calorimetry, is -13.5 kcal/mole, the enthalpy of activation for the back reaction with nitrogen can be calculated to be on the order of 4.3 kcal/mole.

Displacement of hydrogen did not obey simple first order kinetics as shown by the plot of ln(abs) versus time in Figure 5. The kinetic traces appeared to show two limiting reactions, an initial rapid formation of product followed by a slower second reaction. This data was interpreted in terms of the mechanism shown in eqn.(10) below:

$$W_k\text{-}H \underset{k_{-1}}{\overset{k_1}{\rightleftharpoons}} W_k(H_2) \underset{k_{-2}}{\overset{k_2}{\rightleftharpoons}} H_2 + W_k + py \overset{k_3}{\longrightarrow} W_k\text{-}py \qquad (10)$$

The initial rapid reaction is due to the relatively fast loss of H_2 from the molecular hydrogen isomer. On a slower time scale the dihydride complex forms the molecular hydrogen complex which can then dissociate leading to product. Complete kinetic details in this system have been reported for both H_2 and D_2 (9). Data of relevance to this paper are summarized at 25 ^oC: $k_1 = 37$ s^{-1} for H_2 and 33 s^{-1} for D_2; $k_{-1} = 18$ s^{-1} for H_2, $k_2 = 469$ s^{-1} for H_2 and 267 s^{-1} for D_2; $k_{-2} = 2.2 \times 10^6$ $M^{-1}s^{-1}$ for H_2. The enthalpy of activation for the k_1 step, conversion of the dihydride to molecular hydrogen complex is calculated to be 14.4 ± 0.5 kcal/mole. The enthalpy of activation for the k_2 step, loss of molecular hydrogen to form the "agostic" complex is calculated to be 16.9 ± 2.2 kcal/mole.

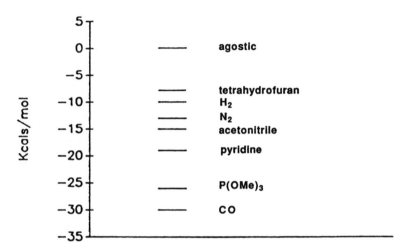

Figure 1. Enthalpies of ligand addition (kcal/mole) to $(P(C_6H_{11})_3)_2W(CO)_3$.

TABLE I

THERMODYNAMIC PARAMETERS FOR BINDING OF H_2 AND N_2

$$M(CO)_3(PCy_3)_2 + L \rightleftharpoons M(CO)_3(PCy_3)_2(L)$$

M	L	ΔH	$\Delta\Delta H$	ΔS	$\Delta\Delta S$	$\Delta\Delta G_{298°}$
Cr	N_2	-9.3 ± 0.2		-35.4 ± 2.3		
Cr	H_2	-7.3 ± 0.1	-2.0 ± 0.3	-25.6 ± 1.7	-9.8 ± 2.6	$+0.9$
Mo	N_2	-9.0 ± 0.6		-32.1 ± 3.2		
Mo	H_2	-6.5 ± 0.2	-2.5 ± 0.8	-23.8 ± 2.1	-8.3 ± 3.9	0.0
W			-4.4 ± 0.4		-13.8 ± 3.5	-0.3

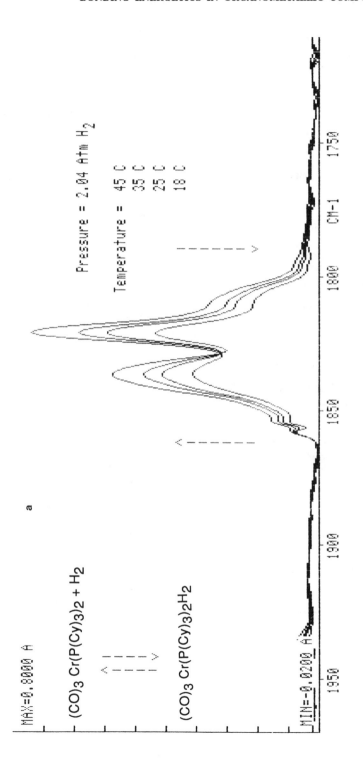

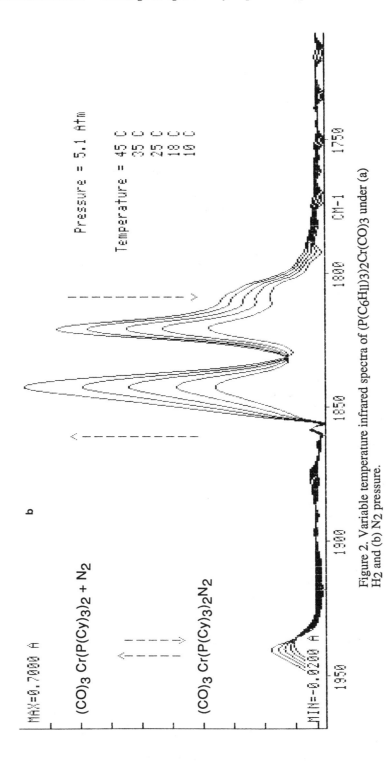

Figure 2. Variable temperature infrared spectra of (P(C6H11)3)2Cr(CO)3 under (a) H2 and (b) N2 pressure.

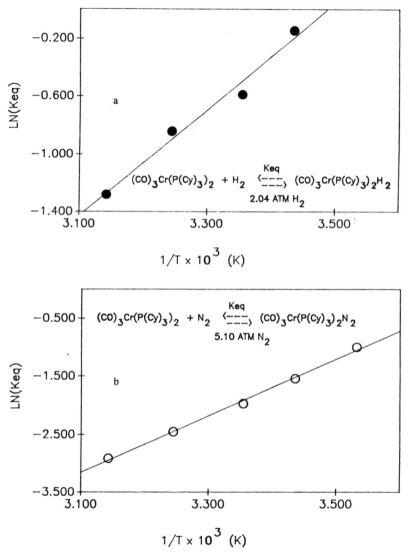

Figure 3. Plots of ln k_{eq} versus l/T for binding of (a) H_2 and (b) N_2 to $(P(C_6H_{11})_3)_2Cr(CO)_3$.

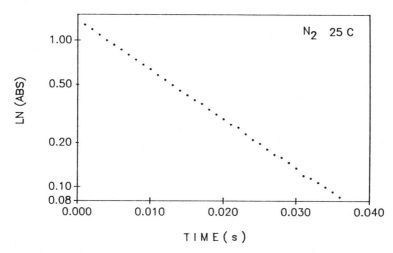

Figure 4. Plot of ln A versus time for the reaction W_k-N_2 + py -> W_k-py + N_2, $W_k = (P(C_6H_{11})_3)_2W(CO)_3$

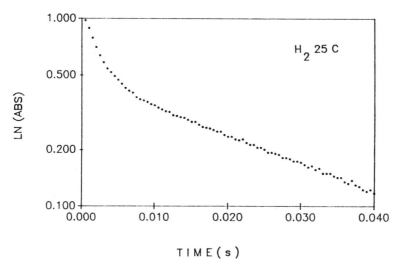

Figure 5. Plot of ln A versus time for the reaction W_k-H_2 + py -> W_k-py + H_2, $W_k = (P(C_6H_{11})_3)_2W(CO)_3$

NMR Measurments of Kinetic and Thermodynamic Parameters of the
Interconversion of the Dihydride and Dihydrogen Forms of $W(CO)_3(PR_3)H_2$
Where R = iso-Propyl and Cyclopentyl.

In parallel to the solution calorimetric and kinetic studies described above, NMR
investigations have been performed on the complexes $(PR_3)_2W(CO)_3$ for R = iso-
propyl (i-Pr) and cyclopentyl (Cyp). Due to its lower solubility, attempts to study
the cyclohexyl complex by NMR methods were not successful. A typical ^{31}P
decoupled ^1H NMR spectrum showing the resonances due to the molecular
hydrogen and dihydride peaks is shown in Figure 6 for the very soluble cyclopentyl
complex. Integration of the peaks allows calculation of the equilibrium constant for
formation of the molecular hydrogen complex from the dihydride as shown in eqn.
(11) below:

$$H_2W(PR_3)_2(CO)_3 \quad \overset{k_1}{\underset{k_{-1}}{\rlap{=}=====}} \quad W(PR_3)_2(CO)_3(H_2) \tag{11}$$

Equilibrium constants for this reaction as a function of temperature are collected
in Table II. Thermodynamic parameters calculated from this data are as follows: for
R = iso-propyl, ΔH = -1.2 $\pm$ 0.6 kcal/mole, ΔS = -1.2 cal/mole deg; for R =
cyclopentyl, ΔH = -1.5 $\pm$ 0.4 kcal/mole, ΔS = -2.4 $\pm$ 1.4 cal/mole deg.
Spin saturation transfer studies of the ^{31}P NMR of the iso-propyl complex yield
data regarding the rate of interconversion of the dihydride and molecular hydrogen
complexes. Thus k_1 is calculated to be 63 s^{-1} (299 K) and k_{-1} is calculated to be 12.4
s^{-1}(300 K). These values are in reasonable agreement with the corresponding values
determined by stopped-flow kinetics for the cyclohexyl complex as described
above; k_1 = 37 s^{-1}(298 K) and k_{-1} = 18 s^{-1} (298 K). Activation parameters for
conversion of the dihydrogen to dihydride reaction (k_{-1} step) were determined based
on analysis of the spectral changes in the temperature range 288-310 K. The values
$\Delta G^{\neq}$= 16.0 $\pm$ 0.2 kcal/mole, $\Delta H^{\neq}$= 10.1 $\pm$ 1.8 kcal/mole and $\Delta S^{\neq}$= -19.9 $\pm$ 6.0
cal/mole deg were determined.

Discussion

Relative and "Absolute" Bond Strengths to $(P(C_6H_{11})_3)_2W(CO)_3$

Enthalpies of ligand addition to $(P(C_6H_{11})_3)_2W(CO)_3$ are summarized in Figure 1.
Relative bond strengths in this system are in the order: "agostic" < H_2 < N_2 <
$NCCH_3$ < py < $P(OMe)_3$ < CO. These values can not be used to estimate "absolute"
bond strengths since the value for the "agostic" bond strength is not known.
Attempts to estimate this value by NMR measurements down to -90 ^0C were
unsuccessful. Recent photoacoustic calorimetric measurements indicate that the
enthalpy of reaction (12) is -13.4 $\pm$ 2.8 kcal/mole (10).

$$W(CO)_5 + \text{heptane} \longrightarrow W(CO)_5(\text{heptane}) \tag{12}$$

The low enthalpy of addition of CO as shown in reaction (13) can be compared to
the more exothermic value for reaction (14) determined in the gas phase by pulsed
laser techniques (11):

$$(PCy_3)_2W(CO)_3 \xoverset{CO}{\longrightarrow} (PCy_3)_2W(CO)_4 \quad \Delta H = -30 \text{ kcal/mole} \tag{13}$$

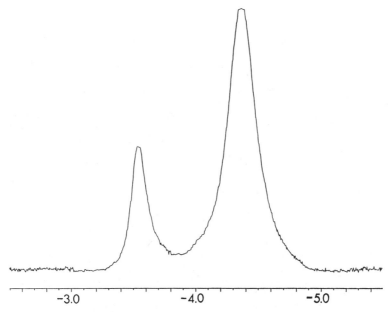

Figure 6. ^{31}P decoupled ^1H NMR spectrum(200 mHz,toluene-d^8, 24 ºC) of (P(Cyp)$_3$)$_2$W(CO)$_3$(H$_2$) showing dihydride(near -3.55 ppm) and molecular hydrogen(near -4.35 ppm) resonances.

TABLE II

EQUILIBRIUM CONSTANTS FOR:
$$WH_2(CO)_3(PR_3)_2 \rightleftharpoons W(CO)_3(PR_3)_2(\eta^2\text{-}H_2)$$

R = i-Pr		R = Cyp	
K_{eq}	T(ºC)	K_{eq}	T(ºC)
3.5	32	3.2	36
4.0	24	3.4	32
4.2	18	3.5	24
4.4	10	3.9	18
4.5	−3	4.0	11
5.0	−8	4.2	5
		4.6	0

Estimated error in K_{eq} values is ±5%.

$$W(CO)_5 \quad \xrightarrow{\text{CO}} \quad W(CO)_6 \quad \Delta H = -46 \text{ kcal/mole} \tag{14}$$

Assuming that the W-CO bond strength is the same in $(PCy_3)_2W(CO)_4$ as in $W(CO)_6$, the lower enthalpy of reaction of the former complex would imply a value of 16 kcal/mole for the W$\cdots$H-C "agostic" bond. While that assumption is probably not warranted, the result is nevertheless in reasonable agreement with the photoacoustic results discussed above. The estimate of -15 ± 5 kcal/mole for this bond strength is reasonable for purposes of comparison to other metals. Adopting this value for the "agostic" bond strength the enthalpy of addition of H_2 and N_2 to "naked" $(P(C_6H_{11})_3)_2W(CO)_3$ would be about -25 and -29 kcal/mole respectively. Since, as discussed below, the molecular hydrogen and dihydride form exist as tautomers with little enthalpic difference between them, the average W-H bond strength for the dihydride form can be estimated as 65 ± 6 kcal/mole.

Enthalpies and Entropies of Binding of N_2 and H_2

Following synthesis of the complex $(P(C_6H_{11})_3)_2Cr(CO)_3$ it was shown that high pressures were needed for quantitative binding of N_2 (1500 psi) and H_2(300 psi) (3). The higher pressure needed for binding of N_2 was surprising since the calorimetric results described above had shown that, with regard to enthalpy of reaction, N_2 was the favored ligand for binding to $(P(C_6H_{11})_3)_2W(CO)_3$. Equilibrium studies were begun on the Cr and Mo complexes expecting to see a cross-over with regard to ligand preference in going from Cr to Mo to W.

The results summarized in Table I show that for all three metals, nitrogen is the preferred ligand with regard to enthalpy of binding, but disfavored with regard to entropy of binding. These data are of interest in reactions in which H_2 and N_2 are competing for the same site on a metal center as shown in eqn.(15):

$$L_nM\text{-}(H_2)(\text{soln}) + N_2(\text{gas}) \longrightarrow L_nM\text{-}(N_2)(\text{soln}) + H_2(\text{gas}) \tag{15}$$

The columns in Table I $\Delta \Delta H, \Delta \Delta S$, and $\Delta \Delta G$ refer to the thermodynamic parameters for reaction (15). In view of the fact that in molecular H_2 complexes, the hydrogen molecule is known to undergo relatively free rotation (1) it was initially considered that the small entropy of binding of hydrogen might be due to less restricted motion of the bound molecule. A simpler explanation is that H_2, due to its small mass and moment of inertia has less entropy to lose on complexation than does N_2. Thus the entropy of reaction (15) is described by eqn.(16) below:

$$\Delta S = S_{L_nM\text{-}(N2)(\text{soln})} - S_{L_nM\text{-}(H2)(\text{soln})} + S_{H2(\text{gas})} - S_{N2(\text{gas})} \tag{16}$$

The entropy change in eqn.(16) corresponds to the term S in Table I. The absolute entropies of H_2 and N_2 are 31.2 and 45.8 cal/mole deg respectively (12). Thus eqn.(16) becomes:

$$\Delta S = -14.6 + S_{L_nM\text{-}(N2)(\text{soln})} - S_{L_nM\text{-}(H2)(\text{soln})} \tag{17}$$

The observed entropy changes as shown in Table I are: Cr(-9.8 ± 2.6), Mo(-8.3 ± 3.9), and W(-13.8 ± 3.5). This indicates that the absolute entropies of the complexes in solution nearly cancel, as might be expected. There is a significant entropic factor(on the order of 10 cal/mole deg) favoring coordination of H_2 over N_2 in eqn.(15). This conclusion is probably general and implies that at high temperatures coordination of hydrogen will be preferred by most complexes. If nitrogen is preferred with respect to enthalpy of binding it will be preferred at low temperatures. The cross-over temprature will be determined by the magnitude of any enthalpic preference for dinitrogen. In the case of the Cr, Mo, and W complexes studied here, the equilibrium constants for reaction (15) have values of 1 at temperatures of about -69, + 28,and + 50 ^{o}C respectively.

In this regard it is also of interest to note that the equilibrium studies discussed earlier for interconversion of the molecular hydrogen and dihydride complexes indicate that the molecular hydrogen complex is favored with regard to enthalpy of formation(circa 1 kcal/mole). The dihydride form appears to be favored with regard to entropy of formation, on the order of 2 cal/mole deg. This should be considered when comparing the tungsten complex to the chromium complex since there is no evidence in the high pressure NMR spectrum of the chromium complex for a stable dihydride form(13). The apparently higher entropy term S for the tungsten complex as shown in Table I could include some contribution of this kind, but experimental errors are too high to warrant further consideration. It is important to note that splitting the H-H bond on the metal does not appear to have a significant entropy of reaction.

Kinetics of Reaction of H$_2$ and (P(C$_6$H$_{11}$)$_3$)$_2$W(CO)$_3$(py)

Displacement of hydrogen by pyridine was studied using stopped flow kinetics as described above. While the reaction was studied in the forward direction shown in eqn.(10),it is instructive to examine the kinetics of this reaction as shown in Figure 7. Half lives for all reactions have been calculated from rate constants at 25 ^{0}C under the pseudo first order conditions [py] = [H$_2$] = 0.01 M, [W] $\leq$ 10^{-4} M.

The first reaction in Figure 7 involves disociation of pyridine which occurs with a half life of about 6 seconds at 25 ^{0}C. This is a fast rate of dissociation since the W-py bond is normally a strong one. Dissociation is facilitated by steric crowding due to the bulky phosphine ligands and concerted formation of a W⋯H-C "agostic" bond as the pyridine ligand is displaced. The "agostic" complex formed can either back react with pyridine(t$_{1/2}$ = 140 microsec) or react with H$_2$ to form the dihydrogen complex(t$_{1/2}$ = 32 microsec). The dihydrogen complex has two reaction channels open to it; oxidative addition to form a dihydride(t$_{1/2}$ = 40 msec) or loss of coordinated dihydrogen to reform the "agostic" complex(t$_{1/2}$ = 1.5 msec). The dihydride complex can isomerize to form the dihydrogen complex(t$_{1/2}$ = 20 msec). The surprising result in this reaction sequence is that, aside from loss of pyridine, oxidative addition of bound H-H is the slowest step in the overall process.

Reaction Profile for Oxidative Addition of H$_2$

The combination of calorimetric, kinetic, and NMR data can be used to generate a fairly complete picture of oxidative addition of H$_2$ in this system. A diagram combining most of these features is shown in Figure 8. The enthalpy of reaction between H$_2$ and the PCy$_3$ "agostic" complex is - 10 $\pm$ 1 kcal/mole. The enthalpy of activation for loss of coordinated H$_2$ is 16.9 $\pm$ 2.2 kcal/mole which implies there is a barrier 6.9 $\pm$ 3.2 kcal/mole for the forward reaction between the tungsten complex and hydrogen.

The second barrier in Figure 8, actual spitting of coordinated H$_2$, is based on thermodynamic parameters determined by NMR studies of the P^{i}Pr$_3$ complex. As discussed below, this data can be compared to kinetic estimates made previously for the PCy$_3$ complex. The ground state enthalpy difference favors formation of the molecular hydrogen complex by 1.2 $\pm$ 0.6 kcal/mole. The enthalpy of activation for cleavage of the H-H bond in this complex was measured to be 10.1 $\pm$ 1.8 kcal/mole. The enthalpy of activation for recombination of the dihydride to regenerate the

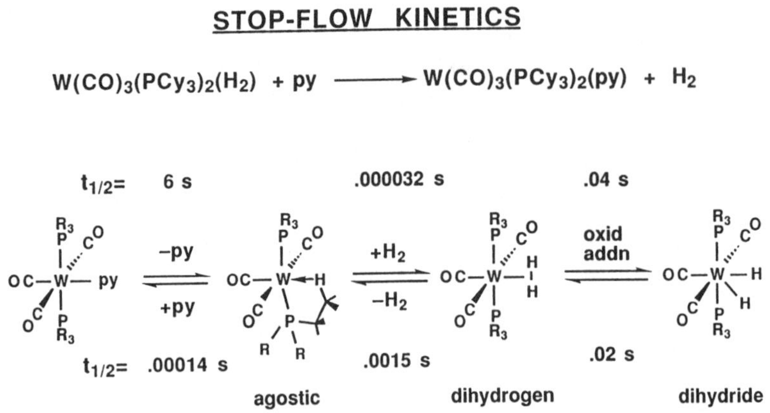

STOP-FLOW KINETICS
W(CO)3(PCy3)2(H2) + py → W(CO)3(PCy3)2(py) + H2
t1/2= 6 s
.000032 s
.04 s
-py
+py
+H2
-H2
oxid addn
t1/2= .00014 s
.0015 s
.02 s
agostic
dihydrogen
dihydride

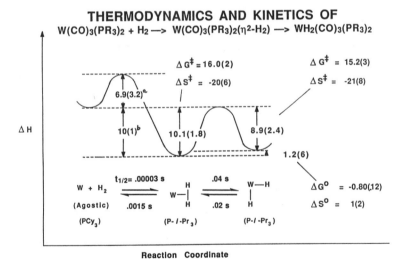

Figure 7. Reaction scheme including half lives under pseudo first order conditions as described in text.

THERMODYNAMICS AND KINETICS OF
W(CO)3(PR3)2 + H2 —> W(CO)3(PR3)2(η2-H2) —> WH2(CO)3(PR3)2
ΔG‡ = 16.0(2)
ΔS‡ = -20(6)
ΔG‡ = 15.2(3)
ΔS‡ = -21(8)
6.9(3.2)a
10(1)b
10.1(1.8)
8.9(2.4)
1.2(6)
ΔH
t1/2= .00003 s
.04 s
W + H2
(Agostic)
.0015 s
.02 s
(PCy3)
(P-i-Pr3)
(P-i-Pr3)
ΔG° = -0.80(.12)
ΔS° = 1(2)
Reaction Coordinate
PARAMETERS DETERMINED BY 1H NMR STUDIES OF DIHYDROGEN–DIHYDRIDE EQUILIBRIUM (Kequil= 0.2-0.3) AND SPIN-SATURATION TRANSFER
a Zhang, Gonzalez, Hoff; J. Am. Chem. Soc. 3627 (1989).
b Gonzalez, Zhang, Nolan, de la Vega, Mukerjee, Hoff, Kubas; Organometallics 2429 (1988).

Figure 8. Reaction profile including enthalpies of activation(kcal/mole) as described in text.

molecular hydrogen complex can then be calculated to be 8.9 ± 2.4 kcal/mole. The kinetic data discussed above for the cyclohexyl complex yielded an enthalpy of activation of 14.4 ± 0.4 kcal/mole for this reaction. These values are in relatively good agreement. It must be considered that they are for different complexes and use completely different approaches to the problem. It is possible that, due to steric factors there is a larger barrier to this reaction for the PCy_3 versus the P^iPr_3 complex. The relatively high experimental errors and the fact that both derivations are based on assumptions regarding mechanism does not allow this conclusion to be firmly held. It seems clear that an energy barrier on the order of 10-15 kcal/mole exists for breaking the H-H bond while it is on the metal. Additional studies on the kinetics and thermodynamics of these and related complexes are in progress.

Acknowledgment

Support of this work by the National Science Foundation (grant # CHE-8618753)(CDH), and the Department of Energy, Division of Chemical Sciences, Office of Basic Energy Sciences(GJK) is gratefully acknowledged.

Literature Cited

1. Kubas, G. J., Acc. Chem. Res.,1988, 21, 120 and references cited therein.
2. Kubas, G. J.; Unkefer, C. J.; Swanson, B. I.; Fukushima, E. J. Amer. Chem. Soc.,1986, 108, 7000.
3. Gonzalez, A. A.; Mukerjee, S. L.; Zhang, K.; Chou, S. J.; Hoff, C. D. J. Amer. Chem. Soc.,1988, 110, 4419.
4. Zhang, K.; Mukerjee, S. L.; Gonzalez, A. A.; Hoff, C. D. manuscript in preparation.
5. Khalsa, G. R. K.; Kubas, G. J.; Unkefer, C. J.; Van Der Sluys, L. S.; Kubat-Martin, K. A., manuscript submitted for publication.
6. Gonzalez, A. A.; Zhang, K.; Nolan, S. P.; de la Vega, R. L.; Mukerjee, S. L.; Hoff, C. D.; Kubas, G. J. Organometal., 1988, 12, 2432.
7. Gonzalez, A. A.; Hoff, C. D. Inorg. Chem.,1989, 28, in press.
8. Gonzalez, A. A.; Zhang, K.; Hoff, C. D. manuscript in preparation.
9. Zhang, K.; Gonzalez, A. A.; Hoff, C. D. J. Amer. Chem. Soc., 1989, 111, 3627.
10. Morse, J. M. Jr.; Parker, G. H.; Burkey, T. J. Organomet.,1989, 12 in press.
11. Lewis, K. E.; Smith, G. P. J. Amer. Chem. Soc., 1984, 106, 4650.
12. Wagman, D. D.; Evans, W. H.; Parker, V. B.; Schumm, R. H.; Halow, I.; Bailey, S. M.; Churney, K. L., Nuttal, R. L. J. Phys. Chem. Ref. Data, 1982, 11, supplement 2.
13. Dr. Istvàn Horvàth, personal communication.

RECEIVED January 9, 1990

Chapter 10

Thermodynamic Studies of the Hydrogenation and Reductive Coupling of Carbon Monoxide by Rhodium(II) Porphyrins

Bradford B. Wayland, Virginia L. Coffin, Alan E. Sherry, and
William R. Brennen

Department of Chemistry, University of Pennsylvania,
Philadelphia, PA 19104–6323

Octaethylporphyrin rhodium II dimer, [(OEP)Rh]$_2$, reacts with H$_2$ and CO to produce an equilibrium distribution of hydride and formyl complexes (Equations 1-3).[1,2] Thermodynamic and kinetic measurements for this system have

$$[(OEP)Rh]_2 + H_2 \rightleftarrows 2 \ (OEP)Rh-H \qquad (1)$$

$$2 \ (OEP)Rh-H + 2 \ CO \rightleftarrows 2 \ (OEP)Rh-CHO \qquad (2)$$

$$[(OEP)Rh]_2 + 2 \ CO + H_2 \rightleftarrows 2(OEP)Rh-CHO \qquad (3)$$

provided estimates for the effective Rh–Rh, Rh–H and Rh–C bond energies: (OEP)Rh–Rh(OEP) (~16.0 kcal mol^{-1}), (OEP)Rh–H (~62 kcal mol^{-1}), (OEP)Rh–CHO (~58 kcal mol^{-1}).[3,4] Approximate bond energy-thermochemical relationships for organometallic reactions (Table I) have subsequently been used in anticipating new types of reactivity for the (OEP)Rh system and in guiding our efforts to modify reactivity through ligand steric and electronic effects. Reductive coupling of CO to form dimetal 1,2-ethanedione complexes[5] and hydrocarbon reactivity[6] are examples of anticipated [(OEP)Rh]$_2$ reactivity. Ligand modifications suggested by thermodynamic and mechanistic considerations have been effectively used in achieving selectivity for CO reductive coupling[7,8] and methane activation.[9] This article summarizes the thermodynamic measurements and reasoning used in organizing and interpreting reactivity studies of rhodium macrocycles relevant to CO reduction.

0097–6156/90/0428–0148$06.00/0

Table I: Thermochemical Estimates for Selected Reactions Relevant to the
Hydrogenation and Reductive Coupling of CO *(Continued on next page)*

Reaction	(I)[a,b] ~$\Delta H°$ kcal mol^{-1}	(II)[c] ~$\Delta G°$ (298K)<0 kcal mol^{-1}
Metallohydrides		
a. M-M + H$_2$ $\rightleftarrows$ 2 M-H	(M-M)-2(M-H)+104	2(M-H)-(M-M)>104
b. 2 M· + H$_2$ $\rightleftarrows$ 2 M-H	-2(M-H)+104	2(M-H)>112
c. M + H$_2$ $\rightleftarrows$ MH$_2$	-2(M-H)+104	2(M-H)>112
Metalloformyls		
d. M-H + CO $\rightleftarrows$ M-CHO	(M-H)-(M-C)-17	(M-H)-(M-C)<9
e. M-M+H$_2$ + 2 CO $\rightleftarrows$ 2 M-CHO	(M-M)-2(M-C)+70	2(M-C)-(M-M)>86
f. 2M·+H$_2$ + 2 CO $\rightleftarrows$ 2 M-CHO	-2(M-C)+70	2(M-C)>94
g. M-M+H$_2$+2CO$\rightleftarrows$2M-H+2CO$\rightleftarrows$2M-CHO		2(M-C)-(M-M)>86 (M-H)-(M-C)<9
h. 2M·+H$_2$+2CO $\rightleftarrows$ 2M-H+2CO$\rightleftarrows$ 2M-CHO		2(M-C)>94 ((M-H)-(M-C)<9) (M-C)>47
Metallohydroxymethyls		
i. M-M+3 H$_2$ + 2 CO$\rightleftarrows$2 M-CH$_2$OH	(M-M)-2(M-C)+42	2(M-C)-(M-M)>74
j. 2M·+3H$_2$ + 2 CO$\rightleftarrows$2 M-CH$_2$OH	-2(M-C)+42	2(M-C)>82
l. M-CHO + H$_2$ $\rightleftarrows$ M-CH$_2$OH[d]	-14	-6
m. M-CH$_2$OH + CO $\rightleftarrows$ M-C(O)CH$_2$OH[d]	-12	-4
Dimetalketones		
n. M-M + CO $\rightleftarrows$ M-C(O)-M	(M-M)-2(M-C)+70	2(M-C)-(M-M)>78
o. 2 M· + CO $\rightleftarrows$ M-C(O)-M	-2(M-C)+70	2(M-C)>86
Dimetal Enediolates		
p. M-M+H$_2$+2 CO$\rightleftarrows$M-OCH=CHO-M	(M-M)-2(M-O)+85	2(M-O)-(M-M)>109
q. 2 M·+H$_2$+2CO$\rightleftarrows$ M-OCH=CHO-M	-2(M-O)+85	2(M-O)>117
r. 2 M-CHO $\rightleftarrows$ M-OCH=CHO-M	2(M-C)-2(M-O)+15	2(M-O)-2(M-C)>23
s. M-M+H$_2$+2CO $\rightleftarrows$ [2M-CHO] $\rightleftarrows$ M-OCH=CHO-M		2(M-O)-(M-M)109 (2(M-O)-2(M-C)>23)
t. 2M·+H$_2$+2CO $\rightleftarrows$ [2M-CHO] $\rightleftarrows$ M-OCH=CHO-M		2(M-O)>117 (2(M-O)-2(M-C)>23)
Dimetal Acetylene Diolates		
u. M-M + 2 CO $\rightleftarrows$ M-OC ≡CO-M	(M-M)-2(M-O)+131	2(M-O)-(M-M)>147
v. 2M· + 2 CO $\rightleftarrows$ M-OC ≡CO-M	-2(M-O)+131	2(M-O)>155
Dimetal 1,2-Ethanediones		
w. M-M+2 CO $\rightleftarrows$ M-C-(O)-C(O)-M	(M-M)-2(M-C)+70	2(M-C)-(M-H)>86
x. 2 M·+2 CO $\rightleftarrows$ M-C(O)-C(O)-M	-2(M-C)+70	2(M-C)>94
y. M-C(O)-M+CO$\rightleftarrows$M-C(O)-C(O)-M[e]	2(M-C)$_k$-2(M-C)$_d$	2(M-C)$_d$-2(M-C)$_k$>8

Table I: *(Continued)* Thermochemical Estimates for Selected Reactions Relevant to the
Hydrogenation and Reductive Coupling of CO

---

[a]   Column I gives the estimated standard enthalpy change ($\Delta H°$) for
a series of reactions in terms of the sum of the effective M-M,
M-H, M-C and M-O bond energies with an enthalpy change for the
organic fragments derived from bond energy data given in b.

[b]   The (C-C), (C-H) and (O-H) bond energies (kcal mol^{-1}) used for
the organic fragments were obtained from data for the organic
molecule that most closely resembles the organic ligand. As an
example, the bond energy parameters for the CH_2OH group of $M-CH_2OH$
were obtained from the parameters for CH_3OH [(C-H)=94, (OH)=104].
The bond energy data for organic oxygenates are found in reference
14, pp. 516   and 517, Table 4.   RC(CO)-H (87), $HOCH_2$-H (94),
HOCH(R)-H (93), HOCH(R)$_2$-H (91), RO(O)C-H (93), RO-H (104), RC(O)O-H
(106), HC(O)-CH_3 (82), $CH_3C(O)CH_3$ (81), CH_3-CH_2R (85), RC(O)O-CH_3
(81).   The bond energy data used for the purely hydrocarbon units
are Laidler   parameters   from   ref.   15,   p.   592,   Table   50.
(C≡C)=183.3, (C=C)=133.0, (C-C)=85.4, (=C-H_2)=101.2, (≡C-H)=104.2,
(C-H)$_p$=98.2, (C-H)$_s$=97.4, (C-H)$_t$=96.5. Bond energy values for the
(C=O) and (C-O) were estimated from $\Delta H°$ values for addition
reactions of CO, CO_2, H_2CO, RCHO, and R_2CO with H_2 and CH_4.[15,16]
(C ≡O)=257, (C=O)=187, (C-O)=100, (C=O)$_{CO_2}$=192, (C(O)-C(O))=70.

[c]   Column II lists the differences in (M-H), (M-C), (M-O) and
(M-M) bond energies that are anticipated to produce a standard free
energy difference of zero at 298 K ($\Delta G°$ (298K)=0). These estimates
are based on using $\Delta S° \sim \Delta n$ (27 cal K^{-1} mol^{-1}) where Δn equals the
difference in the number of product minus the number of reactant
molecules in the respective reactions along with $\Delta H°$ values given
in column I.

[d]   The M-C bond energies in the reactants and products are assumed
equal.

[e]   (M-C)$_k$ designates the M-C bond energy in the dimetal ketone and
(M-C)$_d$ is the M-C bond energy in the dimetal 1,2-ethanedione.

Metallohydride and η^1-Formyl Complexes

Activation of dihydrogen by addition to a metal center is a required step in metal promoted and catalyzed hydrogenation reactions. Guideline criteria for thermodynamically favorable, ($\Delta G°(298K)<0$), addition of H_2 to a M-M bond (Equation 4) are that twice the M-H bond energy minus the M-M bond energy exceed ~104 kcal, and the corresponding criterion for addition to a single metal center (Equation 5) is that the M-H bond energy exceed ~56 kcal mol^{-1} (Table I, entries a-c).

$$M-M + H_2 \rightleftarrows 2\ M-H \qquad (4)$$

$$2\ M + H_2 \rightleftarrows 2\ M-H \qquad (5)$$

Metalloformyl complexes are the most probable first organometallic intermediates in metal complex promoted reactions of H_2 and CO that produce organic oxygenates. Production of large equilibrium concentrations of η^1-carbon bonded formyl complexes from reactions of metal hydrides with CO (Equation 6) requires that the M-H bond

$$MH + CO \rightleftarrows M-CHO \qquad (6)$$

energy not exceed the M-C by more than ~9 kcal mol^{-1} (Table I, entry d). Thermodynamically favorable ($\Delta G°(298K)<0$) production of η^1-metalloformyl species from reactions of metal complexes with H_2 and CO are subject to additional criteria. This reaction for M-M bonded complexes (Equation 7) requires that the M-C bond energy

$$M-M + H_2 + 2\ CO \rightleftarrows 2\ M-CHO \qquad (7)$$

$$2\ M\cdot + H_2 + 2\ CO \rightleftarrows 2\ M-CHO \qquad (8)$$

exceed one half the M-M bond energy plus ~43 kcal and metalloradicals (Equation 8) require that the M-C bond energy exceed ~47 kcal (Table I, entries d-h). At present only rhodium porphyrins[1-4] and rhodium (TMTAA),[10] (TMTAA= dibenzotetramethylaza[14]-annulene dianion), have been demonstrated to produce observable quantities of metalloformyl species from H_2 and CO (Table II), but a wide variety of related macrocycle and chelate complexes can be anticipated to accomplish this reaction of central importance in the hydrogenation of CO.

The standard route proposed for metal promoted and catalyzed conversion of H_2 and CO into organic oxygenates involves the reaction of a metallohydride with CO to form a metalloformyl as the first organometallic intermediate.

Table II: Thermodynamic Values for Selected Reactions Relevant to
the Hydrogenation and Reductive Coupling of CO by Rhodium
Porphyrins

Reaction	$\Delta H°$ kcal mol^{-1}	$\Delta S°$ cal K^{-1} mol^{-1}	$\Delta G°(298K)$ kcal mol^{-1}
[(OEP)Rh]$_2$ + H$_2$ $\rightleftarrows$ 2(OEP)RhH [4]	-3.±1	0.8±2	-3.2
(OEP)RhH + CO $\rightleftarrows$ (OEP)RhCHO [2]	-13±1	-29±4	-3.8
[(OEP)Rh]$_2$ + H$_2$ + CO $\rightleftarrows$ 2(OEP)RhCHO	-29±2	-58±7	-11.3
(TTP)RhH + CO $\rightleftarrows$ (TTP)RhCHO [17]	-10±1	-21±3	-3.7
(OEP)IrH + CO $\rightleftarrows$ (OEP)Ir(H)(CO) [2]	-8.1±1.2	-12±3	-4.5
[(OEP)Rh]$_2$ + CO $\rightleftarrows$ [(OEP)Rh]$_2$CO [5]	-10±1	-26±4	-2.3
[(OEP)Rh]$_2$ + CO $\rightleftarrows$ (OEP)Rh-C(O)-Rh(OEP) [5]	-12±2	-31±5	-2.8
[(OEP)Rh]$_2$+ 2CO $\rightleftarrows$ (OEP)Rh-C(O)-C(O)-Rh(OEP) [5]	-21±2	-62±5	-2.5
(OEP)Rh-C(O)-Rh(OEP) + CO $\rightleftarrows$ (OEP)Rh-C(O)-C(O)-Rh(OEP) [5]	-9±3	-31±7	0.3
[(OEP)Rh]$_2$CO $\rightleftarrows$ (OEP)Rh-C(O)-Rh(OEP) [5]	-2±1	-5±5	-0.5
[(OEP)Rh]$_2$CO+CO $\rightleftarrows$ (OEP)Rh-C(O)-C(O)-Rh(OEP) [5]	-11±1	-36±3	-0.3
2(TMP)Rh-CO $\rightleftarrows$ (TMP)Rh-C(O)-C(O)-Rh(TMP) [8]	-18.5±0.8	(-25)	(-11)

Subsequent carbonyl hydrogenation, CO insertion and reductive elimination steps can produce organic oxygenates. Hydrogenation of M-C(O)-X units and CO insertion reactions are expected to be thermodynamically favorable for virtually any system (Table I, entries 1,m) and thus achieving these steps subsequent to metalloformyl formation is primarily dependent on having appropriate mechanistic pathways.

An alternate route to obtain two carbon products through a metalloformyl intermediate is by formyl coupling to give enediolates (Equation 9). Guideline thermodynamic

$$2 \text{ M-CHO} \rightleftarrows \text{M-OCH=CHO-M} \tag{9}$$

criteria for an η^1-carbon bonded formyl achieving formyl coupling is that the M-O bond energy exceed the M-C by more than ~12 kcal (Table I, entry r). The Rh-C and Rh-O bond energies[3] for the (OEP)Rh system are comparable and thus formyl coupling is neither expected nor observed. Formyl coupling is a prominent reactivity feature for early transition metal, lanthanide and actinide complexes which have particularly strong M-O bonds.[11,12]

CO Reductive Coupling

Reductive coupling of CO to form 1,2-ethanedione or acetylene diolate complexes (Equations 10-11) is a direct route to forming C-C bonded species.

$$\text{M-M} + 2 \text{ CO} \rightleftarrows \text{M-C(O)-C(O)-M} \tag{10}$$

$$\text{M-M} + 2 \text{ CO} \rightleftarrows \text{M-OC} \equiv \text{CO-M} \tag{11}$$

Favorable thermodynamics ($\Delta G^\circ(298K) < 0$) for obtaining acetylene diolates from M-M bonded complexes occurs when $2(\text{M-O}) > (\text{M-M}) + \sim 147$ kcal while single metal units require that the M-O bond energy exceed ~78 kcal (Table I, entries u,v). This limiting type of CO coupling is best known for reactions of alkali metals with CO which form solid ionic acetylene diolate compounds.[13] Rhodium macrocycle complexes have Rh-O bond energies ~50-60 kcal[3] and thus are excluded as potential candidates for acetylene diolate formation.

Reductive coupling of CO that produces α-diketones (1,2-ethanediones) by reaction 10 requires that $2(\text{M-C}) - (\text{M-M})$ exceed ~86 kcal and for metalloradicals the M-C bond energy must exceed ~47 kcal (Table I, entries w,x,y). Recognizing that these M-C and M-M bond energy criteria fall in the range observed for rhodium porphyrins stimulated the search for this unprecedented type of reactivity. Our initial studies involved investigating the equilibria that occur when [(OEP)Rh]$_2$ reacts with CO (P$_{CO}$= 0.1-30 atm) in toluene solution. This study revealed the presence of simultaneous equilibria involving two species with the stoichiometry of two (OEP)Rh units per CO and one species that contained one (OEP)Rh unit per CO. Spectroscopic studies were used in assigning these species to a mono CO adduct, [(OEP)Rh]$_2$CO, a dimetal ketone, (OEP)Rh-C(O)-Rh(OEP), and a dimetal α-diketone, (OEP)Rh-C(O)-C(O)-Rh(OEP), (Equations 12-14).[5]

$$[(OEP)Rh]_2 + CO \rightleftharpoons [(OEP)Rh]_2CO \qquad (12)$$
$$\Delta H_{12}°=-10 \text{ kcal mol}^{-1} \Delta S_{12}°=-26\pm4 \text{ cal K}^{-1} \text{ mol}^{-1},$$
$$\Delta G_{12}°(298K)=-2.3 \text{ kcal mol}^{-1}$$

$$[(OEP)Rh]_2 + CO \rightleftharpoons (OEP)Rh-C(O)-Rh(OEP) \qquad (13)$$
$$\Delta H_{13}°=-12\pm2 \text{ kcal mol}^{-1} \Delta S_{13}°=-31\pm5 \text{ cal K}^{-1} \text{ mol}^{-1}$$
$$\Delta G_{13}°(298K)=-2.8 \text{ kcal mol}^{-1}$$

$$[(OEP)Rh] + 2 CO \rightleftharpoons (OEP)Rh-C(O)-C(O)-Rh(OEP) \qquad (14)$$
$$\Delta H_{14}°=-21\pm2 \text{ kcal mol}^{-1} \Delta S_{14}°=-62\pm5 \text{ cal K}^{-1} \text{ mol}^{-1}$$
$$\Delta G_{14}°(298K)=-2.5 \text{ kcal mol}^{-1}$$

Equilibrium constants determined for reactions 12-14 provide thermodynamic values for both these reactions and the interconversions between the CO containing species (Equations 15-17). Thermodynamic values for reaction 14

$$[(OEP)Rh]_2CO \rightleftharpoons (OEP)Rh-C(O)-Rh(OEP) \qquad (15)$$
$$\Delta H_{15}°=-2\pm1 \text{ kcal mol}^{-1} \Delta S_{15}°=-5\pm5 \text{ cal K}^{-1} \text{ mol}^{-1}$$
$$\Delta G_{15}°(298K)=-0.5 \text{ kcal mol}^{-1}$$

$$[(OEP)Rh]_2CO + CO \rightleftharpoons (OEP)Rh-C(O)-C(O)-Rh(OEP) \qquad (16)$$
$$\Delta H_{16}°=-11\pm1 \text{ kcal mol}^{-1} \Delta S_{16}°=-36\pm3 \text{ cal K}^{-1} \text{ mol}^{-1}$$
$$\Delta G_{16}°(298K)=-0.3 \text{ kcal mol}^{-1}$$

$$(OEP)Rh-C(O)-Rh(OEP) + CO \rightleftharpoons (OEP)Rh-C(O)-C(O)-Rh(OEP) \quad (17)$$
$$\Delta H_{17}°=-9\pm3 \text{ kcal mol}^{-1} \Delta S_{17}°=-31\pm7 \text{ cal K}^{-1} \text{ mol}^{-1}$$
$$\Delta G_{17}°(298K)=0.3 \text{ kcal mol}^{-1}$$

indicate that production of the CO reductive coupling product is inherently favorable ($\Delta G_{14}°(298K)=-2.5$ kcal mol^{-1}), however competition with the other CO containing compounds restricts the α-diketone to a minority species up to relatively high pressures (P$_{CO}$ ~100 atm, T=298K). Reductive coupling reactions like that described in Equation 14 become more thermodynamically favorable when the M-M bond is weakened, and more selective when formation of other CO containing species becomes less favorable. Both the dimetal ketone and the Rh-Rh bonded dimer are sterically restricted and thus sensitive to ligand steric requirements. Increasing the steric bulk of the porphyrin is thus expected to weaken both the Rh-Rh bond and the effective M-C bonding in the dimetal ketone which should make formation of the α-diketone both more thermodynamically favorable and selective.

The general reaction of CO insertion into a dimetal ketone unit is depicted by Equation 18. The dominant

$$M-C(O)-M + CO \rightleftarrows M-C(O)-C(O)-M \qquad (18)$$

enthalpy changes in reaction 18 are associated with forming the C(O)-C(O) bond (-70 kcal mol^{-1}) and the conversion of the CO triple bond to a double bond (+70 kcal mol^{-1}). Reaction 18 is expected to be thermodynamically unfavorable ($\Delta H_{18}° \sim 0$, $\Delta G_{18}°(298K) \sim +8$ kcal) because of the entropy change ($-T\Delta S_{18}°(298K) \sim +8$ kcal mol^{-1}) unless the CO insertion is accompanied by an additional favorable enthalpy term of at least ~8 kcal mol^{-1}. The general process of sequential insertion of two CO units into an M-X bond is thus anticipated to be thermodynamically unfavorable, and examples of this type of reactivity are lacking. Insertion of CO into the rhodium-carbon bond of (OEP)Rh-C(O)-Rh(OEP) is observed and we believe that relief of steric strain present in the dirhodium ketone provides the favorable enthalpy term that overcomes the entropy change associated with this reaction. Using the approximate bond energy relationships for single and double CO insertion (Table I, entries n,u) and the measured enthalpy changes for reactions 8 and 9 provide estimates of 49 kcal mol^{-1} and 54 kcal mol^{-1} for the effective Rh-C bond energies in (OEP)Rh-C(O)-Rh(OEP) and (OEP)Rh-C(O)-C(O)-Rh(OEP), respectively.

Structural models of (OEP)Rh-C(O)-Rh(OEP) suggest that interactions between the porphyrin rings produce an effective weakening of the rhodium-carbon bond. The unfavorable interporphyrin interactions are substantially relieved in the 1,2-ethanedione complex, where the rings are further apart and more nearly parallel. This interpretation suggests that larger substituents on the porphyrin might promote double CO insertion by sterically excluding the single insertion of CO as well as weakening the rhodium-rhodium bond. Tetraphenylporphyrin derivatives are attractive candidates to provide more sterically hindered porphyrins, since the phenyl substituents are oriented perpendicular to the porphyrin ring which inhibits the close approach of two (por)M units that occurs in both the Rh-Rh bonded dimers and dirhodium ketone species.

Rhodium complexes of the bulky tetra-(3,5-dimethyl-phenyl)porphyrin (TXP) ligand and the more sterically demanding tetra-(2,4,6-trimethylphenyl) porphyrin (TMP) were prepared as part of an effort to examine the role of steric influences on promoting CO reductive coupling. The objective was to inhibit formation of the Rh–Rh bonded dimer and dirhodium ketone without seriously affecting the Rh–C bonding in the α-diketone complex. The relative insensitivity of the Rh–C(O)H bonding to steric and electronic effects for this series of porphyrin ligands (OEP, TXP, TMP) is suggested by the remarkably similar IR and NMR parameters for the formyl groups in the corresponding (por)Rh–CHO complexes (Table III).

The anticipated improvement in selectivity for CO reductive coupling is observed for the (TXP)Rh system.[7] (TXP)Rh[II] forms a Rh–Rh bonded dimer, [(TXP)Rh]$_2$, that reacts at mild conditions (P$_{CO}$~1 atm, T=298K) to form (TXP)Rh–C(O)–C(O)–Rh(TXP) as the only ^{1}H NMR observable species in solution (Equation 19). Observation of CO

$$[(TXP)Rh] + 2\ CO \rightleftarrows (TXP)Rh-C(O)-C(O)-Rh(TXP) \quad (19)$$

stretching frequencies of 1778 and 1767 cm^{-1} in (TXP)Rh–C(O)–C(O)–Rh(TXP) illustrates that the dionyl fragment in this dirhodium α-diketone is similar to that in organic analogs where ν$_{CO}$ values of 1720-1740 cm^{-1} are observed. The combined effects of weakening the Rh–Rh bond and virtually excluding the dirhodium ketone give the (TXP)Rh system an appropriate set of thermodynamic factors for selective CO reductive coupling.

Reactivity studies with CO have been extended to the tetramesitylporphyrin (TMP) derivative where ligand steric requirements preclude Rh–Rh bond formation and thus provide a stable rhodium II derivative, (TMP)Rh[II].[8] Reaction of (TMP)Rh· with CO results in equilibria that involve a 17-electron mono-CO complex, (TMP)Rh-ĊO, and the C–C bonded dimer (TMP)Rh–C(O)–C(O)–Rh(TMP) (Equation 20-22). Temperature dependence of the relative EPR intensity

$$(TMP)Rh· + CO \rightleftarrows (TMP)Rh-ĊO \quad (20)$$

$$2\ (TMP)Rh-ĊO \rightleftarrows (TMP)Rh-C(O)-C(O)-Rh(TMP) \quad (21)$$

$$2\ (TMP)Rh· + 2\ CO \rightleftarrows (TMP)Rh-C(O)-C(O)-Rh(TMP) \quad (22)$$

for (TMP)Rh-CO was used in determining the enthalpy change for reaction 21 ($\Delta H_{21}° = -18.5 \pm 0.8$ kcal mol^{-1}).[8] Reversible dimerization of (TMP)Rh-CO through C–C bond formation to produce a 1,2-ethanedionyl bridged complex (Equation 21) can be viewed as being related to acyl radical coupling (2 CH$_3$ĊO → H$_3$C-(O)C-C(O)CH$_3$). Dimerization of (TMP)Rh-ĊO must involve substantially larger electronic and structural rearrangement of the Rh-CO unit compared to that for CH$_3$ĊO because the association energy

Table III: Selected NMR and IR Parameters for the Formyl Group in (por)Rh-CHO Complexes

	OEP	p-TTP	m-TXP	TMP
δ_{CHO} (ppm)	2.82	3.24	3.40	3.78
δ_{CHO} (ppm)	194.4	194.5	194.8	194.3
$J_{103Rh-13CHO}$ (Hz)	29.1	30.0	30.5	30.5
$J_{13C-1HO}$ (Hz)	200	200	197	196
$J_{103Rh-C1HO}$ (Hz)	1.8	1.8	1.9	1.9
$\nu_{C=O}$ (cm^{-1})*	1707 cm^{-1}	1707 cm^{-1}	-	1710 cm^{-1}

*Nujol mull
OEP= octaethylporphyrin
p-TTP= tetra-(4-methylphenyl) porphyrin
m-TXP= tetra-(3,5-dimethylphenyl) porphyrin
TMP= tetra-(2,4,6-trimethylphenyl) porphyrin

($\Delta H_{21}°$=-18.5 kcal mol^{-1}) is much smaller than the expected C(O)-C(O) bond energy (~70 kcal mol^{-1}) in the dionyl complex. This result indicates that a substantial portion (~50 kcal mol^{-1}) of the total energy change for the CO fragment [2(C≡O)-2(C=O)≅140 kcal mol^{-1}) associated with formation of the α-diketone from CO is regained in forming (TMP)Rh-CO. This is equivalent to saying that the CO bond order in (TMP)Rh-CO is greater than two or that the binding of CO with (TMP)Rh· results in only partial rehybridization of CO toward a double bond.

Summary

Rhodium porphyrin complexes have a set of Rh-H (~60 kcal mol^{-1}) and Rh-C(O)X (50-60 kcal mol^{-1}) bond energies appropriate to produce metalloformyl complexes from reactions with H$_2$ and CO and α-diketone complexes by CO reductive coupling. Double insertion of CO into a M-M bond is usually restricted by the thermodynamics for the reaction of the single insertion product with CO. When the M-C bond energies in the dimetal ketone and dimetal α-diketones are equal, unfavorable thermodynamics ($\Delta H°$~0, $\Delta G°$(298K)~+8 kcal mol^{-1}) are anticipated for the second CO insertion step. Recognizing that the effective M-C bond energy in dimetal ketone species is more sensitive to ligand steric requirements than the Rh-C bonding in dimetal α-diketones provides an approach for circumventing this restriction. Application of this strategy to rhodium porphyrin systems results in selective formation of the CO reductive coupling products (Rh-C(O)-C(O)-Rh) for both (TXP)Rh and (TMP)Rh complexes.

Acknowledgment

We gratefully acknowledge support of this work by the National Science Foundation and the Department of Energy, Division of Chemical Sciences, Offices of Basic Energy Sciences, through Grant DE-FG02-86ER13615.

Literature Cited

1)   Wayland, B. B.; Woods, B. A.; Pierce, R. *J. Am.*
      Chem. Soc. **1982**, *104*, 302.
2)   Farnos, M. D.; Woods, B. A.; Wayland, B. B. *J. Am.*
      Chem. Soc. **1986**, *108*, 3659.
3)   Wayland, B. B. *Polyhedron* **1988**, 7, 1545.
4)   Wayland, B. B.; Farnos, M. D.; Coffin, V. L. *Inorg.*
      Chem. **1988**, in press.
5)   Coffin, V. L.; Brennen, W.; Wayland, B. B. *J. Am.*
      Chem.Soc. **1988**, *110*, 6063.
6)   Del Rossi, K. J.; Wayland, B. B. *J. Am. Chem. Soc.*
      1985, *107*, 7941.
7)   Wayland, B. B.; Sherry, A. E.; Coffin, V. L. *J.*
      Chem. Soc., Chem. Commun. **1989**, 662.
8)   Sherry, A. E.; Wayland, B. B. *J. Am. Chem. Soc.*
      1989, *111*, 5010.
9)   Sherry, A. E.; Wayland, B. B. *J. Am. Chem. Soc.*
      1989, in press.
10)   Van Vorhees, S. L.; Wayland, B. B. *Organometallics*,
      1987, *6*, 204.
11)a. Fagan, P. J.; Manriquez, J. M.; Vollmer, S. H.; Day,
      C. S.; Day, V. W.; Marks, T. J. *J. Am. Chem Soc.*
      1981, *103*, 2206-2220.
   b. Fagan, P. J.; Moloy, K. G.; Marks, T. J. *J. Am.*
      Chem. Soc. **1981**, *103*, 6959-6962.
   c. Moloy, K. G.; Fagan, P. J.; Manriquez, J. M.; Marks,
      T. J. *J. Am. Chem. Soc.* **1986**, *108*, 56-67.
   d. Tatsumi, K.; Nakamura, A.; Hofman, P.; Hoffman, R.;
      Moloy, K. G.; Marks, T. J. *J. Am. Chem. Soc.* **1986**,
      108, 4467-4476.
12)a. Manriquez, J. M.; McAllister, D, R.; Sanner, R. D.;
      Bercaw, J. E. *J. Am. Chem. Soc.* **1978**, *100*, 2716-
      2724.
   b. Wolczanski, P. T.; Bercaw, J. E. *Acc. Chem. Res.*
      1980, *13*, 121-127.
   c. Berry, D. H.; Bercaw, J. E.; Jircitano, A. J.;
      Mertes, K. B. *J. Am. Chem. Soc.* **1982**, *104*, 4712-
      4715.
13)a. Weiss, E.; Buchner, W. *Helv. Chem. Acta.* **1963**, *46*,
      1121.
   b. Buchner, W.; Weiss, E. *Helv. Chim. Acta.* **1964**, *17*,
      1415.
14)   McMillen, D. F.; Golden, D. M. *Ann. Rev. Phys. Chem.*
      1982, *33*, 493.
15)   Pedley. J. B.; Naylor, R. D.; Kibrey, S. P.
      Thermochemical Data of Organic Compounds (2nd
      edition). Chapman and Hall, London (1986).
16)   Cox, J. D.; Pilcher, G. *Thermochemistry of Organic*
      and Organometallic Compounds. Academic Press, New
      York (1970).
17)   Wayland, B. B.; Van Vorhees, S. L.; Wilker, C.
      *Inorg. Chem.***1986**, *25*, 4039-4042.

RECEIVED December 6, 1989

Chapter 11

Metal and Ancillary Coordination Effects on Organolanthanide–Ligand Bond Enthalpies

Implications for Bonding and Reaction Patterns

Steven P. Nolan, David Stern, David Hedden, and Tobin J. Marks

Department of Chemistry, Northwestern University, Evanston, IL 60208

This contribution presents a batch titration calorimetric investigation of absolute and relative metal-ligand bond enthalpies in a broad series of organolanthanides of the type L_2LnX, L_2LnR, and L_2LnL' (L = various η^5-cyclopentadienyl-type ligands; Ln = La, Nd, Sm, Eu, Yb; X = Cl, Br, I, alkoxide, dialkylamide; R = $CH(SiMe_3)_2$, η^3-allyl, hydride; $L' = \eta^2$-alkyne). It is found that D(Ln-halide) is a reasonably transferable thermodynamic anchor point from one trivalent lanthanide coordination environment to another. Bond enthalpy trends typical of electropositive metal ions (early transition elements, actinides) are observed in this series and include small D(Ln-H)-D(Ln-alkyl) values, and large D(Ln-halide) and D(Ln-alkoxide) values. Such bond enthalpies are sensitive both to the identity of the lanthanide ion and to the structure of the supporting ligation (e.g., $(Me_5C_5)_2$ vs. $Me_2Si(Me_4C_5)_2$ vs. $Et_2Si(C_5H_4)Me_4C_5)$. While the thermodynamics of $L_2LnR \rightarrow L_2LnR'$ transformations will be relatively insensitive to the identity of Ln, those of binuclear oxidative addition and reductive elimination processes will be highly sensitive to the nature of Ln.

The past decade has witnessed an explosive growth in what we know about the organometallic chemistry of the lanthanide and actinide elements (1-7). A myriad of new metal-ligand linkages, coordination geometries, stoichiometric/catalytic reaction patterns, and insights into metal-ligand bonding have emerged. As this area of chemistry has developed, it has become increasingly apparent that significant departures from reactivity patterns typical of middle and late transition element compounds are pervasive, and that these departures are not simply explained on the basis of kinetic factors. Our desire to understand the origin of these departures as well as to develop guides to predicting new types of transformations thus prompted thermochemical studies aimed at probing f-element metal-ligand bonding energetics (8-11).

0097–6156/90/0428–0159$06.00/0
© 1990 American Chemical Society

Initial calorimetric investigations focussed on determining relative metal-ligand bond disruption enthalpies in tetravalent organoactinide complexes of the type $Cp_2'ThR_2$ ($\underline{12}$), $Cp_2'UR_2$ ($\underline{13}$) ($Cp' = \eta^5\text{-}Me_5C_5$) and Cp_3ThR ($Cp = \eta^5\text{-}C_5H_5$) ($\underline{14}$) using halogenolytic or protonolytic batch titration chemistry (e.g., Equations 1-4). The resulting information is extremely useful for comparing

$$L_nM\text{-}R + X_2 \longrightarrow L_nM\text{-}X + RX + \Delta H_{rx} \tag{1}$$

$$\Delta H_{rx} = D(L_nM\text{-}R) + D(X_2) - D(L_nM\text{-}X) - D(R\text{-}X) \tag{2}$$

$$L_nM\text{-}R + HX \longrightarrow L_nM\text{-}X + RH + \Delta H_{rx} \tag{3}$$

$$\Delta H_{rx} = D(L_nM\text{-}R) + D(H\text{-}X) - D(L_nM\text{-}X) - D(R\text{-}H) \tag{4}$$

$L_nM\text{-}R/L_nM\text{-}R'$ bond enthalpies within a homologous series of compounds or predicting enthalpies of reactions interconnecting members of the series. Converting such relative bond enthalpies to an absolute scale is less straightforward, however, and requires a judiciously chosen "anchor point" (an accurate estimate of $D(L_nM\text{-}X)$). Rigorously absolute values of $D(L_nM\text{-}R)$ are accessible via one-electron sequences as shown in Equations 5-10, provided suitable $L_nM/L_nM\text{-}R/L_nM\text{-}X$ ensembles are available. This approach has recently been applied to an actinide(III)/(IV) and a lanth-

$$L_nM\text{-}R + X_2 \longrightarrow L_nM\text{-}X + RX \tag{5}$$

$$L_nM\text{-}X \longrightarrow L_nM + 1/2\ X_2 \tag{6}$$

$$X\cdot \longrightarrow 1/2\ X_2 \tag{7}$$

$$R\text{-}X \longrightarrow R\cdot + X\cdot \tag{8}$$

$$\overline{L_nM\text{-}R \longrightarrow L_nM + R\cdot} \tag{9}$$

$$D(L_nM\text{-}R) = \Delta H_{rx(1)} + \Delta H_{rx(6)} - 1/2\ D(X_2) + D(R\text{-}X) \tag{10}$$

anide(II)/(III) system: $(Me_3SiC_5H_4)_3U/(Me_3SiC_5H_4)_3U\text{-}R$ ($\underline{15}$, $\underline{16}$) and $Cp_2'Sm/Cp_2'Sm\text{-}R$ ($\underline{17}$). The study of the latter compounds provides the first detailed thermochemical data for any series of organolanthanide complexes. While the results are highly informative, they also raise fascinating additional questions about the metal and ancillary ligation sensitivity of some of the trends as well as their implications for important reaction patterns. In the present contribution, we address these issues with a broader examination of metal-ligand bonding energetics in several classes of organometallic compounds which span the entire lanthanide series.

Experimental

The complexes $Cp_2'Sm$ ($\underline{18}$), $Cp_2'Eu$ ($\underline{19}$, $\underline{20}$), $Cp_2'Yb$ ($\underline{21}$), $Cp_2'Sm(THF)_2$ ($\underline{22}$), $Cp_2'LaCHTMS_2$ (TMS = $SiMe_3$) ($\underline{23}$), $(Cp_2'LaH)_2$ ($\underline{23}$), $Cp_2'NdCHTMS_2$ ($\underline{23}$), $(Cp_2'NdH)_2$ ($\underline{23}$), $Cp_2'NdNMe_2$ ($\underline{24}$), $Cp_2'Nd(\eta^3\text{-}C_3H_5)$ ($\underline{23}$), $Me_2SiCp_2''SmCHTMS_2$ ($\underline{25}$), ($Cp'' = \eta^5\text{-}Me_4C_5$), $(Me_2SiCp_2''SmH)_2$ ($\underline{25}$),

$Cp_2'LuCHTMS_2$ (23), $Me_2SiCp_2''LuCHTMS_2$ (25), $Et_2Si(Cp)Cp''LuCHTMS_2$ (26), and $Cp_2'Yb(\eta^2-Me_2C_2)$ (27) were prepared and purified as described elsewhere. The reagents for titration calorimetry, I_2, t-BuOH, and $(t$-Bu$)_2$CHOH were purified as indicated elsewhere (17). All reactions employed for calorimetry were first investigated by NMR spectroscopy to verify the rapid, quantitative character of the transformations. The anaerobic isoperibol batch titration calorimeter and analytical procedures used in this study have been described in detail elsewhere (17, 28). All heat of solution and heat of reaction measurements were carried out in rigorously purified toluene.

<u>Thermodynamic Anchor Points. $(Me_5C_5)_2Ln$ Thermochemistry.</u>

Titration of the divalent compounds $Cp_2'Sm$, $Cp_2'Eu$, and $Cp_2'Yb$ with iodine proceeds rapidly and quantitatively to yield the corresponding trivalent iodides (Equation 11). The heat of this reaction combined with the known BDE of I_2 yields the $D(Cp_2'Ln-I)$ values set

$$Cp_2'Ln + I_2 \longrightarrow 1/n \ (Cp_2'LnI)_n \tag{11}$$

$$Ln = Sm, \ Eu, \ Yb$$

out in Table I. Structural studies of other $(Cp_2'LnX)_n$ complexes suggest that the above iodides, which are not sufficiently soluble for cryoscopy, are associated in the solid state (e.g., **A** or **B**) (29, 30). Nevertheless, several lines of argument suggest that

A

B

$D(Ln-I)$ and similar parameters for group 16 ligands are not extremely sensitive to the bonding mode (i.e., bridging or terminal). Thus, $D(Sm-I)$ values for $(Cp_2'SmI)_n$ and $Cp_2'Sm(THF)I$ differ marginally: 69.4(2.4) and 72.7(2.9) kcal/mol, respectively (17). Furthermore, $D(Sm^{III}-O)$ values for $(Cp_2'SmO-t$-Bu$)_2$ and $Cp_2'SmOCH(t$-Bu$)_2$ are essentially identical at 82.4(3.5) and 81.3(1.0) kcal/mol, respectively (17, reasonably assuming that $D(t$-BuO-H$) = D[(t$-Bu$)_2$CHO-H$]$). The generality of such trends of course requires additional verification. For more electron-deficient organolanthanide ligation such as alkyl ligands, it appears that differences in bridging versus terminal bonding energetics are somewhat more significant (26).
    In Table II, the present $D(Cp_2'Ln-I)$ data are compared with the corresponding Ln = Sm, Eu, and Yb parameters for the homoleptic lanthanide triiodides, $D_1(LnI_3)$ (31-33). It can be seen that, for both early and late lanthanides, the $D(Ln^{III}-I)$ parameters are

Table I

Enthaplies of Solution, Enthalpies of Reaction of Organolanthanide
Compounds with Iodine in Toluene, and Derived Bond Disruption
Enthalpies[a,b]

Compound	ΔH_{soln}	Titrant	$-\Delta H_{rxn}$[c]	D(Ln-R/X)	R/X
Cp$_2'$Sm	4.7(0.2)	I$_2$	102.4(2.2)	69.4(2.4)	I[d]
Cp$_2'$Eu	3.4(0.4)	I$_2$	80.7(1.8)	58.6(2.0)	I
Cp$_2'$Yb	3.0(0.3)	I$_2$	85.8(1.2)	61.2(1.4)	I
(Cp$_2'$LaH)$_2$	4.6(0.2)	I$_2$	108.1(2.1)	66.6(2.5)	H
Cp$_2'$LaCHTMS$_2$	3.2(0.2)	I$_2$	40.3(1.2)	63.1(1.8)	CHTMS$_2$
(Cp$_2'$NdH)$_2$	5.5(0.3)	I$_2$	109.2(2.2)	56.6(2.5)	H
Cp$_2'$NdCHTMS$_2$	3.1(0.3)	I$_2$	46.4(1.1)	47.5(1.8)	CHTMS$_2$
(Cp$_2'$SmH)$_2$	5.5(0.3)	I$_2$	104.8(1.5)	52.4(2.0)	H[d]
Cp$_2'$SmCHTMS$_2$	3.4(0.2)	I$_2$	40.5(0.9)	46.0(1.8)	CHTMS$_2$[d]
Me$_2$SiCp$_2''$SmCHTMS$_2$	4.1(0.3)	I$_2$	43.2(0.8)	42.7(1.6)	CHTMS$_2$
(Me$_2$SiCp$_2''$SmH)$_2$	4.6(0.2)	I$_2$	100.7(1.8)	52.8(2.2)	H
(Cp$_2'$LuH)$_2$	4.8(0.3)	I$_2$	98.1(2.1)	66.7(2.5)	H
Cp$_2'$LuCHTMS$_2$	3.5(0.3)	I$_2$	31.5(1.1)	67.0(1.8)	CHTMS$_2$
Me$_2$SiCp$_2''$LuCHTMS$_2$	4.2(0.3)	I$_2$	36.1(1.2)	62.4(1.8)	CHTMS$_2$
Et$_2$SiCp''CpLuCHTMS$_2$	4.3(0.3)	I$_2$	43.4(1.2)	55.1(1.8)	CHTMS$_2$

[a] In kcal/mol.

[b] Quantities in parentheses are 95% confidence limits (3σ).

[c] Per mole of titrant.

[d] Reference 17.

Table II

Bond Disruption Enthalpy Data for Organolanthanide Complexes
and the Corresponding Homoleptic Trihalides in Kcal/Mol[a]

$Cp_2'Ln-X$	$D(Cp_2'Ln-X)$	LnX_3	$D_1(LnX_3)$
$Cp_2'SmCl$	$97.1(3.0)$[b]	$SmCl_3$	$102(5)$[c]
$Cp_2'SmBr$	$83.6(1.5)$[b]	$SmBr_3$	$86(5)$[c]
$Cp_2'SmI$	$69.4(2.4)$[b]	SmI_3	$68(5)$[c]
$Cp_2'EuI$	$57.1(2.0)$	EuI_3	$65(5)$[c]
$Cp_2'YbI$	$61.2(1.5)$	YbI_3	$60(5)$[c]

[a]Quantities in parentheses are 95% confidence limits (3σ).

[b]Taken from ref. 17.

[c]References 31-33.

rather similar for Ln constant, indicating that Cp and I ancillary ligations have energetically rather similar effects. Related work also reveals parallel trends in the $(Cp_2'SmBr)_n/SmBr_3$ and $(Cp_2'SmCl)_n/SmCl_3$ pairs (17). These results are combined in Table II and in Figures 1 and 2. Taken together, these data argue strongly that $D(Ln^{III}\text{-halide})$ is a reasonably transferrable thermodynamic anchor point for the lanthanide series. For the actinides, the picture to date appears rather similar in that $D[(Me_3SiC_5H_4)_3U\text{-}I] = 62.4(1.4)$ kcal/mol while $D_1(UI_4) = 66(8)$ kcal/mol (15).

Metal Size and Additional Ancillary Ligand Effects

Excepting effects which are clearly 4f configuration/redox in origin, reactivity trends across the lanthanide series are usually associated with differences in Lewis acidity/electrophilicity and steric interactions. Each of these effects, in turn, is necessarily sensitive to the falling metal ionic radius with increasing atomic number. The quantity $D(Ln M\text{-}H) - D(Ln M\text{-alkyl})$ is a crucial thermodynamic determinant for a number of important homogeneous catalytic reaction components, such as β-H elimination (Equation

$$L_n M \diagdown\!\!\diagup\!\!\diagdown R \rightleftharpoons L_n M\text{-}H + R \diagup\!\!\diagdown \tag{12}$$

12). This parameter was investigated in the pairs $D(Cp_2'Ln\text{-}H)$ versus $D(Cp_2'Ln\text{-}CHTMS_2)$ as shown in Equations 13,14 for the largest, the smallest, and two intermediate lanthanides. To place the bond

$$(Cp_2'LnH)_2 + I_2 \longrightarrow 2/n \ (Cp_2'LnI)_n + H_2 \tag{13}$$

$$Cp_2'LnCHTMS_2 + I_2 \longrightarrow 1/n \ (Cp_2'LnI)_n + ICHTMS_2 \tag{14}$$

Ln = La, Nd, Sm, Lu

enthalpies on an absolute scale, it is assumed that $D(Cp_2'Ln\text{-}I) = D_1(LnI_3)$ as demonstrated above. However, the $D(Cp_2'Ln\text{-}H) - D(Cp_2'Ln\text{-}CHTMS_2)$ difference parameter is not sensitive to this assumption since $D(Cp_2'Ln\text{-}I)$ is constant in the determination for each Ln pair. The results in Table I indicate that this bond enthalpy difference parameter is rather small (0-9 kcal/mol) indicating, in agreement with reactivity observations (5,23,25) that β-H elimination is generally unfavorable for $Cp_2'Ln$-alkyl complexes. As a function of Ln and a 1.160 Å (La(III)) → 0.977 Å (Lu(III)) contraction in eight-coordinate ionic radius (34), the present $D(Cp_2'Ln\text{-}H) - D(Cp_2'Ln\text{-}CHTMS_2)$ parameters vary by only a relatively small amount and in no regular fashion (Table I). Interestingly, for Ln = Sm, this parameter is ca. 4 kcal/mol smaller than for $D(Me_2SiCp_2''SmH) - D(Me_2SiCp_2''SmCHTMS_2)$, indicating that, all other factors being equal, β-H elimination should be more thermodynamically favorable in the latter system. Further comments will be made about such ancillary ligand effects shortly (vide infra).

For pragmatic reasons, the above analyses used $CHTMS_2$ as a

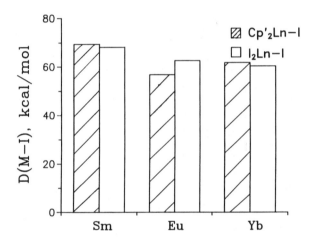

Figure 1. Comparison of metal-iodine bond disruption enthalpies for several $(Me_5C_5)_2LnI$ complexes (Ln = lanthanide) with the published D_1 values for the corresponding lanthanide triiodides.

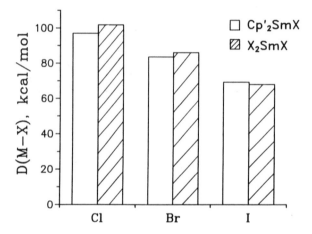

Figure 2. Comparison of samarium-halogen bond disruption enthalpies for several $(Me_5C_5)_2Sm$-halogen complexes with the published D_1 values for the corresponding samarium trihalides.

model alkyl group. In viewing such results, it should be noted that $CHTMS_2$ may be a somewhat atypical alkyl group in that diffraction studies reveal close $Ln\cdots H_3C\text{-}Si$ secondary bonding and a small ⨤ $Ln\text{-}C(\alpha)\text{-}H$ angle (C) ([22,25,26]). In addition, while early lanthanide $(Cp_2'LnH)_2$ compounds have $(\mu\text{-}H)_2$ structure D ([4,23,25]), the lutetium member of the series appears to have $(\mu\text{-}H)H$ structure E ([5]).

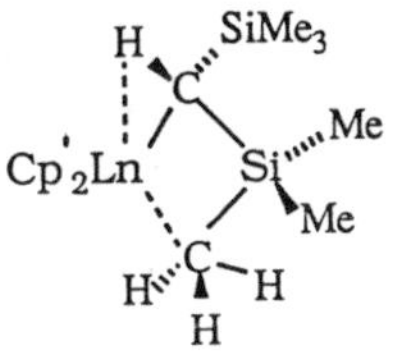

C 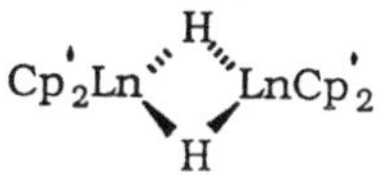   D 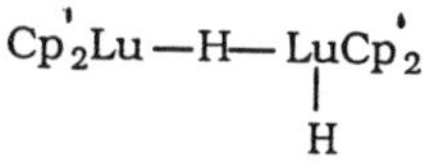   E

It has been previously observed that significant changes in organolanthanide metal-carbon sigma bond reactivity can be effected by the ancillary ligand modifications F → G → H ([23,25,26]). In particular, reactions having large steric demands in the transition state (e.g., olefin insertion) are (not surprisingly) accelerated in the more open coordination spheres of G and H ([23,25,34]).

F    G    H

Curiously, however, sterically rather insensitive Ln-C bond hydrogenolysis is considerably slower in environments G and H ([26]). Iodinolytic thermochemical studies were therefore undertaken on the $Lu\text{-}CHTMS_2$ members of the F, G, H series and $Sm\text{-}CHTMS_2$ members of the F, G series to probe ancillary ligand effects on $D(Ln\text{-}CHTMS_2)$. Assuming that $D(Lu^{III}\text{-}I)$ and $D(Sm^{III}\text{-}I)$ are invariant within each class of compounds (<u>vide</u> <u>supra</u>), the results in Table I are obtained. It can be seen that the $Lu\text{-}CHTMS_2$ bond disruption enthalpy clearly <u>decreases</u> as the ancillary ligation makes the progression F → G → H, and that the $Sm\text{-}CHTMS_2$ data follow an identical trend for F → G. These trends are in opposition to what would be expected from ancillary ligand-$CHTMS_2$ nonbonded interactions. The reasons for these trends appear instead to reflect the nature of the bond dissociation process (e.g., Equation 15) which involves formal reduction of the lanthanide ion. The

$$Cp'_2Ln^{III}\text{-}R \longrightarrow Cp'_2Ln^{II} + R\cdot \tag{15}$$

progression $F \to G \to H$ is one of decreasing electron donation by the ancillary ligands (for example, structural data reveal diminished ring-metal interactions in $Me_2SiCp'_2LnCHTMS_2$ versus $Cp'_2LnCHTMS_2$ (23,25)), hence destabilization of the Ln(III) state and stabilization of Ln(II). Such an effect is expected to lower the bond disruption enthalpy. An alternative explanation would invoke an approximately invariant $D(Ln\text{-}CHTMS_2)$ but increasing $D(Ln\text{-}I)$ through the progression $F \to G \to H$ for steric and/or electronic reasons. While this explanation cannot be rigorously ruled out with the information at hand, it is not supported by the aforementioned $D(Cp'_2Ln\text{-}I)/D_1(LnI_3)$ correlations. That $D(Sm\text{-}H)$ is essentially identical in $(Me_2SiCp''_2SmH)_2$ (52.8(2.2) kcal/mol) and $(Cp'_2SmH)_2$ (52.4(2.0) kcal/mol) (Table I) indicates that either $D(Ln\text{-}I)$ is not stronger in G,H-type complexes (with $D(Ln\text{-}H)$ approximately invariant) or that $D(Ln\text{-}H)$ and $D(Ln\text{-}I)$ change in an almost parallel fashion. Considering the greatly different steric and electronic requirements expected for μ-H and μ-I bonding, the latter contention appears questionable, and additional studies are needed to resolve this issue. The reason for the falling $Ln\text{-}CHTMS_2$ hydrogenolysis reactivity in the $F \to G \to H$ progression appears to reflect diminished charge stabilization at the lanthanide center in the heterolytic, four-center hydrogenolysis transition state (Equation 16) rather than thermodynamic factors. An interesting contrast is presented by $Cp'_2Th(X)R$ compounds where Th-C hydro-

$$Ln\text{-}R \; + H_2 \longrightarrow \; \overset{\delta^-}{\underset{\delta^+}{H}}\cdots\overset{\delta^+}{\underset{}{H}} \quad Ln\cdots R \longrightarrow Ln\text{-}H + R\text{-}H \tag{16}$$

genolysis rates exhibit an approximately opposite dependence on $D(Th\text{-}R)$ (35). In terms of Cp'_2SmR/H interconversions, perhaps the most important result of this study is the observation that the thermodynamics of important transformations such as β-H elimination and metal-carbon hydrogenolysis are sensitive to the cyclopentadienyl ancillary ligation.

Alkoxide, Allyl, Amide, and Alkyne Ligands.

Additional calorimetric studies were undertaken to examine the generality of trends (17) previously noted for Cp'_2SmX compounds and to verify parameter self-consistency within a fairly large series of organolanthanide complexes. Equations 17 and 18 probe Cp'_2Nd-alkoxide bond enthalpies for sterically non-encumbered and encumbered alcohols. From cryoscopy results for the Sm analogues (17),

$$(\text{Cp}_2'\text{NdH})_2 + 2 \ \underline{t}\text{-BuOH} \longrightarrow (\text{Cp}_2'\text{NdO-}\underline{t}\text{-Bu})_2 + 2 \ \text{H}_2 \qquad (17)$$

$$(\text{Cp}_2'\text{NdH})_2 + 2 \ (\underline{t}\text{-Bu})_2\text{CHOH} \longrightarrow 2 \ \text{Cp}_2'\text{NdOCH}(\underline{t}\text{-Bu})_2 + 2 \ \text{H}_2 \qquad (18)$$

we take the molecularities of the products to be dimeric and mono-meric, respectively. The results (Table III) show that the two $D(\text{Cp}_2'\text{Nd-OR})$ values differ insignificantly, arguing as for samarium that such parameters are rather insensitive to the degree of asso-ciation. Equations 19 and 20 probe the strengths of Nd dialkyla-mide and η^3-allyl linkages, respectively. The derived disruption

$$2 \ \text{Cp}_2'\text{NdNMe}_2 + 2 \ \underline{t}\text{-BuOH} \longrightarrow (\text{Cp}_2'\text{NdO-}\underline{t}\text{-Bu})_2 + 2 \ \text{HNMe}_2 \qquad (19)$$

$$2 \ \text{Cp}_2'\text{Nd}(\eta^3\text{-C}_3\text{H}_5) + 2 \ \underline{t}\text{-BuOH} \longrightarrow (\text{Cp}_2'\text{NdO-}\underline{t}\text{-Bu})_2 + 2 \ \text{C}_3\text{H}_6 \qquad (20)$$

enthalpies (Table III) closely mirror those in the Sm series (17) with the lanthanide-to-dialkylamide bond considerably weaker than that to an alkoxide, and the CHTMS$_2$ ligand more strongly bound than the η^3-allyl. The close congruence between the $4f^3$ Cp$_2'$Nd-R and $4f^5$ Cp$_2'$Sm-R bond enthalpy patterns is illustrated in Figure 3.

An additional test of the reliability of the above $D(\text{Cp}_2'\text{Nd-R})$ parameters is provided by an independent determination of $D(\text{Cp}_2'\text{Nd-O-}\underline{t}\text{-Bu})$ as shown in Equation 21. The result, 82.2(2.0) kcal/mol, compares well with that determined via the hydride, 79.8(1.5) kcal/mol (Equation 15; Tables I, III).

$$\text{Cp}_2'\text{NdCHTMS}_2 + \underline{t}\text{-BuOH} \longrightarrow 1/2 \ (\text{Cp}_2'\text{NdO-}\underline{t}\text{-Bu})_2 + \text{H}_2\text{CTMS}_2 \qquad (21)$$

The complex $\text{Cp}_2'\text{Yb}(\eta^2\text{-Me}_2\text{C}_2)$ (**I**) is the only isolable, simple organo-f-element π-alkyne complex (27), and it was of great interest to probe the energetics of the Yb-alkyne bond. The titra-

I

tion of Equation 22 along with the data in Table I yield an alkyne bond disruption enthalpy of 14.0(1.2) kcal/mol. This quantity is rather low, and can be compared to reported transition metal-alkyne

$$\text{Cp}_2'\text{Yb}(\eta^2\text{-Me}_2\text{C}_2) + 1/2\text{I}_2 \longrightarrow 1/\text{n} \ (\text{Cp}_2'\text{YbI})_n + \text{Me}_2\text{C}_2 \qquad (22)$$

bond enthalpies in the range 20-90 kcal/mol (9,10,35,36). For comparison in the f-element series, $D(\text{Cp}_2'\text{Sm-THF}) = 7.3(0.4)$ kcal/mol (17), $D(\text{Cp}_2'\text{Sm(THF)-THF}) = 4.9(1.0)$ kcal/mol (17), and $D((\text{Me}_3\text{SiC}_5\text{H}_4)_3\text{U-CO}) = 10.3(0.2)$ kcal/mol (15). The value for $D(\text{Cp}_2'\text{Zr-benzyne})$ is in excess of 120 kcal/mol (28).

Table III

Enthaplies of Solution, Enthalpies of Reaction of Organolanthanide
Compounds with Various Titrants in Toluene, and Derived
Bond Disruption Enthalpies[a,b]

Compound	ΔH_{soln}	Titrant	$-\Delta H_{rxn}$[c]	$D(Ln-R/X)$	R/X
$(Cp_2'NdH)_2$	5.5(0.3)	$\underline{t}$-BuOH	26.2(1.2)	82.2(2.0)	$O-\underline{t}$-Bu
$Cp_2'NdCHTMS_2$	3.1(0.3)	$\underline{t}$-BuOH	25.7(1.7)	82.2(2.0)	$O-\underline{t}$-Bu
$(Cp_2'NdH)_2$	5.5(0.3)	$(\underline{t}$-Bu$)_2$CHOH	23.0(1.0)	79.8(1.5)	$OCH(\underline{t}$-Bu$)_2$
$Cp_2'NdNMe_2$	1.5(0.2)	$\underline{t}$-BuOH	24.8(1.2)	47.4(1.5)	NMe_2
$Cp_2'Nd(\eta^3-C_3H_5)$	2.5(0.3)	$\underline{t}$-BuOH	25.3(1.3)	40.9(1.7)	$\eta^3-C_3H_5$
$Cp_2'Yb(\eta^2-Me_2C_2)$	4.6(0.2)	I_2	58.2(0.8)	14.0(1.2)	$\eta^2-MeC\equiv CMe$

[a]In kcal/mol.

[b]Quantities in parentheses are 95% confidence limits (3σ).

[c]Per mole of titrant.

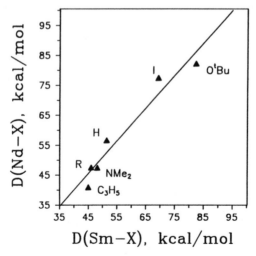

Figure 3. Comparison of $D[(Me_5C_5)_2Nd-R/X]$ data to the correspond-
ing $D[(Me_5C_5)_2Sm-R/X]$ data of ref. 17. $R = CH(SiMe_3)_2$ and $C_3H_5 = \eta^3$-allyl. The line shows a least-squares fit to the data and has
a slope of 1.08.

Discussion

The present study affords the first detailed picture of absolute
and relative organolanthanide bonding energetics for a diverse
selection of metal ions and interesting ligands. At the broadest
level, these data exhibit strong parallels to bond enthalpy
patterns previously identified for actinide and group 4 complexes,
and which appear to be general for this portion of the Periodic
Table. These include small D(M-H)-D(M-alkyl) values, large values
of D(M-halide) and D(M-alkoxide), and smaller values of
D(M-dialkylamide). In all cases, D(M-halide) appears to be rather
insensitive to whether ancillary ligands are halide or π-carbo-
cyclic. The present results indicate that such patterns are rather
general for a series of organolanthanides in which the central 4f
ion can vary considerably in size, $4f^n$ configuration, and redox
characteristics (38). Hence, most $Cp_2'LnR \longrightarrow Cp_2'LnR'$ transformations will have enthalpies which are rather insensitive to the
identity of Ln. As noted previously, many of the aforementioned,
distinctive reaction patterns can be qualitatively understood in
terms of metal and ligand electronegativities (16, 28). The degree
to which the present organolanthanides adhere to this picture is in
good accord with the pronounced electropositive character of the
entire 4f series. The present study reveals in addition that
modification of cyclopentadienyl ancillary ligand structure
significantly affects the enthalpies of some adjacent metal-ligand
sigma bonds. This effect appears also to have some generality.
    As for metal-ligand bond enthalpy differences among compounds
of different lanthanides, the approximate transferability of
D(Ln-halogen) parameters along with fairly general regularities in
D(Ln-halogen) -D(Ln-R/X) values means that unavailable D(Ln-R) data
should be estimable from existing data and should approximately
track D(Ln-halogen) (e.g., Equation 23). Thus organolanthanides

$$D(Ln'-R/X) \approx D(Ln-R/X) - [D(Ln-I)-D(Ln'-I)] \qquad (23)$$

with the weakest metal-halogen bonds are likely to have the weakest
metal-hydride, metal-alkyl, etc. bonds. These lanthanide ions will
be those with the most stable divalent states (e.g., Eu(III),
Yb(III); cf., eq.(15)). Complementary patterns should obtain for
lanthanides with the most stable trivalent states (e.g., La(III),
Ce(III), Pr(III)).
    In terms of interesting organolanthanide chemistry, the above
discussion conveys intriguing implications for trivalent organo-
europium chemistry, the restricted nature of which has generally
been thought to reflect the pronounced stability of the divalent
oxidation state. While complexes such as $(Cp_2'EuH)_2$ and $Cp_2'Eu$-alkyl
are presently unknown, Equation 23 and the data in Tables I and II
suggest that D(Eu-H) $\approx$ 40 kcal/mol and D(Eu-CHTMS$_2$) $\approx$ 35 kcal/mol.
While these bond enthalpies are lower than for other organolanthanides, simple susceptibility to bond homolysis should not be a
major factor in determining the isolability of such complexes (cf.,
isolable cobalt complexes with D(Co-C) $\approx$ 25 kcal/mol (8,11)).
Rather, the kinetic lability of such complexes combined with the
anticipated exothermicity and entropic favorability of many
bimolecular elimination processes (e.g., Equations 24-26) is likely

an important factor. Likewise, even an endothermic β-H elimina-
tion process (e.g., $\Delta H \approx +10 - +30$ kcal/mol ($\underline{17}$)) could be driven

$$(Cp_2'EuH)_2 \longrightarrow 2\ Cp_2'Eu + H_2 \tag{24}$$

$$\Delta H_{calcd} \approx -12 \text{ kcal/mol Eu}$$

$$1/2(Cp_2'EuH)_2 + Cp_2'EuR \longrightarrow 2\ Cp_2'Eu + RH \tag{25}$$

$$\Delta H_{calcd} \approx -12 \text{ kcal/mol Eu}$$

$$2\ Cp_2'EuR \longrightarrow 2\ Cp_2'Eu + R_2 \tag{26}$$

$$\Delta H_{calcd} \approx -8 \text{ kcal/mol Eu}$$

by coupled exergonic follow-up reactions of the resulting hydride.
For example, the sequence of Equations 27, 28 is approximately
thermoneutral, however the $T\Delta S$ contribution to ΔG is expected to be

$$Cp_2'EuR \longrightarrow Cp_2'EuH + \text{alkene} \tag{27}$$

$$\Delta H_{calcd} \approx +25 \text{ kcal/mol}$$

$$Cp_2'EuH + Cp_2'EuR \longrightarrow 2\ Cp_2'Eu + RH \tag{28}$$

overall: $2\ Cp_2'EuR \longrightarrow 2\ Cp_2'Eu + \text{alkene} + RH$ $\quad$ (29)

overall: $\Delta H_{calcd} \approx 0$ kcal/mol Eu

on the order of ca. -10 kcal/mol Eu for two particles forming from
one (39,40). Thus, the intrinsic weakness of the europium-ligand
bonds may promote new types of organolanthanide transformations
which are normally only observed in transition metal systems.
Conversely, many dinuclear addition processes which are exothermic
for other divalent lanthanides ($\underline{17}$) will be endothermic for
europium. An example is dinuclear hydrocarbon activation. The
endothermic process of Equation 30 is estimated to be exothermic
for samarium ($\underline{17}$) yet endothermic for europium. These and a number

$$2\ Cp_2'Ln + \diagup\!\!\!\diagdown \longrightarrow Cp_2'Ln \text{-}\rangle\!\rangle + 1/2\ (Cp_2'LnH)_2 \tag{30}$$

$$\Delta H_{calcd} \approx -12 \text{ kcal/mol, Ln = Sm}$$

$$\Delta H_{calcd} \approx +10 \text{ kcal/mol, Ln = Eu}$$

of other consequences of the data reported herein are currently
under investigation.

Acknowledgments
We thank the National Science Foundation for support of this
research under grant CHE8800813.

Literature Cited

1. Schumann, H.; Albrecht, I.; Gallagher, M.; Hahn, E.; Janiak, C.; Kolax, C.; Loebel, J.; Nickel, S.; Palamedis, E. Polyhedron 1988, 7, 2307-2315, and references therein.
2. Evans, W. J. Polyhedron 1987, 6, 803-835.
3. Marks, T. J.; Fragala, I., Eds. "Fundamental and Technological Aspects of Organo-f-Element Chemistry," D. Reidel: Dordrecht, Holland, 1985.
4. Evans, W. J. Adv. Organomet. Chem. 1985, 24, 131-177.
5. Watson, P. L.; Parshall, G. W. Acc. Chem. Res. 1985, 18, 51-56.
6. Kagan, H. B.; Namy, J. L. In "Handbook on the Physics and Chemistry of Rare Earths"; Gschneider, K. A., Eyring, L. Eds.; Elsevier: Amsterdam, 1984, Chapt. 50.
7. Marks, T. J., Ernst, R. D. In "Comprehensive Organometallic Chemistry"; Wilkinson, G., Stone, F. G. A., Abel, E. W., Eds.; Pergamon Press: Oxford, 1982, Chapt. 21.
8. Marks, T. J., Ed. "Metal-Ligand Bonding Energetics in Organotransition Metal Compounds," Polyhedron Symposium-in-Print, 1988, 7.
9. Pilcher, G.; Skinner, H. A. In "The Chemistry of the Metal-Carbon Bond"; Harley, F. R.; Patai, S., Eds.; Wiley: New York, 1982, pp. 43-90.
10. Connor, J. A. Top. Curr. Chem. 1977, 71, 71-110.
11. Halpern, J. Acc. Chem. Res. 1982, 15, 238-244.
12. Bruno, J. W.; Marks, T. J.; Morss, L. R. J. Am. Chem. Soc. 1983, 105, 6824-6832.
13. Bruno, J. W.; Stecher, H. A.; Morss, L. R.; Sonnenberger, D. C.; Marks, T. J. J. Am. Chem. Soc. 1986, 108, 7275-7280.
14. Sonnenberger, D. C.; Morss, L. R.; Marks, T. J. Organometallics 1985, 4, 352-355.
15. Schock, L. E.; Seyam, A. M.; Marks, T. J., in ref. 8, pp. 1517-1530.
16. Marks, T. J.; Gagné, M. R.; Nolan, S. P.; Schock, L. E.; Seyam, A. M.; Stern, D. Pure Appl. Chem., 1989, 16, 1665-1672.
17. Nolan, S. P.; Stern, D.; Marks, T. J. J. Am. Chem. Soc., 1989, 111, 7844-7853.
18. Evans, W. J.; Chamberlain, L. R.; Ulibarri, T.; Ziller, J. W. J. Am. Chem. Soc. 1988, 110, 6423-6432, and references therein.
19. Evans, W. J.; Hughes, L. A.; Hanusa, T. J. Organometallics 1986, 5, 1285-1291.
20. Tilley, T. D.; Andersen, R. A.; Spencer, B.; Ruben, H.; Zalkin, A.; Templeton, D. H. Inorg. Chem. 1980, 19, 2999-3003.
21. Andersen, R. A.; Boncella, J. M.; Burns, C. J.; Green, J. C.; Hohl, D.; Rösch, N. J. Chem. Soc., Chem. Commun., 1986, 405-407.
22. Evans, W. J.; Grate, J. W.; Chei, H. W.; Bloom, I.; Hunter, W. E.; Atwood, J. L. J. Am. Chem. Soc. 1985, 107, 941-946.
23. Jeske, G.; Lauke, H.; Mauermann, H.; Swepston, P. N.; Schumann, H.; Marks, T. J. J. Am. Chem. Soc. 1985, 107, 8091-8103.
24. Hedden, D.; Marks, T. J. From $(Cp'_2NdH)_2$ + $HNMe_2$ (unpublished results).

25. Jeske, C.; Schock, L. E.; Mauermann, H.; Swepston, P. N.;
    Schumann, H.; Marks, T. J. J. Am. Chem. Soc. 1985, 107,
    8103-8110.
26. Stern, D.; Sabat, M.; Marks, T. J., submitted for publica-
    tion.
27. Burns, C. J.; Andersen, R. A. J. Am. Chem. Soc. 1987, 109,
    941-942.
28. Schock, L. E.; Marks, T. J. J. Am. Chem. Soc. 1988, 110,
    7701-7715.
29. Rausch, M. D.; Moriarity, K. J.; Atwood, J. L.; Weeks, J.
    A.; Hunter, W. E.; Brittain, H. G. Organometallics 1986, 5,
    1281-1283.
30. Evans, W. J.; Brummond, D. K.; Grate, J. W.; Zhang, H.;
    Atwood, J. L. J. Am. Chem. Soc. 1987, 109, 3928-3936.
31. Huheey, J. E. "Inorganic Chemistry," 2nd ed.; Harper and
    Row: New York, 1978; pp. 824-850.
32. Wagman, D. D.; Evans, W. H.; Parker, V. B.; Halow, L.;
    Bailey, S. M.; Schumm, R. H.; Churney, K. L. Natl. Bur.
    Stand. Tech. Note (U.S.), 1971, No. 270-278.
33. Feher, R. C. Los Alamos Report La-3164, 1965.
34. Gagné, M. R.; Marks, T. J., unpublished results.
35. Lin, Z.; Marks, T. J. J. Am. Chem. Soc. 1987, 109, 7979-
    7985.
36. Evans, A.; Mortimer, C. T.; Puddephatt, R. J. J. Organomet.
    Chem. 1975, 96, C58-C60.
37. McNaughton, J. L.; Mortimer, C. T.; Burgess, J.; Hacker, M.
    J.; Kennett, R. D. W. J. Organomet. Chem. 1974, 71, 287-
    290.
38. Morss, L. R. Chem. Rev., 1976, 76, 827-841.
39. Smith, G. M.; Carpenter, J. D.; Marks, T. J. J. Am. Chem.
    Soc. 1986, 108, 6805-6807.
40. Page, M. I. in "The Chemistry of Enzyme Action," Page, M.
    I., Ed., Elsevier: New York, 1984; pp. 1-54.

RECEIVED December 6, 1989

Chapter 12

Novel Extensions of the Electrostatic Covalent Approach and Calorimetric Measurements to Organometallic Systems

Russell S. Drago

Department of Chemistry, University of Florida, Gainesville, FL 32611

Attempts to understand chemical reactivity in organometallic systems often involve comparison of some measured observable with pK_B or other one parameter criteria of sigma donor strength. Deviations from the sigma donor trends are often interpreted in terms of unusual bonding effects in the organometallic system. In this article, the pitfalls associated with selection of a one parameter criteria of donor strength are discussed. An alternative approach based on the ECW model is offered for both the interpretation and design of experiments. Several examples are presented which illustrate both the ways in which the model should be applied and the additional information that can be obtained from the data. Utilization of the approach to reactions in polar solvents and in heterogenous systems is also described.

Most chemists carry out thermodynamic measurements in order to understand trends in chemical reactivity. To accomplish this objective, quantitative criteria are required to provide the basis for what is to be expected under normal circumstances where sigma bonding dominates reactivity. For example, the pK_B scale has been used often to provide the basis for the sigma bond reactivity of organic bases. In the late 1950's, it was recognized that no single scale of sigma donor strength existed ($\underline{1-3}$). As the Lewis acid is varied, changes in the order of donor strength occur.

0097–6156/90/0428–0175$06.00/0
© 1990 American Chemical Society

Some of the most dramatic reversals occur with the acids iodine and phenol ($\underline{3}$). Early, first row transition metal complexes also give rise to orders different from d^8 third row systems ($\underline{1}$).

Qualitative explanations of these reversals in basicity were based on the Mulliken ($\underline{4}$) description of bonding in charge-transfer complexes:

$$\psi^{\circ}{}_{BA} = a\psi^{\circ}{}_{el} + b\psi^{\circ}{}_{cov} \qquad (1)$$

Variations in the importance of covalent, $\psi^{\circ}{}_{cov}$, and electrostatic, $\psi^{\circ}{}_{el}$, bonding were proposed to account for the reversals in strength of oxygen and sulfur donors toward iodine ($R_2S > R_2O$) and phenol ($R_2O > R_2S$). Nitrogen donors, with both a large lone pair dipole moment and low ionization energy, e.g., $(CH_3)_3N$, tend to be stronger than sulfur donors toward I_2 and also stronger than oxygen donors toward phenol. These examples illustrate the qualitative way in which the Mulliken or Pauling ($\underline{5}$) electrostatic-covalent, (EC), model is used to rationalize different donor orders.

If a qualitative rationalization of bonding has any basis in fact, at the very least one should be able to fit bond strengths, e.g. adduct formation enthalpies, to empirical parameters that relate to the effects used in the qualitative description. If additional parameters are needed for a good quantitative fit, then an added effect is also needed for the complete qualitative rationalization. The equation proposed ($\underline{6\text{-}8}$) for the qualitative electrostatic-covalent model is:

$$-\Delta H = E_A E_B + C_A C_B - W \qquad (2)$$

The enthalpy of coordinate bond formation, $-\Delta H$, is given by an electrostatic term, $E_A E_B$, and a covalent term, $C_A C_B$ where A refers to a Lewis acid and B a Lewis base. The W term accommodates ($\underline{8}$) any constant contribution to the enthalpy which for an acid (or base) is independent of the base (or acid) employed. The W term is usually zero but would be finite, for example, for the heat of dissociation of Al_2Cl_6 to form a $B\text{-}AlCl_3$ adduct.

Enthalpy data for reactions of the type:

$$A + B \rightleftharpoons A\text{-}B \qquad (3)$$

in solvents with minimal solvation contributions are used to empirically determine the E and C parameters. The resulting data fit is exceptional. The most recent report (9) of values for these parameters employed 500 enthalpies for 48 bases and 43 acids to solve five hundred equations for 185 unknown parameters. (Seven acids have W values and all other acids and bases have W=0.) The EC parameters that result from the fit represent the tendency of the acid or base to undergo electrostatic or covalent bonding, respectively, when forming an adduct. When the reported (9) empirical parameters for the acid and base are substituted into Equation 2, the calculated enthalpy is found to agree with the experimental result to within 0.1 to 0.2 kcal mole^{-1}. Systems in which steric effects exist and those in which there is metal-ligand π-backbonding show deviations between the calculated and measured values that provide an estimate of the magnitude of these effects.

In contrast to the claim (10) that the ECW model "disguises the relationship between reactivity and periodic elemental properties", elementary application of frontier molecular orbital theory (11) can be used to understand the trends. Using qualitative trends in ionization energies, inductive effects, electronegativities and partial charge/size ratios, one can estimate trends in the HOMO-LUMO separation of the donor and acceptor. Increasing the separation decreases the covalent and increases the electrostatic nature of the interaction. Decreasing the separation has the opposite effect. Trends in the reported acid and base parameters as well as in the $E_A E_B$ and $C_A C_B$ products can be understood in this way.

There have been several attempts in recent inorganic chemistry textbooks to relate the ECW approach to other acid-base theories. Finston and Rychtman (12) have done an outstanding job in their recent text and the reader is referred to this source for this topic. A simple test can be used to judge other approaches. Anyone offering or considering a different qualitative model for donor-acceptor reactivity should express it using an empirical equation, as was done with Equation 2 for the covalent-electrostatic model. The 500 enthalpies compiled (9) should be fit to the proposed equation. If the quantitative expression of the model leads to a poor fit but reproduces the correct trends with parameters consistent with the model imposed then it can be concluded that the qualitative model is acceptable. If the quantitative expression does not reproduce the trends, then the qualitative model is not acceptable. If the new model requires

three or more effects to explain the trends instead of the two of
the EC model it can also be disregarded. Though the above
statements appear obvious, they are mentioned here because the
failure of various qualitative interpretations of chemical
reactivity to fit quantitative data have been justified by the
argument that "it is only a qualitative model."

PHILOSOPHY OF THE ECW MODEL

The ECW model provides a basis for determining what is normal
(Equation 1) in sigma bond, donor-acceptor interactions. As such
it can be used in the study of the coordination chemistry of new
acids or bases to determine the existence, or lack thereof, of
unusual bonding effects. The enthalpy of complexation of a series
of bases (or acids) in the E and C correlation (9) can be studied
toward the new acid (or base) and the series of simultaneous
equations of the form of Equation 2 solved for the two (or three if
W is needed) unknown parameters. A recent study (13) of metal-
metal bonded acids, $M_2(O_2CR)_4$, (where M = Rh(II), Mo(II), Cr(II)
and Ru(II)Ru(III)) illustrates the insight provided by this type of
analysis. An unusually large π-backbond stabilization was observed
in those systems where the π^* orbitals of the metal-metal bond are
occupied. The metals in a metal-metal bond interact in a
synergistic way to enhance the ability of the metal center to π^*-
back donate. The discovery of this synergism has fundamental
implications for understanding the reactivity of metal clusters.
Other examples of these applications have been reviewed. (7-8)
        The utilization of the ECW model need not be restricted to
bond strengths. If one measures the spectral shifts, $\Delta\nu$, of a
series of organometallic base adducts, one can attempt to fit these
to the expression:

$$\Delta\nu = E_A{}^* E_B + C_A{}^* C_B - W^*$$ (4)

The asterisk indicates these are not enthalpy based parameters and
contain conversion units to give $E_A{}^* E_B$ the same units as $\Delta\nu$. Free
energies, infrared, nmr, epr, uv, etc. shifts, rate constants,
contact angles, g.c. retention times, etc., can also be analyzed by
substituting the property, $\Delta\chi$, for $\Delta\nu$ into Equation 3 and a fit
attempted. If the fit is successful, (i.e. the $E_A{}^*$, $C_A{}^*$ and W^*
parameters determined calculate $\Delta\chi$ to experimental error), one can
conclude that the phenomenon is being dominated by normal, sigma

bond coordination chemistry. The more complex the property the greater the chance that the correlation will not work because other factors exist and may make dominant contributions to the observable, $\Delta\chi$. These applications illustrate an important philosophical point about the ECW approach. The ECW parameters are based on solvent minimized enthalpies of adduct formation. Phenol hydrogen bonding shifts in the infrared were included in the fit (9) only after extensive studies showed that they correlated with several known base parameters. Thus, the ECW approach is unique because it is based on the "right stuff". Other empirical approaches use large quantities of more easily measured data to determine the parameters. The philosophy often is the more data the parameters fit, the better the model. The EC parameters would not be changed to fit one-hundred electronic transitions if ten good enthalpies had to be eliminated. Instead, the question would be asked, what else is occurring in the spectroscopy that is not involved in normal sigma bond formation?

The ECW approach can be applied to organometallic spectral and thermodynamic data in polar solvents. If a good correlation results, the phenomenon measured is dominated by donor-acceptor sigma bonding instead of solvation effects, entropies, etc. If a correlation does not result, the phenomenon is being dominated by effects other than normal sigma bond formation. It is not correct to attribute the failure to the EC model and claim other parameters are better in highly coordinating solvents (12). Any scale that has solvation as well as coordination properties included in the same parameter cannot have general applicability. These are different, independent effects that need to be understood and parameterized separately. This is also an important philosophical point that is often not appreciated. The goal of ECW is not to correlate the universe but to provide a tool for understanding it.

PLOTS EMPLOYING REFERENCE ACIDS OR BASES

Next consider the implications of the conclusion that there is no inherent order of donor (or acceptor) ability. The multitude of orders that exist is nicely illustrated by the graphical plot of Cramer and Bopp, (14) who factored the ECW equation (with W=0) to obtain:

$$\frac{-\Delta H}{C_A + E_A} = \frac{C_B + E_B}{2} + \left(\frac{C_B - E_B}{2}\right)\left(\frac{C_A - E_A}{E_A + C_A}\right) \qquad (5)$$

Plotting an enthalpy normalized for acid strength, $-\Delta H/(C_A + E_A)$ versus a "normalized acid covalent bond tendency" $(C_A - E_A)/(E_A + C_A)$ one obtains a straight line for a given base bonding to all sigma acceptors. Select a base from those reported, (9) substitute its E_B and C_B into Equation 5, use any E_A and C_A value you wish, calculate ΔH with Equation 2 and plot this data according to Equation 5. Any values of E_A and C_A chosen, will fall on a straight line for the base. Figure 1 is a plot of such lines for several bases. If you select a value of $(C_A - E_A)/(C_A + E_A)$ on the x-axis and move up parallel to the y-axis, the order of increased donor strength toward this acid results. Every time the line for a given base crosses that for another in Figure 1, the order of donor strength for these two bases will reverse for acids on opposite sides of the intersection. From the large number of intersections in Figure 1, a large number of donor orders is seen to result for the bases plotted as the acid is varied. Thus, it is impossible to find a reference acid that will indicate the order of sigma donor strengths for bases. This shows that it is a fundamental error in chemical reactivity to measure a $\Delta\chi$, plot it vs. pK_B or any other donor scale based on a single reference acid (proton affinity, $SbCl_5$, BF_3, etc.) and interpret deviations from this plot with steric, π-bonding, etc. type arguments. The resulting deviation could result simply because the reference acid chosen is not properly representing the covalent and electrostatic contribution to $\Delta\chi$.

If the C_A/E_A ratio for some base adduct observables, $\Delta\chi$, is the same as that for donor order parameters of a reference acid with the same C_A/E_A ratio, a straight line will result (15) when these base parameters (or enthalpies) are plotted vs. the observable, $\Delta\chi$. Dividing both sides of Equation 2 by E_A produces:

$$-\frac{\Delta H}{E_A} = \left[\frac{C_A}{E_A} C_B + E_B\right] = B_k \qquad (6)$$

For a constant C_A/E_A, the term in brackets is a constant for each base, B_k, where k is the C/E ratio, i.e., when $C_A/E_A = 0.01$, $B_{0.01} = 0.01 C_B + E_B$. We can view these $B_{0.01}$ parameters as a one

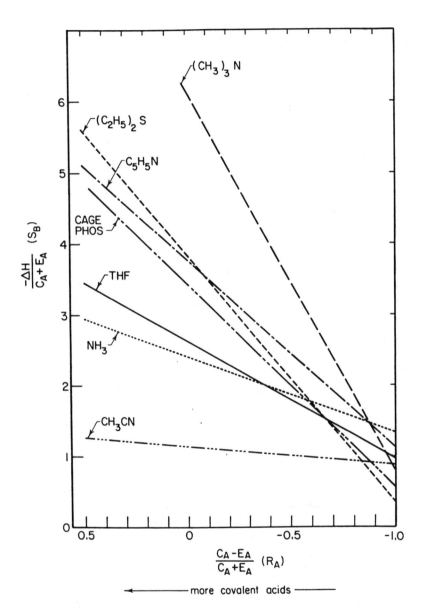

Figure 1. Plot of Equation 5 illustrating the wide variety of donor orders as the acid is varied.

parameter scale of basicity (15b) generated from enthalpies for an acid with $E_A = 1$ that can be used as a scale for other systems whose C_A/E_A ratio is 0.01. If the C/E ratio for $\Delta\chi$ is 0.01, it will plot linearly with $B_{0.01}$ values independent of the E_A^* value of the property examined. Let's generate a set of synthetic $\Delta\chi$ values for an acid whose C/E ratio is 0.1 with E = 2, i.e., $\Delta\chi_{0.1} =$ .2 C_B + 2 E_B. When $\Delta\chi_{0.1}$ is plotted vs. $B_{0.01}$ parameters, Figure 2 results. Certain base points deviate from the line because their covalent contribution to the bonding is greater than that for the $B_{.01}$ reference acid whose C/E = 0.01.

How many times have you seen some property, $\Delta\chi$, for metal-ligand complexes plotted vs. pK_B or a proton affinity or a similar one parameter basicity scale in the literature? How often are the deviations of pyridine and sulfur donors from these plots attributed to π-backbonding? It could just as well be that covalency plays a different role in the coordination chemistry of the complex than in the basicity scale selected. The straight forward approach to the study of organometallic and transition metal complex reactivity is to utilize some of the 48 bases whose E_B and C_B values are known (9) in the experimental design and fit the experimental data to Equation 2. Bases must be selected whose C_B/E_B ratio varies. If pyridine and diethyl sulfide were not used in Figure 2, a good plot would have resulted and one would have concluded incorrectly that the two acids ($\Delta\chi$ and the reference acid used to generate the basicity scale) are similar in their electrostatic-covalent bonding properties. Donor numbers (12) are reported for bases which have a limited range of C_B/E_B ratios. For this reason they often appear to correlate with experimental results. The subtle nature of the above concepts is illustrated in our own research. The change in the OH stretching frequency of phenol upon adduct formation to a series of bases was plotted vs. the enthalpy of adduct formation (16). Sulfur donors did not fall on the line. It was some time later (17) before we realized that the C_A^*/E_A^* ratio for the phenol shift is different than C_A/E_A ratio for the enthalpies. Both sets of data fit Equation 2 very well for all bases.

APPLICATION OF THE ECW APPROACH TO ORGANOMETALLIC SYSTEMS

DIMER CLEAVAGE ENTHALPIES. A variety of applications in which the ECW approach has been used to probe unusual effects in coordination chemistry have been reviewed (7,8,13). Systems in which steric

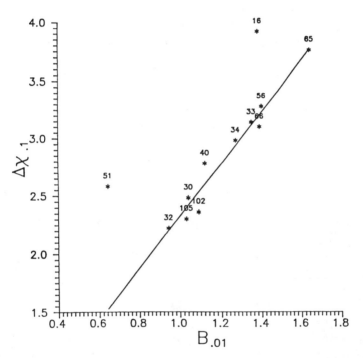

Figure 2. A plot of a measured quantity with a C/E ratio of 0.1, $\Delta_{\chi_{0.1}}$ vs. a basicity order $B_{0.01}$ for acids with a C/E ratio of 0.01. (Reproduced with permission from ref. 8. Copyright 1980 Elsevier Scientific Publishing Company.)

effects exist have been spotted and the magnitude of the effect measured. In other examples, a quantitative measure of the additional stabilization of a metal-ligand bond energy from π-backbonding has been determined and inductive influences on this energy probed (13) Here systems are discussed that provide the basis for future applications of the ECW model to organometallic chemistry.

Table I contains a list of E_B and C_B parameters for bases commonly encountered in organometallic chemistry. Also listed are the $B_{0.01}$ and $B_{0.1}$ parameters. These can be used as reference acids in place of pK_B data to attempt one parameter plots. The reader can consult the literature (9,15b) for a more complete list.

Table I. Parameters for Interpreting Experimental Data

Base	E_B	C_B	$B_{0.01}$	$B_{0.1}$
NH_3	1.48	3.32	1.52	1.81
$(C_2H_5)NH_2$	1.51	5.91	1.58	2.11
$(C_2H_5)_2NH$	1.11	8.59	1.22	1.97
$(C_2H_5)_3N$	1.29	10.83	1.43	2.37
$(CH_2)_5NH$	1.28	9.00	1.39	2.18
$HC(C_2H_4)_3N$	1.14	12.71	1.30	2.61
NCH_3Im	1.12	9.30	1.29	2.05
C_5H_5N	1.30	6.69	1.38	1.96
CH_3CN	0.90	1.34	.917	1.03
$CH_3C(O)CH_3$	1.01	2.38	1.04	1.24
$CH_3C(O)OCH_3$	.92	1.79	.942	1.10
$CH_3C(O)N(CH_3)_2$	1.32	2.48	1.35	1.57
$(C_2H_5)_2O$	1.08	3.08	1.12	1.39
$(CH_2)_4O$	1.06	4.12	1.11	1.49
$(C_2H_5)_2S$	0.55	7.40	.643	1.29
$(CH_2)_4S$	0.58	7.70	.677	1.29
$(CH_3)_2SO$	1.36	2.78	1.40	1.64
C_5H_5NO	1.40	4.40	1.45	1.84
$[(CH_3)_2N]_3PO$	1.52	3.80	1.64	1.88
$(C_2H_5O)_3PO$	1.37	1.84	1.39	1.55
$(CH_3)_3P$	1.11	6.51	1.19	1.76

In the application of these parameters to organometallic systems, the measured quantity that varies as the ligand is changed, $\Delta\chi$, is substituted into Equation 4 along with E_B and C_B leading to a series of simultaneous equations. The least square fit of this data produces E_A^*, C_A^* and W^*.

In the application of the ECW equation to simple adduct formation, Equation 3, W generally has a value of zero. Equations 7 and 8 describe reactions in which there is a constant energy contribution to the reaction enthalpy.

$$AB + B' \longrightarrow AB' + B \qquad (7)$$

$$1/2\ A_2 + B \rightleftharpoons AB \qquad (8)$$

Equation 7 is a base displacement reaction. The enthalpies for a series of different bases B' displacing B can be fit to the ECW equation. In such a fit, W is the enthalpy of adduct formation of AB from A and B, if AB is completely associated in solution. If AB is partially dissociated, it is a fraction of the enthalpy of AB formation corresponding to the fraction of AB that is associated in solution.

Equation 8 is a general reaction for a dimeric acid in which W refers to the heat of dissociation of the dimer. The heat of reaction of a series of bases with $[Rh(COD)Cl]_2$ is measured, leading from Equation 2 and Table I to the following equations in units of kcal mole^{-1} of Rh, i.e. $-\Delta H$ for $B + 1/2[Rh(COD)Cl]_2 \longrightarrow$ BRh(COD)Cl:

Base

C_5H_5N	$7.6 = E_A(1.30)$	$+ C_A(6.69)$	$- W$
$CH_3C_5H_4N$	$8.2 = E_A(1.30)$	$+ C_A(7.42)$	$- W$
$(CH_2)_4S$	$5.3 = E_A(.58)$	$+ C_A(7.70)$	$- W$
$1-CH_3Im$	$9.7 = E_A(1.12)$	$+ C_A(9.30)$	$- W$
$C_5H_{11}N$	$10.1 = E_A(1.28)$	$+ C_A(9.00)$	$- W$
$(CH_3)_2SO$	$3.8 = E_A(1.36)$	$+ C_A(2.78)$	$- W$
$HC(C_2H_4)_3N$	$7.0 = E_A(1.14)$	$+ C_A(12.71)$	$- W$

A least squares solution of these equations for the best fit EC and W parameters leads to $E_A = 4.85$, $C_A = 1.07$, $W = 5.78$ kcal mole^{-1}. W is one half the enthalpy value for cleaving the dimer. This quantity cannot be obtained as easily by any other means. Computer programs are available from the author for the least squares solution of the simultaneous equations. The enthalpy for $[Rh(COD)Cl]_2$ reacting with a whole series of other bases is readily calculated by substituting E_A, C_A and W with E_B and C_B values from Table I into Equation 2. The value for the dissociated monomer $Rh(COD)Cl$ reacting with bases is obtained by setting W = 0 and substituting E_A, C_A, E_B and C_B into Equation 2. Similar studies on $[Rh(CO)_2Cl]_2$ and $[\pi\text{-allyl PdCl}]_2$ lead to E_A = 9.23, C_A = 1.71, W = 11.19 and E_A = 3.2, C_A = 1.0, W = 4.1 respectively. The weaker sigma basicity of CO vs. alkenes is manifested in the greater acidity of $Rh(CO)_2Cl$ vs. $Rh(COD)Cl$.

In the next two sections, the analysis of more complex reactions is illustrated. These will include systems in which a little ingenuity enables one to add or subtract a series of equations which are composed of constant energy processes and adduct formation reactions to produce the equation for the desired reaction.

COBALT-CARBON BOND ENERGIES. The cobalt-carbon bond dissociation energy has been reported (18) for a series of base adducts of alkyl-substituted bis (dimethylglyoximato) cobalt(II),

$$R\text{-}Co(DH)_2\text{-}B \longrightarrow Co(DH)_2\text{-}B + R \qquad (9)$$

(where R is $C_6H_5CH(CH_3)$ and DH is dimethylglyoximato). At first glance it looks as though ECW is not applicable. The cobalt-carbon bond dissociation energy varies when B is changed, but every time B changes we change the ECW of the cobalt center toward R. Furthermore, the cobalt-carbon bond dissociation (or formation if we change the sign of ΔH_D) is not coordinate bond formation but a free radical reaction. However, the formation reaction (i.e., the reverse of Equation 9) can be broken up (9) into the following steps:

$$Co(DH)_2 + \cdot R \longrightarrow Co(DH)_2R \qquad W \qquad (10)$$

$$+ \ BCo(DH)_2 \longrightarrow Co(DH)_2 + B \qquad +\Delta H_{Co} \qquad (11)$$

$$+ \ Co(DH)_2R + B \longrightarrow B\text{-}Co(DH)_2R \qquad -\Delta H_{CoR} \qquad (12)$$

$$\overline{\text{SUM} \quad BCo(DH)_2 + \cdot R \longrightarrow BCo(DH)_2R \qquad -\Delta H_D} \qquad (13)$$

We have rewritten Equation 8 as cobalt-carbon bond formation and expressed it as the sum of a constant energy process (W), an adduct dissociation reaction and an adduct formation reaction to produce the desired reaction, Equation 13. Steps 11 and 12 are the type reaction treated by ECW leading respectively to Equations 14 and 15.

$$+ \Delta H_{Co} = E_{Co}E_B + C_{Co}C_B \tag{14}$$

$$- \Delta H_{CoR} = E_{CoR}E_B + C_{CoR}C_B \tag{15}$$

The heat of cobalt-carbon bond formation (Equation 13) is given by summing W and Equations 14 and 15 to produce:

$$- \Delta H_D = -E_{Co}E_B - C_{Co}C_B + E_{CoR}E_B + C_{CoR}C_B - W$$

or

$$- \Delta H_D = \Delta E_A E_B + \Delta C_A C_B - W \tag{16}$$

where $\Delta E_A = E_{CoR} - E_{Co}$ and $\Delta C_A = C_{CoR} - C_{Co}$. Six bases with different C_B/E_B ratios should have been selected from Table I and their influence on cobalt-carbon bond dissociation studied. The six simultaneous equations of the form of Equation 16 can be solved for three unknowns ΔE_A, ΔC_A and W. We would know if sigma bond formation dominates the Co-R dissociation energy, the relative importance of the base covalent or electrostatic bond forming properties on the stability and, if steric effects exist or if π-backbonding is present. In addition, the cobalt-carbon bond dissociation energy with no base attached is given by W and the cobalt-carbon bond dissociation enthalpy for the 48 adducts of bases in ECW can be calculated. How else can one obtain this much information from experimental data? Unfortunately, the experiment was reported (18) with only four similar bases leading to the four simultaneous equations.

$$-21.2 = 1.37 \, \Delta E_A + 7.99 \, \Delta C_A - W$$
$$-20.1 = 1.33 \, \Delta E_A + 7.24 \, \Delta C_A - W$$
$$-19.5 = 1.30 \, \Delta E_A + 6.69 \, \Delta C_A - W$$
$$-17.9 = 1.14 \, \Delta E_A + 3.89 \, \Delta C_A - W$$

The resulting tentative values from the least squares solutiohn of the four equations are $\Delta E_A = 4.27$, $\Delta C_A = 0.54$ and $W = -10.7$. The quality of the fit is shown in Table II where enthalpies calculated by substituting ΔE_A, ΔC_A and W values into Equation 16 (ΔH_{calc}) are compared to experimental results (ΔH_{meas}).

Table II. Cobalt-Carbon Bond Dissociation Energies,
-ΔH, for R-Co(DH)$_2$-B Adducts

X - C$_5$H$_4$N	-ΔH_{meas}	-ΔH_{calc}
4-NH$_2$	21.2	20.8
4-CH$_3$	20.1	20.3
H	19.5	19.9
4-CN	17.9	17.7

The radical pairing reaction of O$_2$ to cobalt(II) is similar to the reaction of cobalt(II) with an alkyl radical (19). The O$_2$ binding enthalpies have been analyzed (19) with ECW to give ΔE_A = 2.9 ΔC_A = 0.6 and W = -1.5 where Δ refers to CoO$_2$ - Co(II). The ligand system around cobalt in the O$_2$ bonding studies is a porphyrin. It is interesting that binding R· to Co(DH)$_2$ is much more exothermic (-10.7 kcal mole^{-1}) than binding O$_2$ to Co (por) (-.5 kcal mole^{-1}). It is also interesting to note that binding R·to Co(DH)$_2$ has increased the acidity of the cobalt much more than binding O$_2$ to Co (por) (ΔE_A is much larger for Co(DH)$_2$ and ΔC_A comparable). This is consistent with more extensive electron transfer into the alkyl group than into O$_2$, supporting our position (19) that cobalt bound O$_2$ does not resemble ionically bound superoxide. Though the conclusions on Co(DH)$_2$ are tentative because of the experimental design, they are presented here to show the power of an ECW analysis in providing insights into chemical reactivity that cannot be obtained by other means.

2:1 BASE ADDUCTS. The extension of the ECW analysis to 2:1 adducts constitutes another challenge. Consider acids that react in this fashion:

$$A + B \rightleftharpoons AB \qquad (17)$$

$$AB + B \rightleftharpoons AB_2 \qquad (18)$$

The first reaction is a straightforward ECW problem. The second one requires obtaining E_{AB} and C_{AB} values for AB. When B is the same for both steps (Equations 17 and 18), every time we study a different base we generate a new acid for step 18 whose E_{AB} and C_{AB} values are not known. The unknowns mount up faster than the knowns. Usually, we cannot measure enthalpies for AB reacting with several different bases B' to obtain the simultaneous equations needed to solve for E_{AB} and C_{AB}. In the mixed system experiment AB, AB', ABB', AB_2 and AB_2' form. The assumption is made (13) that:

$$E_{AB} = E_A - kE_B \qquad (19)$$
and
$$C_{AB} = C_A - k'C_B \qquad (20)$$

Here k and k' represent the extent to which coordination of B to A modifies the acidity of AB in its subsequent reactivity to form AB_2. It is a measure of the inductive transfer of B's coordinating tendency in modifying the new acid AB. The model has been tested in metal-metal bonding systems involving $M_2(RCO_2)_4$ complexes (13). The proportionality constants measure the effectiveness of the M-M bond to transmit the inductive effect of base coordination at one metal center to the other metal. Substituting Equations 19 and 20 into Equation 2 with W = 0 leads to:

$$-\Delta H_{2:1} = E_A E_B - kE_B^2 + C_A C_B - k'C_B^2 = \Delta H_{1:1} - kE_B^2 - k'C_B^2 \qquad (21)$$

Thus a half dozen enthalpies can be solved for the two unknowns k and k' when $\Delta H_{1:1}$ can be measured separately.

When only the sum of the 1:1 and 2:1 enthalpies can be determined (e.g., $K_2 > K_1$) the data has to be fit to:

$$-\Delta H_T = 2(E_A E_B + C_A C_B) - kE_B^2 - k'C_B^2 \qquad (22)$$

In this case a very extensive set of data is needed to solve for the four unknowns.

REACTIONS OF CATIONS AND ANIONS. Many of the organometallic bond energies in the literature are obtained from the reaction:

$$MX + HY \longrightarrow MY + HX \qquad (23)$$

The enthalpy for

$$M^+ + Y^- \longrightarrow MY \tag{24}$$

can be represented relative to $M^+ + X^- \longrightarrow MX$ being zero. If we had E_B and C_B values for anions, an ECW analysis would lead to absolute energies with W being the value for M-X.

The only place to obtain solvation free energies for ions is the gas phase. However, the ECW equation does not apply to the large enthalpies of interaction associated with gas phase ion-molecule reactions (20,21). The gaseous ion is a strong Lewis acid and when it interacts with a base, a significant contribution to the measured enthalpy of interaction comes from extensive electron transfer to the ion (21). One can view this as a one center energy term in which the cation has regained some fraction of its ionization energy. The magnitude of this effect, for an extreme case, can be appreciated by considering the enthalpy difference in the reaction of two hydrogen atoms to form H_2 with that of H^+ reacting with H^-. Recent work from this laboratory (22) shows that 369 enthalpies can be fit to the equation:

$$-\Delta H = E_A E_B + C_A C_B + T_A T_B \tag{25}$$

where $T_A T_B$ is the transfer term. The significant new finding is that the reported (9) E_B and C_B values from solvent minimized weak adduct enthalpies can be utilized in Equation 25. Twenty three cations including the proton, eighteen new bases and twenty five known bases lead to 148 unknown parameters for the least squares fit. Valuable insights concerning gas phase and solution reactivity result (23). The $T_A T_B$ term dominates gas phase ion chemistry but is largely cancelled out in solution where the reactions are usually displacement. Contributions from this effect must be considered when displacement enthalpies are plotted vs. gas phase bond energies.

The extension of these findings to organometallic displacement reactions, is currently underway. However, the results presently available indicate that variations in the covalent and electrostatic contributions to the bonding occur and cause variations in the trends of sigma bond energies for the different organometallic Lewis acids. Thus, it is quite inappropriate to plot M-X vs. H-X dissociation energies and expect to obtain a linear plot. All of the arguments presented in the section on Plots Employing Reference Acids and Bases apply to the analysis of organometallic bond energies. The CH_3-X, H-X and K-X systems vary substantially in the covalent and electrostatic contributions to the sigma bonds. Plots of M-X bond energies versus all three reference systems may provide some insights. However, deviations from the plots are difficult to interpret and in addition to novel bonding contributions deviations may occur because the covalent-electrostatic contribution to sigma bonding is not properly represented in the reference compound.

HETEROGENOUS CATALYSTS. The coordination of substrate or binding of a small molecule is often involved in the mechanism of heterogeneous catalyzed reactions. Thermodynamic studies which use typical organic bases as ligands to the catalytic metal centers have the potential of characterizing their coordination chemistry and enabling us to compare these heterogeneous systems to conventional coordination compounds. Knowledge of the steric and electronic characteristics (E_B, C_B) of the ligands will aid in the characterization of the acidity of the metal center and in understanding catalyst poisoning. In order to obtain data consistent with the ECW data set, the solid catalyst should be slurried in a poorly solvating solvent and the enthalpy of interaction determined after correcting for the enthalpy of solution of base. When the catalyst support interacts weakly with the base, the calorimetric data also has to be corrected for this contribution. The equation for the reaction of 5% Pd/C reacting with pyridine is thus written as:

$$Pd/C_{(sl)} + Py_{(sol)} \longrightarrow PyPd/C_{(sl)} \tag{26}$$

where sl refers to the solid slurry and sol to solution. The equilibrium constant is written as:

$$K = \frac{PyPd/C}{[Py][n_i^S - PyPd/C]} \tag{27}$$

where n_i^S is the number of active sites per gram of catalyst; PyPd/C the number of grams of pyridine coordinated per gram of catalyst; [Py] is the solution concentration of pyridine in mole liter^{-1}; and K has units of 1 mole^{-1}. The heat liberated in the titration h' is given by (24):

$$\frac{h'}{g\ Pd/C} = \frac{n_i{}^S K_1 [Py]}{1 + K_1 [Py]}\ \Delta H_1 \tag{28}$$

With practical catalysts that often contain multiple acid sites additional terms are added for each site with the subscripts changing from 1 to 2, 3 etc. in each site added.

The information available from a calorimetric titration of a base and slurried catalyst is not sufficient to determine all the unknown quantities in Equation 28. The problem is alleviated by coupling the calorimetric titration with the determination of an absorption isotherm. Equation 26 can be rearranged to give the familiar Laugmuir equation:

$$\frac{[Py]}{PyPd/C} = \frac{1}{n_1{}^S K_1} + \frac{[Py]}{n_1{}^S} \tag{29}$$

For two sites we can write (24):

$$\frac{PyPd/C}{} = \frac{n_1{}^S K_1 [Py]}{1 + K_1 [Py]} + \frac{n_2{}^S K_2 [Py]}{1 + K_2 [Py]} \tag{30}$$

Knowing the amount of pyridine added and measuring the equilibrium concentration in solution (by titration, g.c., uv-visible etc.) we know both PyPd/C and [Py] of Equation 30. The best values of $n_i{}^S$, K_i and ΔH_i that simultaneously fit the data to Equation 30 and 28 (with the added terms for multiple sites) are determined. This analysis groups similar sites into one and gives the minimum number of different types of sites required by the data. Independent confirmation of the conclusion about the number of different sites by surface science techniques is desirable.

For the titration of 5% Pd/C with pyridine, the experiment showed (24) two different sites, n_1 = 2.5 mmole/g and n_2 = 3.2 mmole/g. Values of K_1 = 2.5 x 10^4 M^{-1} and K_2 = 2.9 x 10^2 M^{-1} resulted with ΔH_1 = 13 kcal mole^{-1} and ΔH_2 = 10 kcal mole^{-1}. ESCA studies showed the existence of two distinct sites on the surface, one being Pd(II) and the other Pd(0). Thermogravimetric studies and differential scanning calorimetry were not able to distinguish the two sites but produced average values in agreement with the average found in the titration.

This procedure has many potential applications in catalysis and material science. In the latter area Fowkes and coworkers (25) and Chen (26) have reported several examples where important insights into adhesion at interfaces have been obtained by application of ECW considerations. Verification of some of these conclusions with calorimetric studies is underway.

These diverse examples indicate the philosophy of applying the ECW model. It is a valuable tool for understanding the very complicated area of chemical reactivity because it provides quantitative criteria for what is to be expected when normal sigma bond formation dominates a reaction.

LITERATURE CITED

1. Ahrland, S.; Chatt, J.; Davies, N. R. Chem. Soc., Q. Rev. 1958, 12, 265.
2. Edwards, J. O. J. Am. Chem. Soc. 1956, 78, 1819 and references therein.
3. Drago, R. S.; Meek, D. W.; Longhi, R.; Joesten, M. D. Inorg. Chem. 1963, 2, 1056 and references therein.
4. Mulliken, R. S. J. Am. Chem. Soc. 1952, 74, 811 and references therein.
5. Pauling, L. "Nature of the Chemical Bond," Cornell University Press, Ithaca, New York 1967.
6. Drago, R. S.; Wayland, B. B. J. Am. Chem. Soc. 1965, 87, 375.
7. Drago, R. S. Structure and Bonding 1973, 15, 73.
8. Drago, R. S. Coord. Chem. Rev. 1980, 33, 251.
9. Drago, R. S.; Wong, N.; Bilgrien, C.; Vogel, G. C. Inorg. Chem. 1987, 26, 9.
10. Jensen, W. B. Chem. Rev. 1978, 78, 1.
11. Klopman, G. J. Am. Chem. Soc. 1969, 90, 223 and references therein.
12. Finston, H. L.; Rychtman, A. C. "A New View of Current Acid-Base Theories" Wiley-Interscience, N.Y. 1982.
13. Drago, R. S.; Bilgrien, C. J. Polyhedron 1988, Vol. 7, No. 16/17, 1453.
14. Cramer, R. E.; Bopp, T. T. J. Chem. Ed. 1977, 54, 612.
15. a) Li, M. P.; Drago, R. S.; Pribula, A. J. J. Am. Chem. Soc. 1977, 99, 6901.
    b) Drago, R. S. submitted.
16. Vogel, G. C.; Drago, R. S. J. Am. Chem. Soc. 1970, 92, 5347.
17. Doan, P. E.; Drago, R. S.; J. Am. Chem. Soc. 1982, 104, 4524.
18. Ng, F. T. T.; Rempel, G. L.; Halpern, J. J. J. Am. Chem. Soc. 1982, 104, 621.

19. Drago, R. S.; Corden, B. B. Acc. Chem. Res. 1980, 13, 353 and references therein.
20. Marks, A. P.; Drago, R. S. J. Am. Chem. Soc. 1975, 97, 3324.
21. Kroeger, M. K.; Drago, R. S. J. Am. Chem. Soc. 1981, 103, 3250.
22. Drago, R. S.; Ferris, D.; Wong, N. G. Submitted.
23. Drago, R. S.; Cundari, T. R.; Ferris, D. C. J. Org. Chem. 1989, 54, 1042.
24. Lim, Y. Y.; Drago, R. S.; Babich, M. W.; Wong, N.; Doan, P. E. J. Am. Chem. Soc. 1987, 109, 169.
25. a) Fowkes, F. M.; Tischler, D. O.; Wolfe, J. A.; Lannigan, L. A.; Cedemu-John, C. M.; Halliwel, M. J. J. Polym. Sci-Polym. Chem. Ed. 1984, 22, 547.
    b) Fowkes, F. M.; Mustafa, M. A. Ind. Eng. Chem. Res. Dev. 1978, 17, 3.
    c) Fowkes, F. M.; McCarthy, D. C.; Mustafa, M. A. J. Coll. Interf. Sci. 1980, 78, 200.
26. Chen, F. Macromolecules 1988, 21, 1640.

RECEIVED December 6, 1989

Chapter 13

Metal–Ligand Bond Dissociation Energies in $CpMn(CO)_2L$ and $Cr(CO)_5$(olefin) Complexes

Jane K. Klassen, Matthias Selke, Amy A. Sorensen, and Gilbert K. Yang

Department of Chemistry, University of Southern California, Los Angeles, CA 90089–0744

The metal ligand bond dissociation energies for a variety of $CpMn(CO)_2L$ complexes in heptane solution have been found to be 16.1, 17.4, 24.5 and 29.1 kcal/mol for THF, acetone, cis-cyclooctene and Bu_2S respectively. The reaction of L with $CpMn(CO)_2$ is first-order in [L] with second-order rate constants between 10^6 and 10^7 L/mol-s. Evidence is presented for a Mn-heptane bond strength of 8-9 kcal/mol. Similarly, the $(CO)_5$Cr-olefin bond strengths were determined for cis and $trans$-cyclooctene to be 14.3 and 19.5 kcal/mol respectively. Kinetic data suggest that the difference in bond strengths is predominantly due to the greater ring strain in $trans$-cyclooctene.

Coordinatively unsaturated transition metal complexes are known to bind to a variety of donor molecules. Qualitatively, the enthalpies of these interactions range from very strong dative bonds with ligands such as phosphines and CO to very weak interactions with saturated hydrocarbons. Despite the importance of these interactions in determining structure and reactivity, limited quantitative data exist for metal-ligand bond strengths.

In this work we present bond enthalpies and kinetic information determined by photoacoustic calorimetry for a number of metal-ligand systems. The data presented serve to confirm the accuracy of the photoacoustic technique and illustrate the different types of interactions which might be examined.

Experimental Technique

Time-resolved photoacoustic calorimetry has been used by a number of research groups to determine the enthalpies and quantum yields of reaction in inorganic and organic systems ([1]-[7]). Our experimental technique is similar to those previously described in the literature ([1]). Emission from a N_2 laser (337.1 nm, 1 ns pulse length, 20 μJ/pulse) is passed through a quartz cuvette containing an argon-purged heptane solution at 25 ± 1°C of either a reference compound (ferrocene) or the metal carbonyl being examined. As illustrated in the energy diagram shown in Figure 1, cleavage of the metal-CO bond requires only a portion of the

0097–6156/90/0428–0195$06.00/0

absorbed photon energy. The excess energy is released as heat (q_1, Step B) into the solution. The M-CO bond scission process is very rapid (<1ns) and the sudden deposition of heat into the solution generates a pressure shock wave. The coordinatively unsaturated intermediate "M" then reacts with ligands in solution via a second-order process with a rate constant k_2 (Step C). As the reaction proceeds, the heat of reaction ΔH_2 is released into the solution as q_2. This second heat deposition produces a second shock wave. The two shock waves are simultaneously acoustically detected with a piezoelectric transducer. The reference compound is chosen to only display a "fast" deposition of heat. The transducers are sensitive only to heat generating processes with rate constants near or faster than the transducer frequency and are not sensitive to slower processes. With a 0.5 MHz transducer all "fast" reactions ($k<10^8$ s^{-1}) are detected "in phase" while "slow" reactions ($10^5 s^{-1} < k < 10^7 s^{-1}$) have a phase lag due to the delay in the heat deposition. The phase difference allows deconvolution of the data into the fast (q_1) and slow (q_2) components and the rate constant k_{obs}. An example of the data from an experiment is shown in Figure 2.

Equations 1 and 2, where $E_{hv} = 84.8$ kcal/mol at 337.1 nm and Φ is the quantum yield for the reaction, give the relevant enthalpies of reaction.

$$\Delta H_{BDE} = \frac{1-q_1}{\Phi} \cdot E_{hv} \qquad\qquad \Delta H_2 = \frac{q_2}{\Phi} \cdot E_{hv} \qquad (1,2)$$

For $Cr(CO)_6$ (8) and $CpMn(CO)_3$ (9) the quantum yields have been accurately determined to be 0.67 ± 0.04 and 0.65 ± 0.1 respectively. Quantum yields for CO dissociation have been found to be quite independent of solvent viscosity, wavelength and ligand concentration (8, 9). Relative errors in the quantum yields translate to the same relative error in the enthalpic measurements.

Both species either non-radiatively decay back to the ground state or undergo photodissociation to form the corresponding coordinatively unsaturated intermediate. No evidence has been reported for formation of long-lived intermediates other than the CO loss product or for radiative decay processes.

The measured enthalpies of reaction were all independent of the ligand concentrations used in our experiments suggesting that the quantum yield is also independent of ligand concentration. All reactions were performed in argon-purged n-heptane solutions. The absorbance values of the reference and the metal carbonyl solutions were matched to within 3%.

Metal-ligand Bond Strengths in $CpMn(CO)_2L$

The photochemical ligand substitution reactions of $CpMn(CO)_3$ have been well studied and are synthetically very useful (10, 11). Upon irradiation with light of wavelength less than $ca.$ 400 nm, $CpMn(CO)_3$ readily dissociates one CO ligand (Scheme 1).

In good donor solvents such as THF, the initially formed $CpMn(CO)_2$ is stabilized as the solvated intermediate $CpMn(CO)_2S$ (10, 11). Substitution of the solvent molecule from this intermediate by "good" ligands such as phosphines and olefins occurs readily leading to the corresponding $CpMn(CO)_2L$ complexes. This reaction sequence is ideal for study of the energetics of the Mn-CO bond dissociation process and both the energetics and kinetics of the Mn-ligand bond forming process using photoacoustic calorimetry.

$$CpMn(CO)_3 \xrightarrow[\substack{\Delta H_{Mn-CO} \\ \text{"fast"}}]{337.1nm} CpMn(CO)_2S \xrightarrow[\substack{\Delta H_2, k_2 \\ \text{"slow"}}]{+L} CpMn(CO)_2L$$

Scheme 1

The results of the photoacoustic experiments on $CpMn(CO)_3$ performed in heptane solution are shown in Table I. The ligand concentrations were varied from ~0.1M to ~0.5M. Within this concentration range, the Mn-CO bond dissociation energies (ΔH_{Mn-CO}) were

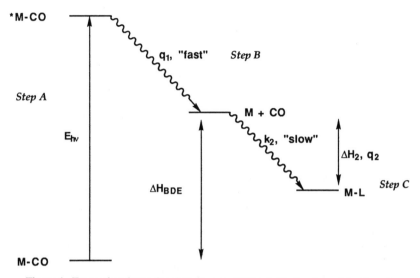

Figure 1.   Energetic scheme for photochemical CO substitution in metal carbonyls.

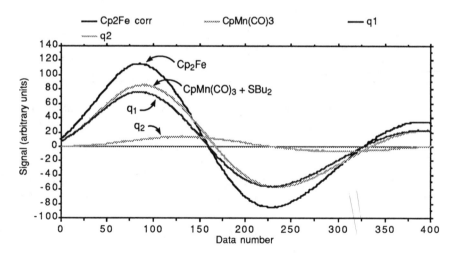

Figure 2.   Plot of photoacoustic data for the photolysis of CpMn(CO)$_3$ in heptane with [SBu$_2$] = 0.10 M both before and after deconvolution into q$_1$ and q$_2$ components. Equations 1 and 2 are used to calculate enthalpies of reaction from q$_1$ and q$_2$.

constant. The ΔH_{Mn-CO} values determined for each of the four ligands agree within experimental error. These observations are consistent with the mechanism shown in Scheme 1 where the CO dissociation step is ligand independent.

Table I. Enthalpic and kinetic data for the reaction of CpMn(CO)$_3$ with L in heptane solution according to Scheme 1 [a]

Ligand	CpMn(CO)$_3$ ΔH_{Mn-CO}	CpMn(CO)$_2$L ΔH_2 (-BDE Mn-L)	k_2 (L/mol-s)
THF	46.2 ± 1.2 kcal/mol	-16.1 ± 1.4 kcal/mol	4.4 ± 0.1 x 10^6
acetone	47.8 ± 1.7	-17.4 ± 1.0	3.6 ± 0.4
cis-cyclooctene	47.6 ± 1.4	-24.5 ± 2.3	2.3 ± 0.4
Bu$_2$S	45.3 ± 1.4	-29.1 ± 2.0	8.1 ± 0.8
P(OMe)$_3$ [b]			7.1 x 10^6
PPh$_3$ [b]			11 x 10^6
Average	46.7 ± 1.7 kcal/mol		

a) Errors are given as one standard deviation of the scatter in the data.
b) See reference 17.

Since the reactions were performed in heptane solution, the bond dissociation energies measured here are the enthalpy of exchange of CO by heptane. The strength of the Mn-CO bond in CpMn(CO)$_3$ is remarkably high when compared to other known metal-CO bond strengths (12, 13). Since is the only report of the Mn-CO bond dissociation energy in this system in either the gas or solution phase, there is little data with which to make comparisons. However, our value is consistent with the observed reluctance of this species to undergo thermal CO substitution reactions (14, 15). Although quantitative data is not readily available, an estimate of Mn-CO can be made from the observations reported by Butler and by Angelici. With the assumption that the CO dissociation reaction has an activation entropy of +20 e.u. we can estimate that the activation enthalpy is *at least* 44 kcal/mol.

The Mn-L bond strengths appear to be reasonable based on the known chemistry of CpMn(CO)$_2$L complexes. The Mn-THF bond is relatively weak in agreement with the observations that CpMn(CO)$_2$THF is thermally very sensitive and is readily substituted by a variety of ligands (10, 11). The cis-cyclooctene and Bu$_2$S ligands are substantially more strongly bound to CpMn(CO)$_2$ and form correspondingly more stable complexes. Complexes of both of these ligands can be isolated by displacement of THF from CpMn(CO)$_2$THF and decompose only above room temperature (14, 16).

The rate constants k_{obs} for the coordination of L to CpMn(CO)$_2$S displayed first-order dependences on the concentration of the incoming ligand. Plots of k_{obs} *vs* [ligand] for cis-cyclooctene and Bu$_2$S are shown in Figure 3. The rate constants are all very large which demonstrates the reactivity of the CpMn(CO)$_2$ intermediate towards donor molecules when weakly solvated by hydrocarbon solvents. In comparison, Poliakoff *et al.* (17) examined the rates of reaction at of P(OMe)$_3$ and PPh$_3$ with CpMn(CO)$_2$ in *n*-heptane solution using time-resolved IR spectroscopy. As shown in Table I, the rate constants they obtained at 22°C for the reaction of CpMn(CO)$_2$ with these two phosphorus ligands are very similar to the k_2 for Bu$_2$S.

Relative Mn-L Bond Dissociation Energies from Equilibrium Studies

As seen from Table I, the heats of formation (ΔH_2) of the Mn-SBu$_2$ bond and the Mn-*cis*-cyclooctene bond were found respectively, to be -29.1 ± 2.0 and -24.5 ± 2.3 kcal/mol. Thus, we expected the Mn-SBu$_2$ bond to be 4.6 ± 3.0 kcal/mol *stronger* than the Mn-*cis*-cyclooctene bond. To confirm this prediction, we examined the equilibrium shown below between the corresponding manganese complexes.

$$CpMn(CO)_2(cis\text{-}C_8H_{14}) + SBu_2 \rightleftharpoons CpMn(CO)_2(SBu_2) + cis\text{-}C_8H_{14}$$
$$\mathbf{1} \qquad\qquad\qquad\qquad\qquad\qquad\qquad \mathbf{2}$$

Both of these complexes are known to undergo thermal substitution near 100°C via clean first-order kinetics (14, 16). Thus we performed the equilibrium experiments between 80 and 120°C. Since the equilibrium constant was expected to heavily favor the SBu_2 complex **2** we ran the experiments with high concentrations of *cis*-cyclooctene, in either neat *cis*-cyclooctene (7.7 M) or 1.2 M solutions of *cis*-cyclooctene in heptane. The results obtained in both solvent systems were identical. Although differential vaporization of the ligands did not appear to be a problem, efforts were made to minimize the gas volume in the reaction vessel. The concentrations of the complexes were monitored by the intensity of the appropriate v_{CO} bands. Identical equilibrium constants were obtained for approach to equilibrium from either direction.

A van't Hoff plot of the equilibrium data obtained between 80 and 120°C (Figure 4) gave $\Delta H = -2.9 \pm 1$ kcal/mol and $\Delta S = 3.6 \pm 2$ eu. This value for ΔH is very close to the -4.6 $\pm$ 3 kcal/mol which we predicted based on the photoacoustic determinations of the individual bond strengths for the *cis*-cyclooctene complex **1** and the SBu_2 complex **2**. The degree to which these two very different techniques agree on this value attests to the accuracy of the photoacoustic technique in determining metal-ligand bond dissociation energies.

Coordination of *cis* and *trans*-cyclooctene to Cr(CO)₅

There are many examples of organometallic complexes which owe their stability, at least in part, to the relief of strain in the organic ligand. Transition-metal benzyne (18-23), cyclohexyne (24) and bicyclic bridgehead olefin (25-29) complexes are among these. The relative stability of transition metal complexes of strained olefins over their unstrained counterparts has long been recognized (25-30) and attempts have been made to determine the magnitude of this stabilization (31). An example of this difference in relative stability is found in the chemistry of *cis* and *trans*-cyclooctene (32, 33). Complexes of both of these olefins can be obtained by photolysis of $Cr(CO)_6$ in hydrocarbon solutions of *cis* or *trans*-cyclooctene (33).

The photochemistry of $Cr(CO)_6$ is similar to that of $CpMn(CO)_3$. Upon photolysis of $Cr(CO)_6$ in a hydrocarbon solution of either *cis* or *trans*-cyclooctene, CO is lost in less than 100 ps (34-36) ("fast" step in Scheme 2) forming the solvated $Cr(CO)_5S$ intermediate (**3**) which subsequently binds olefin in a second-order reaction (k_3 step in Scheme 2) to form either $(CO)_5Cr(cis\text{-}cyclooctene)$ (*cis*-**4**) or $(CO)_5Cr(trans\text{-}cyclooctene)$ (*trans*-**4**) (33). These two steps have rate constants which fit the "fast" and "slow" criteria demanded by the photoacoustic technique allowing us to determine $\Delta H_{Cr\text{-}CO}$, ΔH_3 and k_3. The results are summarized in Table II. The $(CO)_5Cr\text{-}CO$ bond dissociation energy in heptane ($\Delta H_{Cr\text{-}CO}$) of 27.8 $\pm$ 1.7 kcal/mol is in excellent agreement with previous reports (3, 4) and is independent of both the structure and concentration of the cyclooctene used in the experiment.

The heat of reaction, ΔH_3, for the binding of *cis*-cyclooctene to the coordinatively unsaturated intermediate **3** to generate *cis*-**4** was found to be -14.3 $\pm$ 0.9 kcal/mol. This is somewhat more exothermic than the -12.2 kcal/mol olefin-metal bond enthalpy found in (1-hexene)-Cr(CO)₅ (3, 4). Perhaps this stronger bonding interaction can be attributed to the 6 kcal/mol ring strain found in *cis*-cyclooctene (37). A much more dramatic effect is seen in the enthalpy of binding of *trans*-cyclooctene to intermediate **3**. In this case, $\Delta H_3 = -19.5 \pm 2.5$ kcal/mol. Thus, the metal-olefin bond in *trans*-**4** is more than 5 kcal/mol stronger than that in *cis*-**4** i.e., $\Delta\Delta H_3 \approx 5$ kcal/mol. This is consistent with the observation that *trans*-cyclooctene has 9.3 kcal/mol of strain relative to *cis*-cyclooctene (37). Coordination of the olefin relieves a substantial portion of this strain resulting in a greater bond strength (12, 25, 32, 33).

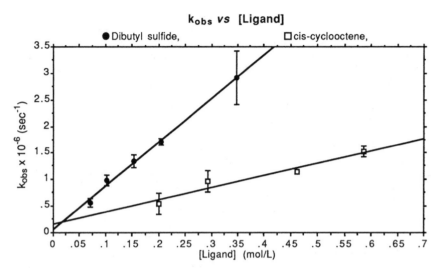

Figure 3. Plot of k_{obs} vs. [Ligand] for the reaction of CpMn(CO)$_2$S with cis-cyclooctene and Bu$_2$S at 25 ± 1°C in n-heptane. For cis-cyclooctene, r^2=0.93; for Bu$_2$S, r^2=0.996. Other data is given in Table I. Error bars are the greater of the standard deviation calculated from the scatter in the data and ± 10%.

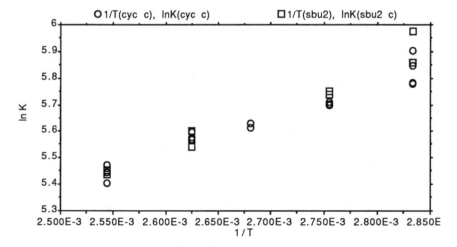

Figure 4. van't Hoff plot for the equilibrium CpMn(CO)$_2$(cis-C$_8$H$_{14}$) + SBu$_2$ $\rightleftharpoons$ CpMn(CO)$_2$(SBu$_2$) + cis-C$_8$H$_{14}$ in neat cyclooctene (7.7 M). T = 80-120°C; ΔH=-2.9 ± 1 kcal/mol and ΔS=3.6 ± 2 eu.

Different steric interactions may also contribute to $\Delta\Delta H_3$, although we believe that this effect is minimal (*vide infra*).

Table II. Enthalpic and kinetic data for substitution of $Cr(CO)_6$ with olefins in heptane solution according to Scheme 2 [a]

Ligand	$\Delta H_{Cr\text{-}CO}$ kcal/mol	ΔH_3 kcal/mol	$k_3 \times 10^7$ (L/mol-s)
cis-cyclooctene	27.8 ± 1.6	-14.3 ± 0.9	1.4 ± 0.3
trans-cyclooctene	27.7 ± 1.9	-19.5 ± 2.5	1.7 ± 0.2
1-hexene[b]	27.0 ± 1.2	-12.2 ± 1.2	3.0 ± 1.5

[a] Errors are given as one standard deviation of the scatter in the data.
[b] Data from reference 3, 4.

Muhs and Weiss ([31]) have determined that the equilibrium constant for formation of *trans*-cyclooctene-Ag^+ (K_{trans}) at 40°C is at least 70 times greater than the equilibrium constant for formation of *cis*-cyclooctene-Ag^+ (K_{cis}). Assuming that there is little difference in ΔS for these reactions and that the $\Delta\Delta H_3$ we determined for the cyclooctene-Cr bond can be applied here, we estimate that the ratio K_{trans}/K_{cis} is close to 3000.

The rate of substitution of solvent in intermediate 3 by both *cis* and *trans*-cyclooctene exhibited an apparent first-order dependence on the concentration of the cyclooctene. The data is plotted in Figure 5. Assuming an overall second-order rate law we found that the rates of these two substitution processes are nearly identical with second-order rate constants k_3 for *cis* and *trans*-cyclooctene respectively of $1.4 \pm 0.3 \times 10^7$ and $1.7 \pm 0.2 \times 10^7$ L/mol-s (Table II). The rate constants were obtained from the slopes of the lines. These rate constants are very similar to previously reported rate constants for reaction of intermediate 3 with a number of different ligands ([3], [4], [38]). It is possible that a residual reaction at zero ligand concentration is responsible for the non-zero intercept (e.g. reaction with water) but since there is no dependence of q_2 on ligand concentration the energetics obtained from these experiments is still valid. However, given the scatter in the data, we cannot conclude that the non-zero intercept seen in both lines is physically significant.

The larger olefin-metal bond strength in *trans*-4 than in *cis*-4 could be due to either steric or electronic factors or both. From comparisons of the rates and enthalpies of reaction of α-methyl substituted and unsubstituted pyridine and THF ligands with $Cr(CO)_5S$ in heptane solution, it is apparent that steric influences which have an effect on the enthalpy of reaction also have a substantial effect on the rate of reaction. For example, in heptane solution, pyridine reacts with $Cr(CO)_5S$ more exothermically (2.3 kcal/mol) and faster (10 times) than does 2-picoline ([4]). The lack of a substantial difference in the rate constants k_3 for *cis* and *trans*-cyclooctene supports the conclusion that strain relief and not steric bulk, is the primary component of $\Delta\Delta H_3$.

Discussion and Conclusions

In both the $CpMn(CO)_3$ and the $Cr(CO)_6$ systems we have obtained bond enthalpy data which, until now, has been unavailable. From the reactive $CpMn(CO)_2THF$ complex to the substitutionally inert $CpMn(CO)_3$ the Mn-L bond strengths agree with empirically expected trends. The rate constants for the reactions of ligands with either $CpMn(CO)_2$ or $Cr(CO)_5$ appear to bear little correlation to the exothermicity of the reaction. This observation has been noted previously ([3], [4]) and has been attributed to steric effects playing a dominant role in determining the rate constant for the reaction. The much greater reactivity of $Cr(CO)_5$ compared to $CpMn(CO)_2$ may be due to greater steric demands imposed by the cyclopentadienyl ligand in comparison to 3 CO ligands. Experiments are in progress to help elucidate this.

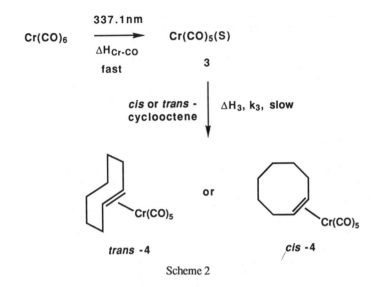

Scheme 2

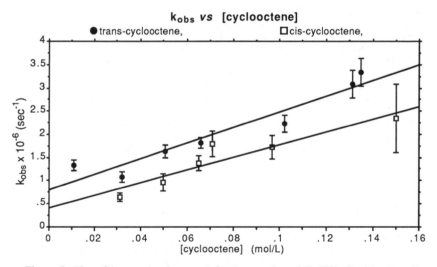

Figure 5. Plot of k_{obs} vs [cyclooctene] for the reaction of $Cr(CO)_5S$ with *cis* and *trans*-cyclooctene in *n*-heptane. For *cis*-cyclooctene, $r^2 = 0.92$. For *trans*-cyclooctene, $r^2 = 0.87$. Other data given in Table II. Error bars are the greater of the standard deviation calculated from the scatter in each point or ± 10%.

An interesting question remains unanswered. How large is the heptane-Mn interaction? Angelici and Loewen (14) have studied the ligand substitution of the *cis*-cyclooctene complex **1** by PPh$_3$ in methylcyclohexane solution. Similarly, Butler and Sawai (16) have studied the ligand substitution of the SBu$_2$ complex **2** by P(O-*n*-Bu)$_3$ in methylcyclohexane solution.

$$CpMn(CO)_2(cis\text{-}C_8H_{14}) + PPh_3 \longrightarrow CpMn(CO)_2(PPh_3) + cis\text{-}C_8H_{14}$$
$$\mathbf{1} \qquad \Delta H^\ddagger = 34.9 \pm 0.7 \text{ kcal/mol}; \quad \Delta S^\ddagger = 27.5 \pm 2.0 \text{ eu}$$

$$CpMn(CO)_2(SBu_2) + P(O\text{-}n\text{-}Bu)_3 \longrightarrow CpMn(CO)_2(P(O\text{-}n\text{-}Bu)_3) + SBu_2$$
$$\mathbf{2} \qquad \Delta H^\ddagger = 36 \pm 1 \text{ kcal/mol}; \quad \Delta S^\ddagger = 22 \pm 1 \text{ eu}$$

The activation enthalpies for these reactions are respectively 10 and 7 kcal/mol *greater* than the corresponding Mn-L bond strengths in heptane. Assuming that methylcyclohexane and heptane are similar as solvents and that both reactions proceed via dissociative pathways as proposed, this implies that the Mn-heptane interaction is close to 8 or 9 kcal/mol. This is close to the 10 kcal/mol Cr-heptane interaction in (CO)$_5$Cr-heptane (3, 4) and the 9.6 kcal/mol W-ethane interaction in (CO)$_5$W-ethane (39). We can also calculate the gas phase Mn-CO bond dissociation energy in CpMn(CO)$_3$ to be close to 55 kcal/mol. Further experiments are necessary to confirm the magnitude of the Mn-heptane interaction.

Acknowledgments

The authors would like to acknowledge the National Institutes of Health (GM-42704-01), the University of Southern California Faculty Research and Innovation Fund and the donors of The Petroleum Research Fund, administered by the ACS for financial support of this research.

Literature Cited

1. Rudzki, J.E.; Goodman, J.L.; Peters, K.S. *J. Am. Chem. Soc.* **1985,** *107,* 7849-7854.
2. Burkey, T. J.; Majewski, M.; Griller, D. *J. Am. Chem. Soc.* **1986,** *108,* 2218-2221.
3. Yang, G. K.; Peters, K. S.; Vaida, V. *Chem. Phys. Lett.* **1986,** *125,* 566-568.
4. Yang, G. K.; Vaida, V.; Peters, K. S. *Polyhedron* **1988,** *7,* 1619-1622.
5. Lynch, D.; Endicott, J. F. *Inorg. Chem.* **1988,** *27,* 2181-2184.
6. Gould, I. R.; Moser, J. E.; Armitage, B.; Farid, S.; Goodman, J. L.; Herman, M. S. *J. Am. Chem. Soc.* **1989,** *111,* 1917-1919.
7. Herman, M. S.; Goodman, J. L. *J. Am. Chem. Soc.* **1988,** *110,* 2681-2683 and references cited therein.
8. Nasielski, J.; Colas, A. *Inorg. Chem.* **1978,** *17,* 237-240.
9. Giordano, P. J.; Wrighton, M. S. *Inorg. Chem.* **1977,** *16,* 160-166.
10. Strohmeier, W. *Angew. Chem. internat. Edit.* **1964,** *3,* 730-737.
11. Caulton, K. G. *Coord. Chem. Rev.* **1981,** *38,* 1-43.
12. Collman, J.P.; Hegedus, L.S.; Norton, J.R.; Finke, R.G. *Principles and Applications of Organotransition Metal Chemistry;* University Science Books: Mill Valley, California, **1987,** p37.
13. Lewis, K. E.; Golden, D. M.; Smith, G. *J. Am. Chem. Soc.* **1984,** *106,* 3905-3912.
14. Angelici, R. J.; Loewen, W. *Inorg. Chem.* **1967,** *6,* 682.
15. Fenster, A. E.; Butler, I. S. *Inorg. Chem.* **1974,** *13,* 915-920.
16. Butler, I. S.; Sawai, T. *Inorg. Chem.* **1975,** *14,* 2703.
17. Creaven, B. S.; Dixon, A. J.; Kelly, J. M.; Long, C.; Poliakoff, M. *Organometallics* **1987,** *6,* 2600-2605.
18. Gowling, E. W.; Kettle, S. F. A.; Sharples, G. M. *J. Chem. Soc., Chem. Commun.* **1968,** 21-22.

19. McLain, S. J.; Schrock, R. R.; Sharp, P. R.; Churchill, M. R.; Youngs, W. J. *J. Am. Chem. Soc.* **1979**, *101*, 262, 265.
20. Bennett, M. A.; Hambley, T. W.; Roberts, N. K. Robertson, G. B. *Organometallics* **1985**, *4*, 1992-2000.
21. Buchwald, S. L.; Watson, B. T.; Huffman, J. C. *J. Am. Chem. Soc.* **1986**, 108, 7411-7413.
22. Arnold, J.; Wilkinson, G.; Hussain, B.; Hursthouse M. B. *J. Chem. Soc., Chem. Commun.* **1988**, 704-705.
23. Hartwig, J. F.; Andersen, R. A.; Bergman, R. G. *J. Am. Chem. Soc.* **1989**, *111*, 2717-2719.
24. Bennett, M. A.; Yoshida, T. *J. Am. Chem. Soc.* **1978**, *100*, 1750-1759.
25. Jason, M. E.; McGinnety, J. A.; Wiberg, K. B. *J. Am. Chem. Soc.* **1974**, *96*, 6531-6532.
26. Bly, R. S.; Hossain, M. M.; Lebioda, L. *J. Am. Chem. Soc.* **1985**, *107*, 5549-5550; Bly, R. S.
27. Bly, R. K.; Hossain, M. M.; Lebioda, L.; Raja, M. *J. Am. Chem. Soc.* **1988**, *110*, 7723-7730.
28. Bly, R. S.; Silverman, G. S.; Bly, R. K. *J. Am. Chem. Soc.* **1988**, *110*, 7730-7737.
29. Godleski, S. A.; Gundlach, K. B.; Valpey, R. S. *Organometallics* **1985**, *4*, 296-302.
30. Herberhold, M. *Transition Metal π-Complexes, vol. II;* Elsevier: New York, **1974**.
31. Muhs, M. A.; Weiss, F. T. *J. Am. Chem. Soc.* **1962**, *84*, 4697-4705.
32. von Büren, von M.; Hansen, H.-J. *Helv. Chim. Acta* **1977**, *60*, 2717-2722.
33. Grevels, F.-W.; Skibbe, V. *J. Chem. Soc., Chem. Commun.* **1984**, 681-683.
34. Simon, J. D.; Peters, K. S. *Chem. Phys. Lett.* **1983**, *98*, 53.
35. Simon, J. D.; Xie, X. *J. Phys. Chem.* **1986**, *90*, 6751-6753.
36. Wang, L.; Zhu, X.; Spears, K. G. *J. Am. Chem. Soc.* **1988**, *110*, 8659-8696.
37. Benson, S. W. *Thermochemical Kinetics, 2nd ed.*; Wiley: New York, **1976**.
38. Kelly, J. M.; Bent, D. V.; Hermann, H. Shulte-Frohlinde, D.; Koerner von Gustorf, E. *J. Organomet. Chem.* **1974**, *69*, 259-269.
39. Ishikawa, Y.; Brown, C. E.; Hackett, P. A.; Raynor, D. M. *Chem. Phys. Lett.* **1988**, *150*, 506-510.

RECEIVED December 18, 1989

Chapter 14

Metal–Carbon and Metal–Hydrogen Bond Dissociation Enthalpies from Classical and Nonclassical Calorimetric Studies

A. R. Dias[1], H. P. Diogo[1], D. Griller[2], M. E. Minas da Piedade[1], and J. A. Martinho Simões[1]

[1]Centro de Química Estrutural, Complexo I, Instituto Superior Técnico, 1096 Lisboa Codex, Portugal
[2]Division of Chemistry, National Research Council of Canada, Ottawa, Ontario K1A 0R6, Canada

The application of classical and non-classical (photoacoustic) reaction-solution calorimetry to probe the energetics of metal-ligand bonds in organometallic systems is briefly analysed and illustrated by thermochemical results involving two families of compounds. The classical reaction-solution studies enabled the discussion of the systematics of metal-carbon bond enthalpies in several complexes $M(\eta^5-C_5H_5)_2L_2$. The photoacoustic studies addressed the effect of phenyl goups on the energetics of silicon-hydrogen bonds in phenyl-substituted silanes.

Classical and Non-Classical Reaction-Solution Calorimetry

One of the limitations of "classical" reaction-solution calorimetry (either relying on temperature change or heat flux measurements) (1) as a tool for obtaining information on the energetics of chemical bonds, stems from the fact that reactions seldom involve the formation or the cleavage of an individual bond. Therefore, the enthalpy of reaction usually reflects a *difference* between bond dissociation enthalpies. A second drawback of the method is that the bond enthalpy balance can be masked by solvation effects.

In organometallic reactions it is often possible to make measurements or reasonable estimates of vaporization and solution enthalpies, in order to derive the enthalpy of reaction in the gas phase. Yet, the situation where only one bond dissociation enthalpy is unknown is rather uncommon. Despite these shortcomings, classical reaction-solution calorimetry, being a well tested, inexpensive, reliable technique, has provided an outstanding contribution to our present knowledge in the area of transition metal organometallic thermochemistry.

A "non-classical" type of reaction-solution calorimetry has been developed in recent years (2,3). In this new method, the enthalpy of a reaction is not measured with thermometers or

0097–6156/90/0428–0205$06.00/0

thermopiles but determined by using piezoelectric transducers. As these transducers have very short response times (ca. 10^{-6}-10^{-7} s), only very rapid reactions can be examined – the case, for example, of many radical-molecule reactions. Another difference between classical and non-classical calorimetry concerns the way reactions are initiated. As illustrated by Reactions 1 and 2, in the new method the products or one of the reactants are directly produced by photolysis ($\underline{4}$).

$$AA(soln) \xrightarrow{h\nu} 2A^{\cdot}(soln) \tag{1}$$

$$BB(soln) \xrightarrow{h\nu} 2B^{\cdot}(soln) \tag{2a}$$

$$B^{\cdot}(soln) + XH(soln) \longrightarrow X^{\cdot}(soln) + BH(soln) \tag{2b}$$

Net: $$BB(soln) + 2XH(soln) \longrightarrow 2X^{\cdot}(soln) + 2BH(soln) \tag{2c}$$

The remarkable advantage of the non-classical "photoacoustic" technique over classical calorimetry is that it can provide an accurate value of a bond dissociation enthalpy, D(A-A) or D(X-H), since D(B-B) and D(B-H) are usually available. Although that value is determined in solution, there is a good deal of experimental evidence indicating that solvation enthalpies play a minor role in the reaction energetics (particularly in process 2c) when AA, BB and XH are organic or even main group organometallic molecules ($\underline{5}$). Significant solvation effects can, however, be expected when A$^{\cdot}$ or B$^{\cdot}$ are transition metal coordinatively unsaturated complexes, so that the measured enthalpy of reaction may no longer enable to derive an absolute value of a bond dissociation enthalpy. This and other complicating factors (e.g. the scarcity of reliable values for quantum yields in the case of process 1 and also the frequent possibility of competing reactions) have hindered an extensive use of photoacoustic calorimetry to study transition metal systems ($\underline{6}$).

Results obtained by classical and non-classical reaction-solution calorimetry are described in the present paper. The former deal with the thermochemistry of several $Mo(Cp)_2R_2$ (R= Me, Et, Bu, C_2H_4, C_2Ph_2) and $Nb(Cp)_2Me_2$, and provided values for molybdenum-carbon and and niobium-methyl bond enthalpies. The latter concern the energetics of silicon-hydrogen bonds in phenyl-substituted silanes.

Molybdenum-Carbon and Niobium-Carbon Bond Dissociation Enthalpies

Despite intense current efforts to probe the energetics of transition metal-ligand bonds, there are still few examples of families of organometallic complexes for which the "strengths" of those bonds are known for a variety of ligands. In the case of $Mo(Cp)_2L_n$ complexes, the systematics of Mo-L bond enthalpies is fairly well established for a series of ligands, including hydrogen, methyl, halogen, benzoate, thiolates, and azobenzene ($\underline{7}$). In an effort to investigate the effect of the size of the n-alkyl chain on Mo-C bond enthalpies and to obtain information on the energetics of Mo-alkene and Mo-alkyne bonds, the compounds $Mo(Cp)_2Et_2$, $Mo(Cp)_2Bu_2$, $Mo(Cp)_2(C_2H_4)$, and $Mo(Cp)_2(C_2Ph_2)$ were studied by classical reaction-solution calorimetry ($\underline{8}$).

Another illustration of the same technique is provided by the the complex $Nb(Cp)_2Me_2$. In this case, the goal was to enlarge our knowledge on the systematics of metal-methyl bond enthalpies. The calorimetric set-up used in Lisbon has been described in the literature ([7]), and details of the experiments will be published elsewhere.

The enthalpies of formation of the crystalline complexes $Mo(Cp)_2Et_2$, $Mo(Cp)_2Bu_2$, and $Nb(Cp)_2Me_2$ rely on measurements of the enthalpies of their reactions with HCl in isopropyl ether solutions, Equation 3. The enthalpies of formation of $Mo(Cp)_2(C_2H_4)$ and $Mo(Cp)_2(C_2Ph_2)$ were determined from the measurements of the enthalpies of reaction of these complexes with toluene solutions of iodine, Equation 4.

$$M(Cp)_2R_2(c) + 2HCl(soln) \longrightarrow M(Cp)_2Cl_2(c/soln) + 2RH(g/soln) \quad (3)$$

$$M = Mo, \; R = Et, \; Bu; \; M = Nb, \; R = Me$$

$$Mo(Cp)_2L(c) + I_2(soln) \longrightarrow Mo(Cp)_2I_2(c) + L(g/soln) \quad (4)$$

$$L = C_2H_4, \; C_2Ph_2$$

Several measurements of enthalpies of solution were also needed in order to calculate the final results shown in Table I. In addition, it was necessary to determine the enthalpies of formation of HCl in the ether solutions, a task which proved to be more difficult than anticipated and which is still being pursued, aiming to achieve a better accuracy. This problem, together with the uncertainties associated with the enthalpies of solution of the alkanes, particularly ethane and butane, may affect the values reported for the alkyl complexes by about ±15 kJ/mol. However, it can be seen from the discussion below that this inaccuracy is reduced by half when bond enthalpies are derived.

The measured value for the enthalpy of sublimation, ΔH_s^0, of the ethyl complex, and the values estimated for the remaining molecules are also included in Table I. While it is obvious that this exercise involved speculation, its effect on the bond enthalpy data will (hopefully!) be covered by the assigned uncertainties. Anyhow, it is stressed that the experimental values for the enthalpies of formation of the crystalline complexes enable to make reliable estimates of reaction enthalpies in solution or in standard reference states.

The method used to derive metal-ligand bond enthalpy terms, $E(M-L)$, and mean bond dissociation enthalpies, $\bar{D}(M-L)$, has been discussed in a recent review ([7]) and only a brief account is given here. The enthalpies of formation of the complexes $M(Cp)_2L$ and $M(Cp)_2Cl_2$ in the gas phase afford the enthalpy of Reaction 5 which

$$M(Cp)_2L_n(g) + 2Cl(g) \longrightarrow M(Cp)_2Cl_2(g) + nL(g) \quad (5)$$

$$\Delta H_r(5) = n\bar{D}(M-L) - 2\bar{D}(M-Cl) \quad (6)$$

in turn is equal to the difference between M-L and M-Cl mean bond dissociation enthalpies. When n = 1, $\bar{D}(M-L)$ is of course replaced by

Table I.   Standard Enthalpies of Formation of Several
Bis(cyclopentadienyl) Metal Complexes
Data in kJ/mol

Complex	$\Delta H_f^0(c)$	ΔH_s^0 [a]	$\Delta H_f^0(g)$
$Nb(Cp)_2Me_2$	(364.8 ± 2.0) [b]	(25 ± 10) [b]	(339.8 ± 10.2) [b]
$Mo(Cp)_2Cl_2$ [c]	-95.8 ± 2.5	(100.4 ± 4.2)	4.6 ± 4.9
$Mo(Cp)_2Me_2$ [c]	262.4 ± 4.0	70.4 ± 4.2	332.8 ± 5.8
$Mo(Cp)_2Et_2$	200.3 ± 4.6	93.6 ± 1.8	293.9 ± 4.9
$Mo(Cp)_2Bu_2$	106.4 ± 6.1	(103 ± 10)	209.4 ± 11.7
$Mo(Cp)_2(C_2H_4)$	285.0 ± 8.2	(77 ± 10)	362.0 ± 12.9
$Mo(Cp)_2(C_2Ph_2)$	511.4 ± 8.0	(150 ± 20) [d]	661 ± 22

[a]Estimated values in parentheses.   [b]The enthalpies of formation
and sublimation of the niobium dichloride complex are not
available. The values represent the differences between the
enthalpies of formation or sublimation of the dimethyl and the
dichloride complexes.   [c]Reference 7.   [d]See text.

M-L bond dissociation enthalpy, D(M-L). The calculation of either
one of these quantities requires a value for $\bar{D}$(M-Cl), which is not
available. While the experimentally derived value for the difference
is itself rather useful, it is often possible to predict a tentative
value for $\bar{D}$(M-Cl). In any case, even if the estimate is
unreasonable, the trend in $\bar{D}$(M-L) or D(M-L) for a series of ligands
will not be affected by the error.
    Different methods have been used in attempts to assign
reasonable numbers to $\bar{D}$(M-Cl) (7,9). In one of those methods (7),
the M-Cl "bond strength" in the complex $M(Cp)_2Cl_2$ is identified with
the M-Cl bond strength in the homoleptic molecule MCl_m, on the basis
of similar bond lengths. As bond dissociation enthalpies are often
poor measures of the intrinsic strengths of bonds, since they include
the relaxation energies of the fragments, a new parameter had to be
introduced. This quantity, called bond enthalpy term, E(M-L), was
defined so that it does not contain the above relaxation energies and
therefore can be "transferred" from one molecule to another.
    Once E(M-Cl) in the complex is identified with E(M-Cl) =
$\bar{D}$(M-Cl) in MCl_m, it has to be tied to $\bar{D}$(M-Cl) in $M(Cp)_2Cl_2$. The
bridge is the relaxation energy of the fragment MCp_2 from the
dichloride complex (7).
    The bond enthalpy term for the metal-ligand bond, E(M-L), can
also be obtained by subtracting the relaxation energies of the
fragments MCp_2 and L, formed by cleavage of M-L bond(s), from $\bar{D}$(M-L).
    Although the method just described is essentially

semi-quantitative, as it relies e.g. on extended Hückel calculations to derive relaxation energies (7), it has at least the merit of emphasizing some cautions that should be taken when transferring bond enthalpy data. On the other hand, bond enthalpy terms resemble Laidler terms, tabulated for organic compounds (1), and in fact can be used to estimate new enthalpies of formation of organometallic complexes (7).

The bond dissociation enthalpies shown in Table II are based on E(Nb-Cl) = 407.1±2.1 kJ/mol (from $NbCl_5$) and on E(Mo-Cl) = 303.8±7.1 kJ/mol (from $MoCl_6$). The similarity between $\bar{D}$(Nb-Cl)-$\bar{D}$(Nb-Me) and $\bar{D}$(Mo-Cl)-$\bar{D}$(Mo-Me) and the trend in molybdenum-carbon bond enthalpies are independent of those assignments.

Table II. Molybdenum- and Niobium-Carbon Bond Enthalpy Data
Data in kJ/mol

Complex	D or $\bar{D}$(M-L)[a]
$Nb(Cp)_2Cl_2$	407.1±2.1[b]
$Nb(Cp)_2Me_2$	263±6
$Mo(Cp)_2Cl_2$	303.8±7.1[b]
$Mo(Cp)_2Me_2$	166±8
$Mo(Cp)_2Et_2$	156±9
$Mo(Cp)_2Bu_2$	154±12
$Mo(Cp)_2(C_2H_4)$	(59±20)[c,d]
$Mo(Cp)_2(C_2Ph_2)$	(120±27)[c,d]

[a] Mean bond dissociation enthalpies unless indicated otherwise.
[b] See text. Value for Mo from ref. 7 and for Nb from Chase, Jr.,M. W.; Davies, C. A.; Downey, Jr., J. R.; Frurip, D. J.; McDonald, R. A.; Syverud, A. N. *J. Phys. Chem. Ref. Data* 1985, 14, Suppl. No. 1.
[c] Bond dissociation enthalpy. [d] See text.

The molybdenum-alkyl mean bond dissociation enthalpies are in the expected order and in good agreement e.g. with the trend reported for $Th(Cp^*)_2R_2$ complexes ($Cp^* = \eta^5-C_5Me_5$) (9a). On the other hand, it is interesting to note that the group additivity rule seems to apply to the Mo-alkyl family, as shown by the excellent linear relationships 7 and 8, where N is the number of carbon atoms in the alkyl chain.

$$\Delta H_f^o[Mo(Cp)_2R_2, c] = -(51.3\pm3.8)N + (309.4\pm9.9) \qquad (7)$$

$$\Delta H_f^o[Mo(Cp)_2R_2, g] = -(41.3\pm0.8)N + (375.1\pm2.2) \qquad (8)$$

A comparison of the results in Table II with other literature
data shows that $\bar{D}(M-Me)-\bar{D}(M-Et) \simeq (15\pm5)$ kJ/mol is nearly independent
of the metal [21 kJ/mol for $Th(Cp^*)_2R_2$ ($\underline{9a},\underline{10}$); 19 kJ/mol for
$Th(Cp^*)_2(OBu-t)R$ ($\underline{9a},\underline{10}$); 10 kJ/mol for $Mo(Cp)_2R_2$ ($\underline{7},\underline{10},\underline{11}$); 18 kJ/mol
for $Mo(Cp)(CO)_3R$ ($\underline{10},\underline{12},\underline{13}$) and 10 kJ/mol for $Ir(Cl)(CO)(PMe_3)_2(I)R$
($\underline{10},\underline{14},\underline{15}$)]. This difference is also in the range of the average
value for main group elements ca. 20 kJ/mol ($\underline{16}$), suggesting that
$D(M-Me)-D(M-Et)$ is approximately constant along the Periodic Table.
We expect this conclusion to hold for any other differences between
metal-σ-alkyl bond dissociation enthalpies in the absence of relevant
steric effects.

The comparison between $\bar{D}(Mo-Me)$, $\bar{D}(Nb-Me)$, and $\bar{D}(Zr-Me) = 285\pm2$
kJ/mol in $Zr(Cp)_2Me_2$ ($\underline{9b}$) is of course hindered by the unknown
accuracies in the anchor values, $D(M-Cl)$. Nevertheless, the
expected trend, $\bar{D}(Zr-Me)>\bar{D}(Nb-Me)>\bar{D}(Mo-Me)$ is observed.

$D(Mo-C_2H_4)$ is small as compared to $D(Mo-C_2Ph_2)$. The estimated
value for the enthalpy of sublimation of $Mo(Cp)_2(C_2Ph_2)$
may be uncertain ($\underline{11}$), thus implying an adjustment of $D(Mo-C_2Ph_2)$.
However, the enthalpy of Reaction 9, $\Delta H_r(9) = 56.6\pm2.5$ kJ/mol,

$$Mo(Cp)_2(C_2Ph_2)(c) + C_2H_4(g/soln) \longrightarrow$$
$$\longrightarrow Mo(Cp)_2(C_2H_4)(c) + C_2Ph_2(soln) \qquad (9)$$

directly obtained from the experimental data, confirms that trend (the enthalpies of solution of
the complexes should cancel within ±10 kJ/mol). In addition, the energetics of
molybdenum–carbon bonds in these complexes, where the ethylene and diphenylacetylene
fragments are η^2-bonded to the metal (17) follows the expected order
$D[Mo-C(sp)]>D[Mo-C(sp^2)]$. This is in line, for example, with the trend observed in several
scandium σ-alkyl complexes, $Sc(Cp^*)_2R$ ($R=C\equiv CCMe_3$, Ph, $CH_2CH_2CH_2C_5Me_4$),
$D[Sc-C(sp)]> D(Sc-aryl)>D[Sc-C(sp^3)]$ (18).

Silicon-Hydrogen Bond Dissociation Enthalpies in Methyl- and Silyl-silanes. A Critical Summary

One of the most interesting properties of methyl-substituted
silanes, Me_nSiH_{4-n} (n = 0-3), is the constancy of silicon-hydrogen
bond dissociation enthalpies, with an average value of 376±2 kJ/mol
($\underline{19}$). The corresponding carbon-hydrogen bond dissociation enthalpies
span a range of about 43 kJ/mol. Although this striking difference
between silanes and alkanes has been questioned ($\underline{20}$), there seems to
be no reason to doubt the accuracy of the experimental values of
$D(Si-H)$.

The negligible effect of methyl groups on silicon-hydrogen bond
dissociation enthalpies in methylsilanes is supported by simple
correlations which only involve enthalpy of formation data. For
example, when $\Delta H_f^\circ(Me_nSiH_{3-n}, g)$ are plotted against $\Delta H_f^\circ(Me_nSiH_{4-n}, g)$
(n = 0-3), the excellent linear fitting, Equation 10, implies
that the enthalpy of Reaction 11 is constant, which in turn

$$\Delta H_f^o(Me_{n+1}SiH_{3-n},g) = (1.034\pm0.004)\Delta H_f^o(Me_n SiH_{4-n},g) - (64.7\pm0.4) \quad (10)$$

$$Me_n SiH_{4-n},(g) + CH_4(g) \longrightarrow Me_{n+1}SiH_{3-n},(g) + H_2(g) \quad (11)$$

implies that $D(Me_n SiH_{3-n}-H) - D(Me_n SiH_{3-n}-Me)$ are also constant or, more likely, that silicon-hydrogen and silicon-carbon bond dissociation enthalpies are both little affected by the number of methyl groups bonded to the silicon atom. No similar correlations are observed for the corresponding hydrocarbons.

Another fact in support of the constancy of $D(Si-H)$ in methylsilanes, is that the enthalpies of formation of these molecules can be predicted by using only three "bond enthalpy terms", $E(Si-H)$, $E(Si-C)$, and $E(C-H)$, whereas five terms are needed for the hydrocarbons, including $E(C-H)$, $E(C-H)_p$, $E(C-H)_s$, ($\underline{1}$). The subscripts designate primary, secondary, and tertiary C-H bonds, respectively.

A recent analysis of thermochemical data based on relationships between enthalpies of formation and the unshielded core potential of a moiety X, V_x ($\underline{21}$), throws some light into the behavior of silicon-hydrogen bonds in methylsilanes. Consider first the case of carbon-hydrogen bonds. The enthalpy of Reaction 12, which is shown to be a linear function of V_x ($\underline{21}$) can be expressed in terms of bond dissociation enthalpies, Equation 13.

$$MeX(g) + t\text{-}Bu(g) \longrightarrow t\text{-}BuX(g) + Me(g) \quad (12)$$

$$D(Me-X) - D(t\text{-}Bu-X) = 69.0 - 9.62V_x \quad (13)$$

Differences $D(Me-X)-D(t\text{-}Bu-X)$ for several X are displayed in Table III. Besides a very good agreement between calculated and experimental data, it is observed that $D(Me-X)-D(t\text{-}Bu-X)$ becomes negative when the electronegativity of X (as measured by V_x) exceeds a given value, $V_x = 7.17$. It can be expected that this "critical" value will be significantly lower when the central carbon is replaced by a much less electronegative element, as happens with silicon. Indeed, the same exercise applied to silanes leads to Equations 14a,b and to the results collected in Table IV ($\underline{22}$).

$$D(H_3 Si-X) - D(Me_3 Si-X) = -67.8 + 24.9V_x \quad (14a)$$

$$X = H, Me, SiH_3$$

$$D(H_3 Si-X) - D(Me_3 Si-X) = -39.8 + 2.1V_x \quad (14b)$$

$$X = halogen, OH, SH, NH_2$$

Although the trend in Table IV is affected by the participation of silicon d orbitals in Si-X bond (Equation 14b) ($\underline{22b}$), it is seen that the critical value, $V_x = 2.72$ (Equation 14a), is quite low and close to the unshielded core potential of hydrogen atom. It is also noted that $D(H_3 Si-X)-D(Me_3 Si-X)$ increase with V_x (within each family), whereas the corresponding difference for hydrocarbons decrease for higher electronegativity moieties.

Table III.   Carbon-X Bond Dissociation Enthalpies as a
Function of the Electronegativity of X

X	V_x [a]	[D(Me-X) - D(t-Bu-X)] (kJ/mol)	
		calc.	exp.
H	2.70	43	43
SH	5.77	13	16
Cl	7.04	1	3
OH	8.11	-9	-8

[a] Data from Reference 22b.

Table IV.   Silicon-X Bond Dissociation Enthalpies as a Function
of the Electronegativity of X

X	V_x [a]	[D(H_3Si-X) - D(Me_3Si-X)] (kJ/mol)	
		calc.	exp.
H	2.70	-1	0
SiH_3	3.41	17	
SH	5.77	-28	
Cl	7.04	-25	-21
OH	8.11	-23	

[a] Data from Reference 22b.

The thermochemical correlations, the bond terms, and the unshielded core potential method all support a negligible effect of methyl groups on silicon-hydrogen bond dissociation enthalpies. This constancy is also in keeping with the hydridic nature of Si-H bonds in silanes (23). The effect of other substituents is, by comparison, poorly defined (19), although some results have recently become available. For example, it has been shown that silylation destabilizes Si-H bonds: D(H_3SiSiH$_2$-H) = 361±8 kJ/mol (19), D(Me_3SiSiMe$_2$-H) = 357±8 kJ/mol, and D[(Me_3Si)$_3$Si-H] = 331±8 kJ/mol (24). The large difference between D(Si-H) in tris(trimethysilyl)silane and silane, 47 kJ/mol, led to the suggestion, later confirmed (25), that this molecule would be an efficient reducing agent. The silylation effect has been supported by *ab initio* calculations, which also addressed the influence of other substituents on D(Si-H) (26). According to these calculations, the difference $D(H_3Si-H)-D(H_2XSi-H)$ depends in a large extent on the electronegativity of X and on the low electronegativity of Si: positive differences were predicted for elements or moieties that are more electropositive than hydrogen and inversely for very electronegative X.

Silicon-Hydrogen Bond Dissociation Enthalpies in Phenylsilanes

If, in silane, one hydrogen is replaced by a phenyl group, the silicon-hydrogen bond dissociation enthalpy is only reduced by about 9±7 kJ/mol (19). This is in sharp contrast with the decrease observed for the carbon analogues, where the large stabilization energy of benzyl radical makes $D(Me-H)-D(PhCH_2-H) = 71±4$ kJ/mol. The effect of a second phenyl group in the hydrocarbon family is also known, $D(Ph_2CH-H) = 340±8$ kJ/mol (27,28) and there is a recent estimate for $D(Ph_3C-H)$, 338±13 kJ/mol (28).

The small effect induced by one phenyl group on D(Si-H) raises the question of whether these bond dissociation enthalpies in phenylsilanes will follow a pattern similar to the one observed in methylsilanes. In order to tackle this problem, photoacoustic calorimetry experiments on several phenyl- and methylsilanes were carried out at NRCC.

The photoacoustic set-up and the standard experimental procedure have been described elsewhere (3), and details of the studies involving the phenylsilanes will be reported in a separate publication (29). As in the case of the non-classical studies, only a few selected points will be addressed here.

In a typical photoacoustic experiment, a benzene solution of the silane containing a given amount of di-*tert*-butyl peroxide was irradiated in a quartz cuvette with a pulsed nitrogen laser ($\lambda = 337.1$ nm). The energy of a photon (354.9 kJ/mol) was used, in part, to cleave the oxygen-oxygen bond in the peroxide, producing two *tert*-butoxyl radicals, which then abstracted hydrogen from silane molecules, yielding silyl radicals and *tert*-butyl alcohol. The reaction sequence is described by Equations 2a,b, by identifying B˙ with *t*-BuO˙ and XH with the silane.

The amount of heat deposited in solution by a laser shot, ΔH_{obs}, is therefore the difference between the photon energy and the net reaction enthalpy (ΔH_r for Reaction 2c) multiplied by the quantum yield, Φ, for peroxide dissociation, Equation 15.

$$\Delta H_{obs} = 354.9 - \Phi \Delta H_r \qquad (15)$$

The sudden heat release produced a shock wave which was detected by the piezoelectric transducer that was clamped to the side of the cuvette, amplified and stored in an oscilloscope. The amplitude of that wave was proportional to ΔH_{obs} and to the optical density of the solution, so that the measurement of these two quantities led to the value of ΔH_{obs}, Equation 16.

$$\Delta H_{obs} = 354.9 - \Phi a_r/a_s \qquad (16)$$

Here, a_r is the slope of a plot of the wave amplitudes versus the optical densities of the solutions at different concentrations of di-*tert*-butyl peroxide; the calibration constant a_s is the slope of a similar plot involving solutions containing variable amounts of *o*-hydroxybenzophenone in place of di-*tert*-butyl peroxide. That substance converts all of the absorved radiation into heat in a few nanoseconds (3).

As mentioned above, a crucial requirement in the photoacoustic experiment is that the net reaction 2c must be fast when compared to the response of the tranducers. The rate constants for the slower reaction, 2b, were available for most of the silanes studied and allowed the concentrations of those substrates to be calculated so as to meet the above condition. Another important issue concerns the possible addition of silyl radicals to solvent molecules. Since this reaction is predicted to be exothermic by ca. 60 kJ/mol, its occurence would have a significant impact on the results. However, parallel experiments involving triethylsilane, carried out in benzene and also using the silane as solvent, led to similar ΔH_r values, indicating that the addition to benzene was irrelevant on the timescale of the photoacoustic experiment.

The calculation of silicon-hydrogen bond dissociation enthalpies follows the determination of ΔH_r and relies on literature values for $D(t\text{-BuO-OBu-}t)$ and $D(t\text{-BuO-H})$ (27), Equation 17.

$$D(Si-H) = \Delta H_r/2 - D(t\text{-BuO-OBu-}t)/2 + D(t\text{-BuO-H})$$

$$= \Delta H_r/2 + (360.2\pm1.7) \qquad (17)$$

The values obtained for $D(Si-H)$ are shown in Table V, together

Table V.　Silicon-Hydrogen and Carbon-Hydrogen Bond Dissociation
Enthalpies in Phenylsilanes and Phenylmethanes
Data in kJ/mol

Silane	$D(Si-H)$	Hydrocarbon	$D(C-H)$
$PhSiH_3$	369 ± 5^a	$PhCH_3$	368 ± 4^b
$PhMe_2SiH$	358 ± 7^c	$PhMe_2CH$	353 ± 6^b
Ph_2SiH_2	360 ± 6^c	Ph_2CH_2	340 ± 4^d
Ph_2MeSiH	342 ± 10^c	Ph_2MeCH	339 ± 8^b
Ph_3SiH	352 ± 2^c	Ph_3CH	338 ± 13^e

[a]Reference 19.　[b]Reference 27.　[c]Photoacoustic result.　The error is twice the standard deviation of the mean.　[d]Average literature value. See Reference 28.　[e]Reference 28.

with data for PhSiH$_3$ and the carbon analogues. Although the uncertainties are relatively large, in particular for diphenylmethylsilane, it is clear that the replacement of hydrogen or methyl groups by phenyl groups decreases D(Si-H). The low value obtained for Ph$_3$SiH suggested that it could be used as a useful reducing agent for alkyl bromides. This was confirmed by additional kinetic experiments which showed that the rate constant of Reaction 18, the rate-determing step of the reaction chain, is about 10^4 M^{-1} s^{-1} at 90°C (30).

$$R^{\cdot} + Ph_3SiH \longrightarrow Ph_3Si^{\cdot} + RH \qquad\qquad (18)$$

$$Ph_3Si^{\cdot} + RBr \longrightarrow Ph_3SiBr + R^{\cdot} \qquad\qquad (19)$$

Conclusions

The Mo-alkyl mean bond dissociation enthalpies decrease with the increase of the n-alkyl chain length, in a way closely resembling the D(C-H) trend in the alkanes. The results suggest that, in the absence of important steric effects, further increases of the chain length will not significantly affect the Mo-alkyl bond dissociation enthalpy.

The trend $\bar{D}$(Zr-Me)>$\bar{D}$(Nb-Me)>$\bar{D}$(Mo-Me) is observed accompanying the decrease of the electronegativity difference between the metal and the methyl ligand.

D(Mo-C$_2$H$_4$) is small as compared to D(Mo-C$_2$Ph$_2$).

The constancy of silicon-hydrogen bond dissociation enthalpies in methyl substituted silanes is supported by a thermochemical analysis which does not rely on the enthalpies of formation of the corresponding silyl radicals.

The replacement of hydrogen or methyl groups by phenyl groups induces a considerable decrease in D(Si-H). The use of Ph$_3$SiH as a reducing agent for alkyl bromides was suggested by the low silicon-hydrogen bond dissociation enthalpy in this molecule.

Literature Cited

1. Cox, J. D.; Pilcher, G. Thermochemistry of Organic and Organometallic Compounds; Academic Press: London and New York, 1970.
2. Peters, K. S.; Snyder, G. J. Science 1988, 241, 1053 and references cited therein.
3. (a) Burkey, T. J.; Majewski, M.; Griller, D. J. Am. Chem. Soc. 1986, 108, 2218. (b) Kanabus-Kaminska, J. M.; Gilbert, B. C.; Griller, D. J. Am. Chem. Soc. 1989, 111, 3311.
4. In classical photocalorimetry the processes are also light-induced but as this technique relies on temperature or heat flux measurements, only reactions involving stable, long-lived species, can be examined. See, e.g. Harel, Y.; Adamson, A. W. J. Phys. Chem. 1986, 90, 6693 and references cited therein.

5. The evidence is mainly provided by the agreement between photoacoustic and gas-phase bond dissociation enthalpies. See also Reference 3b.

6. Examples of application of photoacoustic calorimetry to study transition metal complexes are listed in Reference 2. See also Yang, G. K.; Vaida, V.; Peters, K. S. Polyhedron 1988, 7, 1619.

7. Dias, A. R.; Martinho Simões, J. A. Polyhedron 1988, 7, 1531 and references cited therein.

8. Preliminary results for the first three compounds were given in Reference 7.

9. See, for example: (a) Bruno, J. W.; Marks, T. J.; Morss, L. R. J. Am. Chem. Soc. 1983, 105, 6824. (b) Schock, L. E.; Marks, T. J. J. Am. Chem. Soc. 1988, 110, 7701. The "reference" bond can of course be other than M–Cl.

10. Martinho Simões, J. A.; Beauchamp, J. L. Chem. Rev. to be published.

11. The measured enthalpy of sublimation of C_2Ph_2 is 99.9±1.5 kJ/mol (Minas da Piedade, M. E. Tese de Doutôramento, Instituto Superior Técnico, Lisboa, 1988). This indicates that the enthalpy of sublimation of the diphenylacetylene complex is at least ca. 100 kJ/mol.

12. Nolan, S. P.; Lopez de la Vega, R.; Hoff, C. D. J. Organometal. Chem. 1986, 315, 187.

13. Nolan, S. P.; Lopez de la Vega, R.; Mukerjee, S. L.; Hoff, C. D. Inorg. Chem. 1986, 25, 1160.

14. Yoneda, G.; Blake, D. M. J. Organometal. Chem. 1980, 190, C71.

15. Yoneda, G.; Blake, D. M. Inorg. Chem. 1981, 20, 67.

16. Pilcher, G.; Skinner, H. A. in The Chemistry of the Metal-Carbon Bond; Hartley, F. R.; Patai, S., Eds.; Wiley: New York, 1982; Chapter 2.

17. (a) De Cian, A.; Colin, J.; Shappacher, N.; Ricard, L.; Wiess, R. J. Am. Chem. Soc. 1981, 103, 1850. (b) Thomas, J. L. J. Am. Chem. Soc. 1973, 95, 1838. (c) Benfield, F. W. S.; Green, M. L. H. J. Chem. Soc. Dalton Trans. 1974, 1324.

18. Bulls, A. R.; Bercaw, J. E.; Manriquez, J. M.; Thompson, M. E. Polyhedron 1988, 7, 1409.

19. Walsh, R. In The Chemistry of Organo Silicon Compounds; Patai, S.; Rappoport, Z., Eds.; Wiley: New York, 1988; Chapter 5.

20. McKean, D. C. J. Molecular Structure 1984, 113, 251.

21. Luo, Y. R.; Benson, S. W. J. Phys. Chem. 1988, 92, 5255.

22. Correlations between enthalpies of formation and V_x from: (a) Luo, Y. R.; Benson, S. W. J. Phys. Chem. 1989, 93, 4643. (b) Luo, Y. R.; Benson, S. W. J. Phys. Chem. 1989, 93, 3791.

23. (a) Fritz, G.; Matern, E. Carbosilanes; Springer-Verlag: Berlin, 1986. (b) Pawlenko, S. Organosilicon Chemistry; de Gruyter: Belin, 1986.

24. Kanabus-Kaminska, J. M.; Hawari, J. A.; Griller, D.; Chatgilialoglu, C. J. Am. Chem. Soc. 1987, 109, 5267.

25. Chatgilialoglu, C.; Griller, D.; Lesage, M. J. Org. Chem. 1988, 53, 3641.

26. Coolidge, M. B.; Borden, W. T. J. Am. Chem. Soc. 1988, 110, 2298.

27. McMillen, D. F.; Golden, D. M. Ann. Rev. Phys. Chem. 1982, 33, 493.

28. Bordwell, F. G.; Cheng, J. P.; Harrelson, J. A., Jr. J. Am. Chem. Soc. 1988, 110, 1229.
29. Griller, D.; Kanabus-Kaminska, J. M.; Martinho Simões, J. A. submitted for publication.
30. Griller, D.; Lesage, M.; Martinho Simões, J. A. to be published.

RECEIVED January 22, 1990

Chapter 15

Adsorbate-Induced Restructuring of Surfaces

Surface Thermodynamic Puzzle

G. A. Somorjai

Department of Chemistry, University of California, Berkeley, CA 94720 and Center for Advanced Materials, Lawrence Berkeley Laboratory, Berkeley, CA 94720

Most solid surfaces restructure when clean. Absorbates that chemically bind cause additional marked restructuring of metal substrates that is well demonstrated by recent low energy electron diffraction (LEED)-surface crystallography studies. The rearrangement of atoms in the solid substrate may be local or long range and can occur on time scales of adsorption (less than $<10^{-3}$ seconds) surface reactions (seconds) or atom transport (larger than $>10^{+3}$ seconds). Atomically rough surfaces rearrange more readily and at lower temperatures and are also more chemically active in breaking bonds. Surface thermodynamic data are missing and very much needed to elucidate the nature and reasons for surface restructuring. Adsorbate induced restructuring can explain the need for thermal activation of chemical bond breaking, the dominant role of rough surfaces in dissociative chemisorption and catalytic activity and the structure insensitivity of a class of catalyzed reactions.

In the beginning of the century, one of the major thermodynamic puzzles, the discrepancy in the equilibrium constant of ammonia as determined by Haber and Nernst has induced the rapid development of physical chemistry(1). The data points reported by these two giants of chemistry are shown in Figure 1. The equilibrium constant as determined in their laboratories was very different and the search for the reasons of the discrepency lead to the development of low temperature heat capacity measurements, commerical high pressure reactor technology and to the accelerated development of catalytic research and catalyst based chemical technologies. The use of catalysts and the establishment of high pressure reactor technology was needed to shift the equilibrium between nitrogen, hydrogen and ammonia.

0097–6156/90/0428–0218$06.25/0
© 1990 American Chemical Society

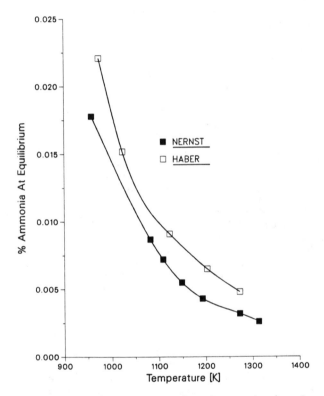

Figure 1.   The equilibrium constant for the synthesis of ammonia from nitrogen and hydrogen as determined by Nernst and Haber as a function of temperature.

Modern surface chemistry and heterogeneous catalysis have many
puzzles that await solution. Even the use of single crystal
surfaces shown in Figure 2 and characterization techniques that
provide information about the atomic structure, composition and
oxidation state of surface atoms on the molecular level have not
provided solutions to these puzzles as yet. One of the most
interesting puzzles is the need for 1) thermal activation to break
large binding energy chemical bonds. Discovery of this phenomenon
was in the 1930's when dinitrogen was adsorbed on iron and its low
heat of adsorption was measured. As the temperature was increased
above around 130K a large exotherm was observed indicating
dissociative adsorption of dinitrogen(2). That is, at this
temperature molecular nitrogen dissociated to make iron–nitrogen
bonds as the atomic nitrogen chemisorbed. Similar observation for
the thermal activation of chemisorbed hydrocarbons is shown in
Figure 3. Ethylene, propylene and higher olefins adsorbed on
platinum decompose upon heating(3). Hydrogen desorbs, leaving a
partially dehydrogenated organometallic fragment on the metal
surface. The decomposition occurs sequentially at well-defined
temperatures. Why strong 100-200 kcal chemical bonds break by
changing the temperature a few degrees for a given chemisorbed
molecule on a given surface is one of the puzzles of modern surface
chemistry. While the molecular structure before and after the
thermal decomposition is well characterized (as shown for ethylene
in Figure 4), the reasons for this transition and the mechanism for
its occurrence are not understood.

     2) Rough Surfaces Do Chemistry. Another puzzle is the way
rough surfaces do chemistry, both bond breaking and catalysis. The
rougher, more open the surface is the more likely that it breaks
chemical bonds and at a lower temperature. Figure 5 shows three
crystal faces of platinum, the flat, stepped and the kinked
surface. Figure 6 shows the stepwise decomposition of ethylene on
the flat (111) surface of nickel as compared to the stepped surface
of nickel(4). The same decomposition pathway is found on both of
these surfaces. However, on the stepped surface decomposition
occurs at a much lower temperature, below 150K, as compared to the
nickel (111) face which shows chemical bond breaking only at about
230K. The activity of rough surfaces can perhaps be best
demonstrated by studying the behavior of hydrogen on flat, stepped
and kinked surfaces of platinum. When hydrogen is adsorbed on these
surfaces and then a temperature programmed thermal desorption
experiment is carried out, it is found that hydrogen desorbs at the
maximum rate at the lowest temperature from the flat surface while
from stepped and kink surfaces desorption occurs at higher
temperatures (Figure 7). The higher temperature desorption peaks
are associated with the hydrogen desorption from steps and kinks
respectively(5). Similarly, the catalytic H_2/D_2 exchange also
occurs by at least an order of magnitude higher reaction probability
at steps as compared to the flat surface(6). This can be studied by
a molecular beam surface scattering process where a mixed H_2+D_2
beam is scattered from a stepped platinum surface or a flat platinum
surface and the resulting HD that forms upon single scattering is
measured. The presence of HD clearly indicates dissociative
adsorption and reaction of hydrogen on the metal surface. As Figure

Figure 2. Small area (1 cm^2) single crystal surface that is used in surface science studies as well as a model heterogeneous catalyst. (Reproduced with permission from Lawrence Berkeley Laboratory.)

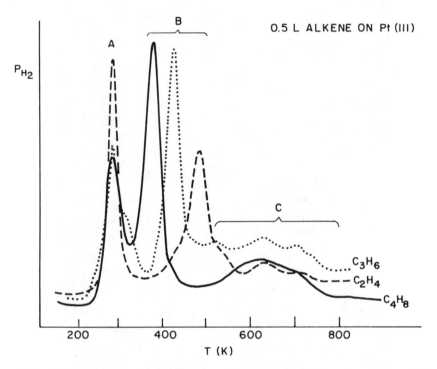

Figure 3. Thermal desorption spectra for small alkenes adsorbed on the platinum (111) surface showing sequential hydrogen evolution and decomposition. (Reproduced from ref. 3. Copyright 1982 American Chemical Society.)

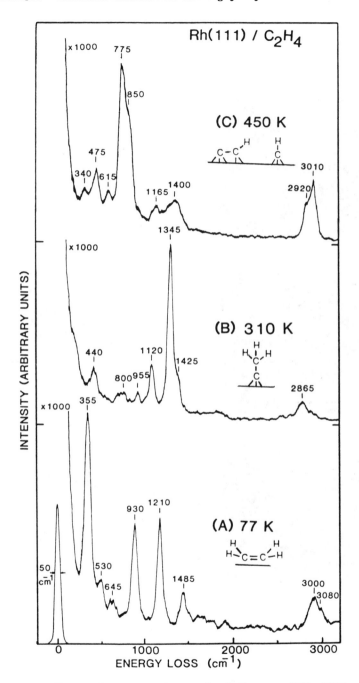

Figure 4.   The surface structures of ethylene at 77K, 310K and 450K as determined by high resolution electron energy loss spectroscopy and low energy electron diffraction.

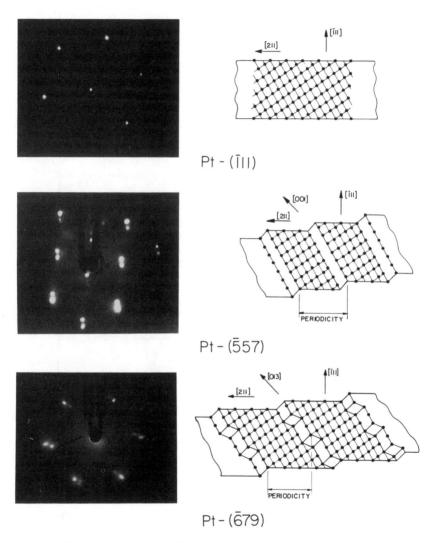

Figure 5. The flat stepped and kinked surfaces of platinum. (Photo credit: Lawrence Berkeley Laboratories. Reproduced with permission from ref. 5. Copyright 1980 Elsevier Science Publishers.)

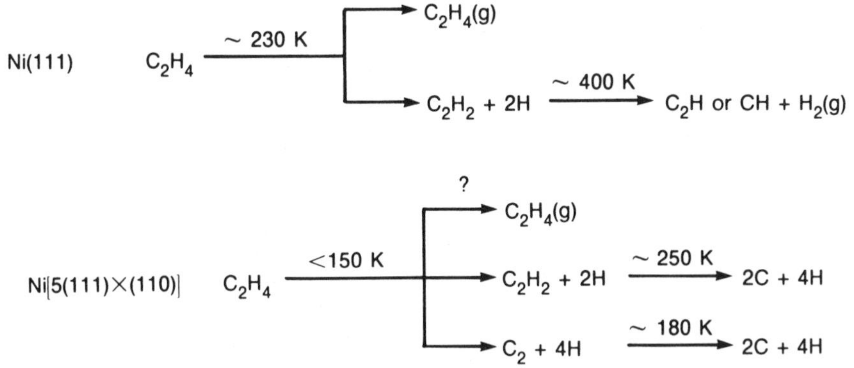

Ni(111)
C2H4
~ 230 K
C2H4(g)
C2H2 + 2H
~ 400 K
C2H or CH + H2(g)
?
C2H4(g)
Ni[5(111)×(110)]
C2H4
<150 K
C2H2 + 2H
~ 250 K
2C + 4H
C2 + 4H
~ 180 K
2C + 4H

Figure 6. Scheme of the decomposition pathway of ethylene on the nickel (111) and the stepped nickel [5(111)x(110)] single crystal surfaces.

Pt (12, 9, 8)
50 L
0.5 L
1 L
Pt (557)
Pt (111)
H2 Partial Pressure (a.u.)
0
200
400
Temperature (°C)

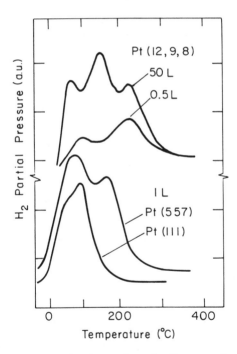

Figure 7. Thermal desorption of hydrogen from the flat stepped and kinked platinum single crystal surfaces. (Reproduced with permission from ref. 5. Copyright 1980 Elsevier Science Publishers.)

8 indicates, the reaction probability is nearly 90% from a stepped surface, especially when the beam impinges on the open step edges. While from the (111) face the reaction probability is at least an order of magnitude lower, recent studies by Comsa(7) and his co-workers indicate that the reaction probability from the flat (111) platinum surface can be three orders of magnitude lower, once the uncontrolled defects are removed from the surface. Thus, the reactivity of the (111) face, if any, is due to uncontrolled defects and not to the ordered (111) crystal face. This reaction, of course, is catalytic and as it is shown in Figure 9 catalytic reactions require many turnovers. Long surface residence times would be detrimental to catalytic activity. Yet, we find that hydrogen chemisorption yields more strongly bound hydrogen atomic species at stepped surfaces of platinum which are the same sites that also carry out catalytic reactions. This is a contradiction since the strongly bound hydrogen has long surface residence times and therefore should have low turnover rates. This is another puzzle of surface chemistry, why rough surfaces do both strong chemisorption of species of high binding energy as well as exhibit high catalytic activity.

The Rigid Lattice Model; Its Inadequacy and the Reasons for it. The atomic model of a surface that was frequently utilized in the past is shown in Figure 10. In this model, the surface is pictured as rough and the various surface sites are distinguishable by the number of their nearest neighbors(8). There are steps, kinks and terrace sites along with ad-atoms and vacancies. There is experimental evidence for the existence of all of these species. However, the model assumes that the lattice is rigid. That is, all the atoms are in their bulk-like equilibrium positions that could be obtained by extrapolating the bulk structure to that surface. As we shall demonstrate, the surface restructures readily and this rigid lattice model is incorrect. When we carry out studies of the structure of adsorbates and their surface chemical bonds, we tacitly assume that the substrate structure remains unchanged during adsorption. The usual reason for the assumption of the rigid lattice under chemisorption conditions is that most of the surface science techniques that provide us with molecular level information about the structure of the surface are sensitive only to the top monolayer and provide little or no information about the second layer under the surface. As a result, it has been most comforting to assume that the second layer or the substrate remains unchanged during chemisorption in the absence of experimental proof to the contrary. However, recently there are many reports of low energy electron diffraction surface crystallography studies that can determine the bond distance and bond angles down to three atomic layers under the surface. These studies provide clear-cut evidence for the restructuring of surfaces under chemisorption conditions.

To demonstrate the insensitivity of some of the most modern surface science techniques to the structure of the second layer under the adsorbed molecule, Figure 11a shows a scanning tunneling microscope picture of a monolayer of graphite adsorbed on a platinum (111) surface. The graphite structure is clearly visible, but there is no information on the platinum atoms underneath the graphite.

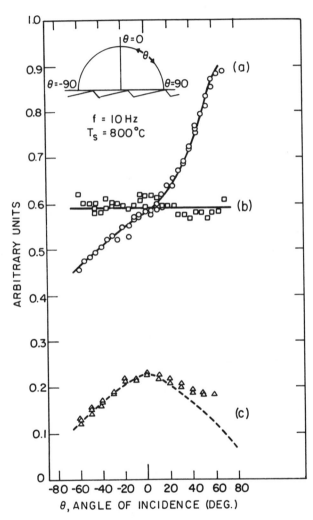

Figure 8. The hydrogen–deuterium exchange on the stepped and flat platinum surfaces. The reaction probability is much higher on the stepped metal surface. (Reproduced with permission from ref. 20. Copyright 1977 American Physical Society.)

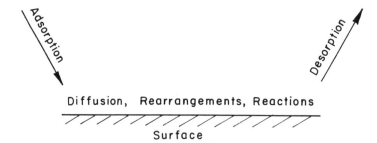

Figure 9. A scheme of catalytic reaction that require many turnovers.

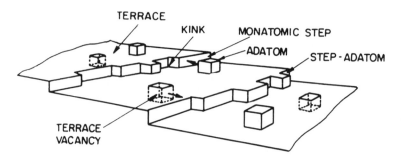

Figure 10. Model of a heterogeneous solid surface depicting different surface sites. These sites are distinguishable by the number of nearest neighbors. (Reproduced with permission from ref. 5. Copyright 1980 Elsevier Science Publishers.)

When a low energy electron diffraction-surface crystallography study
is carried out on the same system, a second layer of carbon can be
detected under the graphitic overlayer (Figure 11b). In this second
layer the carbon atoms are in the troughs and atomic crevices
provided by the metallic substrate and the metal surface atoms also
occupy positions different from their equilibrium bulk-like
positions.

Over the past twenty years over 250 surface structures have
been solved, most of them by low energy electron diffraction-surface
crystallography(9). From these studies an entirely different model
of the structure of surfaces emerged on the atomic scale. This
model indicates a dynamic structure, a structure where the atomic
positions of the surface atoms are different from that predicted by
the rigid lattice model when the surface is clean and change again
when chemisorption occurs.

The Dynamic Lattice Model of Surfaces. The Experimental Facts. Let
us now review what is known about the structure of clean surfaces
without adsorbates. There are three phenomena that have been
identified; relaxation, reconstruction and the presence of steps and
kinks.

Relaxation and Reconstruction at Clean Surfaces. The surface atoms
are pulled inward at the clean metal surface providing an interlayer
spacing between the first and second layer that is shorter than the
interlayer distances between subsequent layers. This inward
relaxation is increased greatly with increasing surface roughness.
In the work by Marcus Jona(10) roughness was defined as one over the
packing density. It is clear that very large relaxations occur at
open and rough surfaces. The first layer relaxation is always
inward while the second layer could be either inward or outward. In
each case, however, the interlayer spacing between the first layer
and second layer of atoms is much shorter than subsequent interlayer
spacings. Large relaxation leads to reconstruction and a few
reconstructed surfaces are shown in Figures 12 (a) and (b). Figure
12a shows the restructured silicon (100) surface(11) that shows
silicon atoms as dimer like species on the surface separated by
large troughs. Figure 12b shows the missing row reconstruction of
the Iridium (110), face where a whole row of atoms is periodically
missing(12).

Stepped Surfaces. Scanning tunneling microscope studies show
readily the presence of steps even on flat surfaces. Surface
crystallography studies by low energy electron diffraction show that
instead of the rigid step model there are large relaxations of step
atoms both perpendicular and parallel to the surface that leads to
the formation of clusters of atoms at the step edges. It appears
that, perhaps, cracks may propagate into the solid from such step
edges.

Adsorbate Induced Restructuring. Perhaps, the most striking
observation of recent years is the adsorbate induced restructuring
of surfaces. This can be demonstrated by the restructuring of the
nickel (100) face(13) in the presence of half a monolayer of carbon

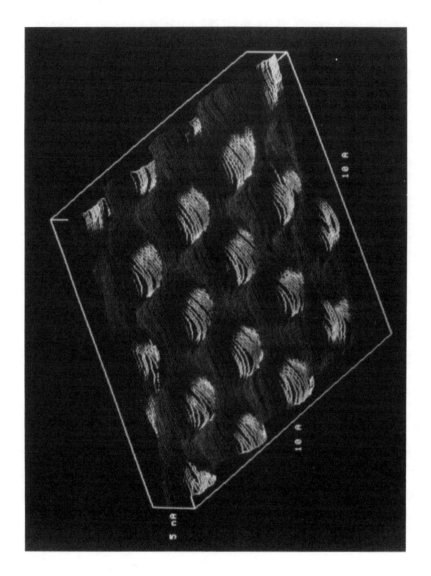

Figure 11a. A scanning tunneling microscope image of a graphite adsorbed layer on the platinum (111) crystal face. (Reproduced with permission from Lawrence Berkeley Laboratory.)

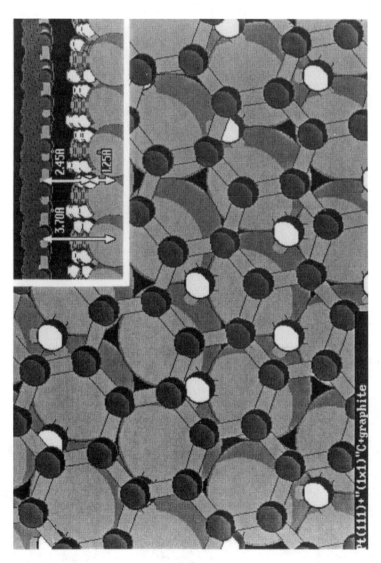

Figure 11b. Low energy electron diffraction surface crystallography determination of the structure of a graphite adsorbed layer on the platinum (111) crystal face. The presence of two layers of carbon is clearly visible. (Reproduced with permission from Lawrence Berkeley Laboratory.)

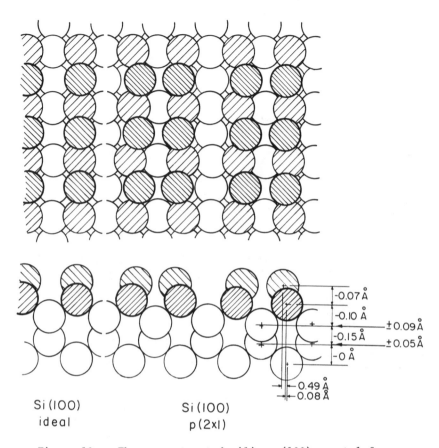

Si (100)
ideal

Si (100)
p (2x1)

Figure 12a.   The reconstructed silicon (100) crystal face.

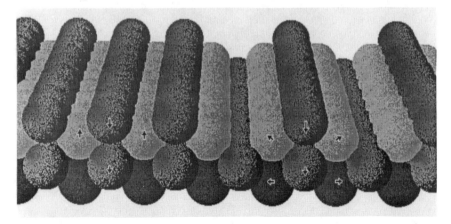

Figure 12b.   The reconstructed iridium (110) crystal face exhibiting missing row restructuring. (Reproduced with permission from Lawrence Berkeley Laboratory.)

(Figure 13). When carbon adsorbs it occupies a four-fold site. In addition, the nearest neighbor metal atoms are displaced outward around the carbon atom providing a possibility for the carbon atom to sink more deeply into the surface and make bonds to the second layer of nickel atoms. This expansion leads to a large compression between nearest neighbor and next nearest neighbor nickel atoms which can be relieved by a small rotation of the nickel atoms around the carbon atoms. This is shown in Figure 13. Such an expansion and rotation of the nickel atoms are associated with the chemisorption of carbon and are induced by it. When the carbon is removed by using hydrogen to form methane, the metal atoms rotate back to their clean four-fold lattice positions. Such restructuring occurs within the time scale of adsorption, milliseconds or less and this is the phenomenon we call adsorbate induced restructuring.

Another adsorbate induced restructuring may be demonstrated during the adsorption of sulfur on the iron (110) surface(14). (Figure 14) In this case the iron atoms cluster around the sulphur atoms providing larger iron-iron distances away from the sulfur atom, among the next nearest neighbors. This clustering may be viewed as producing Fe_4S species. The removal of sulfur re-establishes the original clean iron (110) surface structure.

The missing row structure can be stabilized by the presence of a small amount of sulfur on the iridium (110) surface where the sulfur atoms are placed in bridge sites on the side of the troughs created by the missing row(13) of iridium in the top layer (Figure 15). Thus, chemisorption restructures the substrate drastically. This is not suprising since the chemisorption bond is as strong as the metal-metal bonds. As a result, the restructuring may be induced by substituting the adsorbate-substrate bond for the strong bonds between the substrate atoms.

The adsorbate induced restructuring processes that occur upon chemisorption ($\sim 10^{-3}$ sec), on time scales equal to that of catalytic reactions ($\sim$seconds) or longer atom transport controlled time scales ($\sim 10^3$ sec) indicate that adsorbate induced restructuring is perhaps one of the key phenomenon in surface science that holds the promise to explain many of the thermodynamic puzzles of surface chemistry. It is proposed that the thermally activated bond breaking of adsorbed molecules may be induced by a phase transformation, perhaps first order, that causes the restructuring of the surface which induces bond breaking. In turn the bond formation upon adsorption induces restructuring. It is proposed that rough surfaces are more reactive because they restructure more readily when chemisorption occurs. Having fewer nearest neighbor their relaxation and reconstruction when clean and in the presence of adsorbates in more drastic and more facile. Adsorbate induced restructuring or surface restructuring induced bond breaking and catalytic processes is proposed to be one of the missing links in our understanding of many phenomena in surface science and catalysis.

Catalytic Reactions

Adsorbate induced restructuring can also explain some of the important properties of catalytic reactions. The classification of

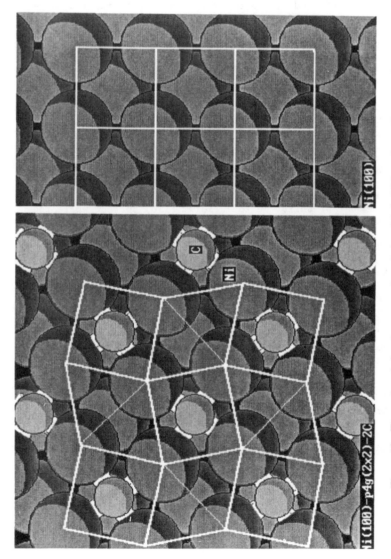

Figure 13. The carbon induced restructuring of the nickel (100) surface. (Reproduced with permission from Lawrence Berkeley Laboratory.)

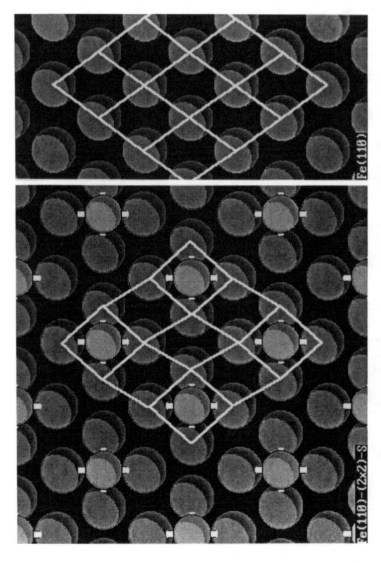

Figure 14. The sulfur induced restructuring of the iron (110) surface. (Reproduced with permission from Lawrence Berkeley Laboratory.)

Figure 15. The sulfur induced restructuring of the iridium (110) surface. (Reproduced with permission from Lawrence Berkeley Laboratory.)

catalytic reactions into two groups, structure sensitive and structure insensitive reactions is one of the most widespread and useful distinctions (15-16). Ammonia synthesis that has also been carried out on single crystal surfaces clearly belongs to the structure sensitive reaction class(17). The (111) and (211) surfaces of iron are orders of magnitude more active for producing ammonia from nitrogen and hydrogen then the close packed (110) iron surface. While the structure of the active site is not well established, it is clear that it can be correlated with the structure of the clean metal surface before the reaction.

Ethylene hydrogenation is an example of a structure insensitive reaction(18). This is much more difficult to understand from surface science studies. It is known that the catalytically active surface is covered with a monolayer of ethylidyne C_2H_3, species which do not participate directly in the reaction(19). It is thought that the ethylidyne adsorption restructures the metal surface by providing perhaps somewhat larger intermetallic distance among platinum atoms in between the ethylidyne metal cluster moiety. These adsorbate created metal sites are the active centers where ethylene hydrogenation occurs with a high turnover without the direct participation of the strongly chemisorbed ethylidyne species. The metal surface creates it own active sites upon the chemisorption of the first monolayer of reactants. These sites did not exist on the uncovered clean metal surface. The reaction is structure insensitive as the active site concentration that is created by chemisorption depends on the coverage of the adsorbates and not on the catalyst surface structure prior to the chemical reaction. It appears that surface restructuring induced bond breaking and catalytic processes are likely to be at the root of many of the puzzles, thermodynamic and dynamic, of surface chemical processes.

Acknowledgment

This work was supported by the director, Office of Energy Research, Office of Basic Energy Sciences, Materials Sciences Division, of the U.S. Department of Energy under Contract No. DE-AC03-76SF00098.

Literature Cited

1.   Topham, S. A. In Catalysis; Anderson, J. R.; Boudart, M., Eds.; Springer Verlag Berlin 1985; Vol. 7, Chpt. 1.
2.   Emmett, P. H. In The Physical Basis for Heterogeneous Catalysis; Drauglis, E.; Jaffee, R. I., Eds.; Phenum, New York 1975; p.3.
3.   Salmeron, M.; Somorjai, G. A. J. Phys. Chem. 1982, 86, 341.
4.   Lehwald, S.; Ibach, H. Surf. Sci. 1984, 89, 42.
5.   Davis, S. M.; Somorjai, G. A. Surf. Sci. 1980, 91, 73.
6.   Salmeron, M.; Gale, R. J.; Somorjai, G. A. J. Chem. Phys. 1979, 70 (6), 2807.
7.   Poelsema, B.; Brown, L. S.; Lenz, K.; Verheji, L. K.; Comsa, G. Surf. Sci. 1986, 171, L395.
8.   Somorjai, G. A. Chemistry in Two Dimensions: Surfaces; Cornell University Press, 1981.

9. MacLaren, J. M.; Pendry, J. B.; Rous, P. J.; Saldin, D. K.; Van
    Hove, M. A.; Wedensky, D. D. A Handbook of Surface Structures;
    Reidel, D. Publ. Co., Holland 1987.
10. Jona, F.; Marcus, P. M. In The Structure of Surfaces II; van
    der Veen, J. F.; Van Hove, M. A., eds.; Springer, Heidelberg
    1988.
11. Holland, B. W.; Duke, C. B.; Paton, A. Surf. Sci. 1984, 140,
    L269.
12. Chan, C. M.; Van Hove, M. A. Surf. Sci. 1987, 183, 303.
13. Onuferko, J. H.; Woodruff, D. P.; Holland, B. W. Surf. Sci.
    1979, 87, 357.
14. Shih, H. D.; Jona, F.; Jepsen, D. W.; Marcus, P. M. Phys. Rev.
    Letters, 1981, 46, 731.
15. Somorjai, G. A. In The Building of Catalysts: A Molecular
    Surface Science Approach, in Catalyst Design, Progress and
    Perspectives; John Wiley & Sons, New York, 1987, Chapter 2
    11-69.
16. Nix, R. M.; Somorjai, G. A. In Perspectives in Quantum
    Chemistry; Jortner/Pullman, 1988.
17. Spencer, N. D.; Schoonmaker, R. C.; Somorjai, G. A. J. Catal.
    1982, 74, 129.
18. Zaera, F.; Somorjai, G. A. J. Am. Chem. Soc. 1984, 106 (8)
    2288.
19. Godbey, D.; Zaera, F.; Yeates, R.; Somorjai, G. A. Surf. Sci.
    1986, 167, 150-166.
20. Gale, R.J.; Salmeron, M.; Somorjai, G.A.; Phys. Rev. Letters, 1977, 38, 1027.

RECEIVED December 7, 1989

Chapter 16

Acid Sites on Chemically Modified Molybdenum Surfaces

Peter C. Stair

Department of Chemistry, Northwestern University, Evanston, IL 60208

Modification of metal surfaces by the introduction of
electronegative atomic adsorbates forms Lewis acid sites
associated with electron deficient metal atoms at the
surface. The strength and hard/soft character of the acid
sites depend on the coverage and identity of the electronegative modifier. These effects are evident in surface
electron spectroscopic measurements and in the experimental results from a variety of surface/adsorbate systems.
They are shown most clearly by quantitative thermal
desorption measurements of NH_3, PH_3, CH_3OCH_3, $CH_2=CH_2$, and
$CH_2=CHCH_3$ adsorbed on C, O, and S modified Mo(100) surfaces. Trends in desorption activation energies are
consistent with increasing Lewis acid strength with the
coverage of the modifier and with the electronegativity of
the modifier at a fixed coverage. The effect of steric
blocking by sulfur is also clearly revealed. The ordering
in desorption activation energies among the probe molecules indicates the acid sites are hard (not soft) in
character despite the metallic nature of the adsorbing
surface. Product selectivities in methylcyclopropane
hydrogenolysis are consistent with the conclusions based
on physical and chemisorption experments.

Much of surface science research to date has focussed on the physical and chemical properties of clean metal surfaces, a state of
matter that is only obtainable under ultrahigh vacuum. However,
under practical, real world conditions most metals are covered by an
oxide layer or take the form of various compounds, eg. sulfide,
carbide, etc. For the last several years my research group has
investigated the properties of "chemically modified" molybdenum
surfaces which serve as models for the surface of molybdenum compounds. Surfaces that are models for the oxides, carbides, sul-
fides, and borides of molybdenum are fabricated by the reaction,

0097–6156/90/0428–0239$06.00/0

respectively, of oxygen, ethylene, H_2S, and B_2H_6 with a clean surface in ultrahigh vacuum followed by annealing to order the surface
and remove hydrogen.

The geometric and electronic structure of the metal surface are
modified by the introduction of electronegative elements such as
oxygen, carbon, and sulfur. Geometric modifications may involve
surface reconstruction and/or steric blocking of metal atoms. Electronic structure changes include transfer of valence electron
density between surface atoms, shifts in the electron population
among the atomic orbitals of a given surface atom, as well as
modifications in the energy and spatial distribution of HOMO and
LUMO states. The chemical consequences of the electronic structure
changes are described using the traditional chemical concept of
Lewis acids and bases as discussed in reference ([1]). These physical
and chemical effects of surface modification can be codified into a
series of qualitative rules based upon physical and chemical intui-
tion.

1. Electronegative modifiers produce electron deficient metal
atoms (cations) and electron rich adatoms (anions).

2. The degree of electron transfer away from surface metal
atoms increases with modifier coverage and/or
electronegativity.

3. The surface polarizability decreases with increasing
modifer coverage as the surface region transforms from metal to
metal compound.

4. The metal cations are Lewis acid sites, and the adatom
anions are Lewis base sites.

5. The strength of acid sites increases with modifier coverage
and/or electronegativity. The strength of base sites decreases
with modifer coverage.

6. The acid sites become harder in character with increasing
modifer coverage.

The first three rules describe the changes in surface electronic
structure which are responsible for the appearance of acidic and
basic sites. The identity of these sites, their strength as a
function of modifer coverage and their hard/soft character are
governed by the last three rules.

The present paper reviews the physical and chemical evidence
for the above rules obtained over the last several years from
ultrahigh vacuum surface science studies of molybdenum single
crystals chemically modified by O, C, S, and B. Additionally, the
results of recent studies of methylcyclopropane hydrogenolysis will
be presented which illustrate the influence of surface acid/base
sites on catalytic hydrocarbon conversions. The surface coverage of
each modifier was determined by quantitative Auger electron spectroscopy or x-ray photoelectron spectroscopy (XPS). The atomic structure of oxygen, carbon, and sulfur adlayers below one monolayer (ML)

coverage was established by low-energy ion scattering spectroscopy (2,3). The effective oxidation state of surface molybdenum atoms on oxygen, carbon, boron and a mixed carbon and oxygen modified surface was assigned by angle resolved XPS measurements of Mo(3d) core level shifts (4,5). The acid/base properties of chemically modified Mo(100) surfaces were investigated by temperature programmed desorption of a series of chemisorbed Lewis bases (6,7). The molecules used were hard Lewis bases, ammonia - NH_3 and dimethylether - CH_3OCH_3, and soft Lewis bases, phosphine - PH_3, ethylene - $CH_2=CH_2$, and propene - $CH_3CH=CH_2$.

Surface Geometry

At coverages less than or equal to $10^{15}/cm^2$, O and C adatoms are located in the four-fold hollow of an essentially unreconstructed Mo(100) surface at a distance of .3 - .5 Å above the top layer of molybdenum atoms (2). Above $10^{15}/cm^2$ of atomic oxygen the coverage is in excess of monolayer capacity. The structure is unknown but must involve formation of a three dimensional surface phase. The geometry of sulfur overlayers on Mo(100) has been discussed by Clarke (8) and by Salmeron et. al. (9). At coverages below 0.5 ML sulfur adatoms are located on four-fold hollow sites as was found for carbon and oxygen. The position perpendicular to the surface is higher than for the smaller adatoms, ~1 Å. Above 0.5 ML the situation is less clear. Both four-fold hollow and two-fold bridge positions have been proposed. However, the important structural parameter needed to interpret chemisorption and catalytic results on sulfur modified surfaces is the height of the adlayer. In either structural model the perpendicular distance between the sulfur layer and the top molybdenum layer is larger than the corresponding distance determined for carbon and oxygen adlayers.

Effective Molybdenum Oxidation State

XPS measurements of chemically modified metal surfaces contain contributions from both surface and bulk metal atoms. The component due to surface molybdenum atoms was determined using a procedure established by Citrin et. al. (10) which is based on the changing contribution of surface and bulk components to the measured signal as a function of photoemission takeoff angle. The surface

TABLE I $Mo(3d_{5/2})$ Surface Binding Energies and Oxidation States

Modifier	Coverage $10^{15}/cm^2$	BE (eV)	Oxidation State
Clean		227.5	0.0
B	1.1	227.6	0.4
C	1.0	227.7	0.9
β-CO	1.0	227.8	1.3
O	0.3	227.8	1.3
O	0.8	227.8	1.3
O	1.0	227.9	1.8
O	1.2	228.0	2.2
O	1.4	228.4	4.0

molybdenum atom core levels shift to higher binding energy with
respect to the clean surface for all adlayers studied as shown by
the data in Table I. The surface designated β-CO was a 50:50
mixture of atomic carbon and oxygen produced by the dissociative
adsorption of CO. Three aspects of the results are of significance
for understanding the chemistry of the chemically modified molybdenum surfaces. First, the surface Mo core level binding energy
increases monotonically with increasing oxygen coverage indicative
of progressive molybdenum oxidation. Second, it is evident that the
most rapid increase in molybdenum binding energy occurs between 1.0
and 1.4 x 10^{15}/cm^2 oxygen coverage as indicated by the plot in
Figure 1. This is due to formation of a thin MoO$_2$ layer on the
surface (11).

The final point of interest concerns the effective oxidation
states assigned to molybdenum atoms on the chemically modified
surfaces. The adlayer at 1.4 ML oxygen coverage is MoO$_2$ (see below), therefore, the molybdenum atoms in the layer are assigned an
oxidation state corresponding to this oxide, namely +4. Effective
oxidation states are assigned on the rest of the surfaces by assuming a linear relationship between surface binding energy shifts and
oxidation state. An important observation in connection with the
assigned oxidation states for different adlayers at constant 1 ML
coverage, is the linear increase in surface oxidation state with
increasing modifier electronegativity (see Figure 2). Linearity
implies a direct relation between the degree of electron transfer
from metal to adsorbate and the adsorbate electronegativity. This
relationship can be understood as a consequence of the electronegativity equalization postulate of Sanderson (12) which states that
when two or more atoms bond to each other electron transfer adjusts
their electronegativities to some common intermediate value.

Surface Polarizability

The shift in measured by XPS core level binding energies between a
rare gas such as xenon in the gas phase and adsorbed on a surface
results from a combination of chemical shift, local potential at the
site of the adsorbate, and stabilization of the photoionization core
hole by polarization of the substrate electron density. As discussed in reference (13), the contribution due to substrate polarization is related to the surface electronic polarizability and can be
isolated from the other contributions to a good approximation by
measurements of the xenon gas phase and adsorbed phase "Auger
parameter", α. α is defined as the difference between the (jkl)
core level Auger electron kinetic energy, $K^{Auger}(jkl)$, and the (j)
core level photoelectron kinetic energy, $K^{PE}(j)$.

$$\alpha \equiv K^{Auger}(jkl) - K^{PE}(j)$$

The difference in measured values of α for xenon in the gas phase
and on the surface is equal to twice the polarization relaxation
energy of the substrate, E_R^{PE}. The details of this argument can be
found in reference (13).

$$\Delta\alpha = \alpha(ads) - \alpha(gas) = 2E_R^{PE}$$

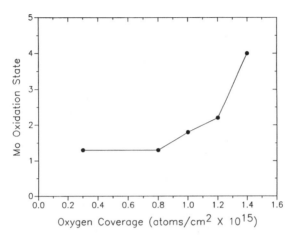

Figure 1. Effective Mo oxidation state vs. oxygen coverage.

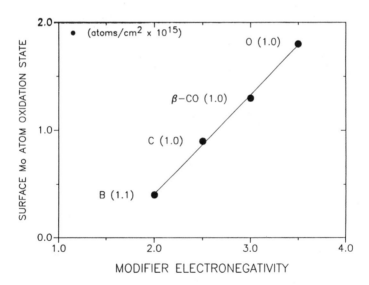

Figure 2. Effective Mo oxidation state vs. modifier electronegativity.

Measured values of E_R^{PE} for xenon adsorbed on the Mo(100) surface with varying coverages of atomic oxygen are plotted in Figure
3. The relaxation energy and hence the surface polarizability
decreases monotonically with increasing oxygen coverage with a
pronounced break in the slope at a coverage corresponding to one
monolayer. At 1.4 to 1.5×10^{15} atoms/cm^2, the surface relaxation
energy has reached the value measured for xenon adsorbed on bulk
MoO$_2$. The data is interpreted as evidence for 1) the presence of a
chemisorbed oxygen phase below 10^{15} atoms/cm^2 with a surface polarizability characteristic of a metal and 2) the formation of oxide
above 10^{15} atoms/cm^2 which is completed at an oxygen coverage of
1.5×10^{15} atoms/cm^2.

Acid Properties of Chemically Modified Mo(100)

Temperature programmed desorption measurements of desorption activation energies were performed for ammonia, dimethylether, phosphine,
propene, and ethene adsorbed on the five surfaces listed in Table
II. In each case the dose of molecules was selected such that the
desorption spectrum contained only a single desorption peak. Longer
doses resulted in a variety of lower temperature desorption peaks.
Thus the reported activation energies correspond only to the most
strongly chemisorbed surface species. With few exceptions the
molecules adsorbed associatively on these surfaces (no decomposition). For the exceptions, (molecule/surface) PH$_3$/1.0×10^{15} O,
PH$_3$/1.0×10^{15} C, NH$_3$/1.5×10^{15} O, NH$_3$/1.0×10^{15} O, only the parent molecule desorption spectrum was analyzed to obtain an activation
energy.

Thermal desorption spectra were measured at four heating rates
ranging from 3 K/sec to 150 K/sec. The spectra were analyzed by
finding desorption rates and surface temperatures at several common

TABLE II   Mo(100) surfaces used for chemisorption experiments

Surface	Adatom Positions	Mo Oxdn. State	Comments
1.0 ML O	4-fold hollow .3 Å above Mo	1.8	Chemisorbed Phase
1.2 ML O	-	2.2	Mix of Chem. Phase and Oxide
1.5 ML O	disordered	4.0	Oxide
1.0 ML C	4-fold hollow .3 Å above Mo	0.9	Chemisorbed Phase
0.8 ML S	4-fold hollow + Bridged 1 Å above Mo	-	Chemisorbed Phase

coverages on each desorption curve. This data was used to construct
an Arrhenius plot of log[desorption rate] vs. 1/T at each coverage
(14). Desorption activation energies and prefactors were calculated
from the slopes and intercepts of these plots. No significant
coverage dependence of the desorption kinetic parameters was observed. Therefore, a single value of the activation energy is
sufficient for purposes of comparing the various molecule/surface
combinations.

The measured desorption activation energies are summarized in
Figure 4. Both hard and soft Lewis bases, NH_3, CH_3OCH_3, PH_3,
CH_3CHCH_2, and CH_2CH_2, exhibit increasing desorption activation energies with increasing oxygen coverage. This trend reflects an increase in the Lewis acidity of the surface molybdenum atoms with
increasing oxygen coverage and is consistent with the oxidation
state changes measured by XPS. Desorption activation energies
measured for all the molecules adsorbed on the sulfur modified
surface are much lower than observed for any of the other surfaces.
This effect cannot be explained by electron transfer from molybdenum
to sulfur else a similar effect would be observed with the oxygen
modified surfaces. Rather, the weak bonding to sulfur modified
Mo(100) must be due to steric blocking of the molybdenum atoms by
the sulfur overlayer. The greater influence of steric blocking with
sulfur compared to the carbon and oxygen modifiers can be attributed
to its increased size and the larger sulfur to molybdenum layer
spacing (see Table II). Furthermore, this result suggests that
bonding of all the molecules to the modified surfaces involves
coordination to molybdenum rather than to the modifier adatoms.
Otherwise, bonding to the sulfur modified surface would be enhanced
by the greater accessibility to the sulfur layer.

The hard bases are more strongly bonded to the oxygen modified
surfaces than the soft bases. In fact, the desorption activation
energies through the series of molecules studied show a linear
correlation with the measured proton affinities (15) as shown in
Figure 5. This suggests that the Lewis acid sites on all of the
oxygen modified surfaces should be classified as hard, ie. the empty
acceptor orbitals associated with the molybdenum cations are spa-
tially localized and nonpolarizable. The slope of the correlation
line in Figure 5 decreases with decreasing oxygen coverage indicating that the surface electronic structure softens at lower oxygen
coverages. This chemical trend agrees with the physical measurements of surface polarizability discussed above.

Comparison of desorption activation energies for the molecules
adsorbed on 1.0×10^{15} atoms/cm^2 carbon and oxygen modified surfaces
provides a test for the effect of surface electronic structure
changes with little or no change in the surface geometry. The
activation energies are lower for desorption from the carbon modified surface than from the oxygen modified surface. This trend
supports the model for surface Lewis acidity and is in agreement
with the measured molybdenum oxidation states. The corresponding
activation energy increase observed for PH_3 between O/Mo(100) and
C/Mo(100) is not consistent with the acid/base bonding model but may
be an artifact of the partial decomposition which accompanies PH_3
adsorption on these surfaces, ie. the surface from which desorption
occurs has an altered composition.

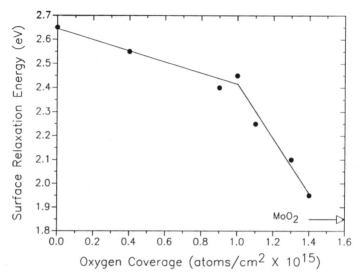

Figure 3.   Surface relaxation energy vs. oxygen coverage.   The arrow indicates the relaxation energy measured for bulk MoO₂.

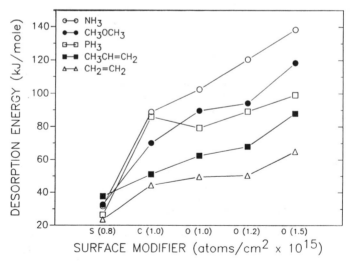

Figure 4.   Desorption activation energies for various probe Lewis bases on chemically modified Mo surfaces.

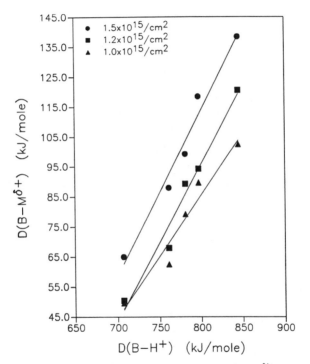

Figure 5. Desorption activation energies, D(B-Mo$^{\delta+}$), vs. proton affinities, D(B-H^{+}).

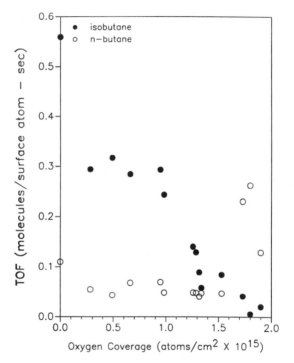

Figure 6. Turnover frequency for isobutane formation (filled circles) and n-butane formation (open circles) vs. oxygen coverage.

Methylcyclopropane Hydrogenolysis

The physical and chemical properties of the oxygen modified molybdenum surfaces described above indicate the formation of acidic
sites with variable strength and hard/soft character as a function
of oxygen coverage. The hydrogenolysis of methylcyclopropane (MCP)
was investigated to probe the catalytic properties of these surfaces. A full account of this study will appear elsewhere (Touvell,
M. S.; Stair, P. C. J. Catal. submitted).
     MCP may undergo single hydrogenolysis (ring opening) to form
n-butane or isobutane or double hydrogenolysis to form either equal
amounts of methane and propane or two molecules of ethane. The
formation of isobutane is taken as evidence for reaction catalyzed
by metal sites whereas n-butane formation is characteristic of
acidic sites (16). n-butane product is not favored by metal catalyzed hydrogenolysis due to the electron donating methyl substituent
which strengthens the adjacent bonds and weakens the bond opposite
the substituent (17). Production of isobutane by an acid catalyzed
reaction is unfavorable due to the required formation of a primary
carbenium ion intermediate (18).
     MCP hydrogenolysis was carried out over oxygen modified Mo(111)
surfaces with varying oxygen coverages. The reaction conditions (5
torr, MCP; 755 torr, H_2; 100°C) were similar to those reported for
MCP hydrogenolysis catalyzed by low valent Mo supported on alumina
(16). The turnover frequencies per surface Mo atom as a function of
oxygen coverage are plotted in Figure 6. Note that for oxygen
coverages below $10^{15}/cm^2$ isobutane is the major product while above
$10^{15}/cm^2$ the major product is n-butane. These results are indicative of catalytic function which is predominantly metallic at low
oxygen coverage and acidic at coverages greater than $10^{15}/cm^2$.
     The variation in catalytic function with increasing oxygen
coverage is in excellent agreement with surface characterization by
molybdenum core level shifts and surface polarizability measurements. Clearly the physical measurements were performed on the
actual catalytically active sites, and the nature of the active
sites has been established. The catalytic results combined with the
spectroscopic data also serve to calibrate the minimum molybdenum
oxidation state required for acid catalyzed C-C bond breaking and
formation of a secondary carbenium ion, namely, Mo(IV).
     Finally, it should be noted that the present work calls into
question a previous report by Holl et. al. (19) that zero valent
molybdenum catalysts possess acidic function. In light of the
present results and the fact that molybdenum has a very high affinity for oxygen, it is likely that the molybdenum particles in
their catalyst were covered by a thin oxide layer imparted the
acidic function. Similar observations of acid function on other
zero valent metal catalysts may also be spurious due to the presence
of oxide overcoatings.

Literature Cited

1.   Stair, P. C. J. Am. Chem. Soc. 1982, 104, 4044.
2.   Overbury, S. H.; Stair, P. C. J. Vac. Sci. Technol., 1983, A1,
     1055.

3.   DeKoven, B. M.; Overbury, S. H.; Stair, P. C. Phys. Rev. Lett.,
     1984, 53, 481.
4.   Grant, J. L.; Fryberger, T. B.; Stair, P. C. Surf. Sci., 1985,
     159, 333.
5.   Grant, J. L.; Fryberger, T. B.; Stair, P. C. Appl. Surf. Sci.,
     1986, 26, 472.
6.   Deffeyes, J. E.; Smith, A. H.; Stair, P. C. Surf. Sci., 1985,
     163, 79.
7.   Deffeyes, J. E.; Smith, A. H.; Stair, P. C. Appl. Surf. Sci.,
     1986, 26, 517.
8.   Clarke, L. J. Surf. Sci., 1981, 102, 331.
9.   Salmeron, M.; Somorjai, G. A.; Chianelli, R. R. Surf. Sci., 1983,
     127, 526.
10. Citrin, P. H.; Wertheim, C. K.; Baer, Y. Phys. Rev., 1983, B28,
     3160.
11. Fryberger T. B.; Stair, P. C. Chem. Phys. Lett., 1982, 93, 151.
12. Sanderson, R. T. J. Am. Chem. Soc., 1952, 74, 272.
13. Stair, P. C. Israel J. Chem. 1982, 22, 380-85.
14. Taylor J. L.; Weinberg, W. H. Surf. Sci., 1978, 78, 259.
15. Aue D. H.; Bowers, M. T. in Gas Phase Ion Chemistry; Bowers, M.
     T., Ed.; Academic Press, New York, 1979; Vol. 2; p. 1.
16. Chung, J.-S.; Burwell, R. L., Jr. J. Catal. 1989, 116, 519.
17. Günther, H. Tetrahedron Lett. 1970, 5173.
18. Brouwer, D. M. in NATO Adv. Study Instit. on Chemistry and
     Chemical Engineering of Catalytic Processes; Prins, R.; Schuit,
     G. C. A., Eds.; Sijthoff & Noordhoff, 1980; p. 137.
19. Holl, Y.; Garin, F.; Maire, G. J. Catal., 1988, 113, 569.

RECEIVED December 6, 1989

Chapter 17

Absolute Electronegativity, Hardness, and Bond Energies

Ralph G. Pearson

Department of Chemistry, University of California, Santa Barbara, CA 93106

The absolute electronegativity, (I+A)/2, determines whether a molecule, atom, radical or ion, is a net electron donor or acceptor. The chemical hardness, (I-A)/2, determines how easily the electron number can be changed. The two properties are related to bond energies in several different ways. An equation for calculating the amount of electron transfer between two groups can be related to bond energies, in some cases. The principle of hard and soft acids and bases can be used, in conjunction with tables of values of hardness, to predict the sign of ΔH for exchange reactions. The effect of auxiliary ligands on bond energies can be predicted.

The concept of hard and soft acids and bases was introduced in 1963 to help rationalize bond dissociation energies in Lewis acid-base complexes. Organometallic compounds can be viewed as such complexes, with the metal atom acting (usually) as the Lewis acid and the various ligands as bases.

Recently there have been important new developments, both with respect to acid-base interactions and with the concept of hardness (1). These developments come from density functional theory (DFT), a branch of quantum mechanics which focuses on the one-electron density function of a molecule, rather than its wave function (2).

According to DFT every atom, molecule, radical or ion is characterized by two properties: the electronic chemical potential, μ, which measures the escaping tendency of electrons in the system; the hardness, η, which measures how difficult it is to change μ. The softness, σ, is the inverse of the hardness, $\sigma = 1/\eta$.

It turns out that good approximations to these properties are $\mu = -(I+A)/2$ and $\eta = (I-A)/2$, where I and A are the ionization potential and electron affinity. Since (I+A)/2 is simply the Mulliken electronegativity, χ, it is convenient to use χ and η, and to call χ the absolute electronegativity. Note that χ is now quite

0097–6156/90/0428–0251$06.00/0

different from Pauling electronegativity, which is a property of atoms only. As noted above, χ refers to atoms, molecules, radicals or ions.

In any chemical system χ, or μ, must be constant everywhere at equilibrium, whereas η can have local values. The average value is given by (I-A)/2. The local values determine which parts of a molecule are most reactive. The most reactive parts of a molecule are the softest.

If two molecules, C and D, are brought close enough to inter- act, there must be a flow of electron density from the one of

$$C + D \rightarrow C - D \tag{1}$$

higher μ to that of lower μ until a single average value exists everywhere. The fractional number of electrons transferred is given by ($\underline{3}$)

$$\Delta N = \frac{(\mu_D - \mu_C)}{2(\eta_C + \eta_D)} = \frac{(\chi_C - \chi_D)}{2(\eta_C + \eta_D)} \tag{2}$$

The difference in the original electronegativities drives the electron transfer, and the sum of the hardness parameters acts as a resistance. Equation 2 refers only to the initial interaction between the two molecules and gives no direct information about the ionic and covalent bonding that ensues. Nevertheless, a large value of ΔN obviously means a strong interaction between the two molecules. This gives rise to a stronger bond being formed in some instances, or to a lowering of the activation energy for reaction in others ($\underline{4},\underline{5}$).

While DFT does not require orbitals, it is not inconsistent to consider molecular orbitals in the form of frontier orbital theory. This means the electrons must come from definite occupied orbitals in D, and go into definite empty orbitals in C. Usually there is some electron transfer in both directions, as in $\sigma+\pi$-bonding. The nature of these interacting orbitals is of great importance in determining the interaction between C and D.

According to Koopmans' theorem, the frontier orbital energies are given by

$$-\epsilon_{HOMO} = I \text{ and } -\epsilon_{LUMO} = A \tag{3}$$

The value of χ falls on the midpoint between the HOMO and LUMO. The energy gap between the HOMO and the LUMO is equal to 2η. Thus hard molecules have a large HOMO-LUMO gap, and soft molecules have a small gap. ($\underline{6}$)

Table I contains values of χ and η for a number of atoms and common molecules. These are calculated from experimental values of I and A, many of which are now available. Most common molecules have negative electron affinities, meaning that kinetic energy must be supplied to an electron before the molecule will accept it. The table only gives a sampling of more extensive lists that have been compiled for molecules, atoms, free radicals and monatomic cations. ($\underline{3}$)

The entries in Table I are arranged in order of decreasing values of χ, so that Lewis acids or electrophiles, start the list and bases, or nucleophiles, are at the bottom. The ordering must not be taken as one of decreasing acid strength, but of inherent tendency to gain or lose electrons. The acid strength depends heavily on two other factors, the charge or dipole moment of the acid, and the empty orbital which accepts the electrons. The same properties of each base must also be considered.

Table I. Experimental Values for Some Molecules in eV

Molecule	χ	η		χ	η
SO_3	7.2	5.5	HCl	4.7	8.0
N_2	7.0	8.6	C_2H_4	4.4	6.2
Cl_2	7.0	4.6	Ni	4.4	3.3
SO_2	6.7	5.6	C_6H_6	4.0	5.2
H_2	6.7	8.7	PH_3	4.0	6.0
O_2	6.3	5.9	H_2O	3.1	9.5
PF_3	6.3	5.3	V	3.6	3.1
BF_3	6.2	9.7	NH_3	2.9	7.9
CO	6.1	7.9	$(CH_3)_3As$	2.8	5.5
I_2	6.0	3.4	$(CH_3)_3P$	2.8	5.9
Pt	5.6	3.5	$(CH_3)_2S$	2.7	6.0
CH_3I	4.9	4.7	CH_4	2.5	10.3

In principle a molecule in Table I should donate electron density to any molecule above it. But for certain kinds of orbitals, there may be little evidence that this occurs. Figure 1 shows the HOMO, the LUMO and χ for the molecules CO and NH_3, and the Pd atom. It can be seen that Pd is a net electron donor to CO, but a net electron acceptor from NH_3. While it appears that NH_3 should be a good donor to CO, there actually will be little interaction. The accepting orbital of CO is of π-type, whereas NH_3 is a σ-donor. Unlike the favorable case of Pd reacting with CO, there will be no synergistic $\sigma+\pi$-bonding for NH_3 and CO.

The most detailed use of Tables of χ and η would be the application of Equation 2 to a series of related molecules, all reacting with a common substrate. This restriction is necessary to ensure that the interacting orbitals remain fairly constant. In certain cases we can assume that ΔN will be proportional to the strength of the coordinate bond formed in Equation 1.

The requirements are that electron transfer in one direction be more important than in the other, but that ΔN be small. If ΔN is too large, then the bond will be very ionic. Its strength will depend mainly on size factors and ΔN will not be a good criterion.

A suitable example is provided by Tolman's data on the equilibrium constants in benzene for

$$NiL_3 + olefin = NiL_2(olefin) + L \quad K_{eq} \qquad (4)$$

where L is a phosphite (*7*). Values of K_{eq} are given in Table II for various olefins, along with their χ and η values. The assumption is made that the unknown values of χ and η for NiL$_2$ are about the same as for low spin Ni atom in the $\underline{d}^{10}$ configuration.

Table II.   Electron Transfer in Reaction of Olefins
with Low-Spin Nickel Atoms

Olefin	Keq[a]	χ	η	Δ N
maleic anhydride	4×10^8	6.3	4.7	0.20
trans-NCCH=CHCN	1.6×10^8	6.2	5.6	0.17
CH$_2$=CHCN	4.0×10^4	5.4	5.6	0.12
C$_2$H$_4$	250	4.4	6.2	0.053
CH$_2$=CHF	90	4.2	6.1	0.042
styrene	10	4.1	4.4	0.045
H$_3$CH=CH$_2$	0.5	3.9	5.9	0.024
trans-2-butene	2.7×10^{-3}	3.5	5.6	0.000
cyclohexene	3.5×10^{-4}	3.4	5.5	-0.006
(CH$_3$)$_2$C=CHCH$_3$	3.0×10^{-4}	3.3	5.5	-0.013

[a] Reference 7

    The calculated values of ΔN correlate very well with the equilibrium constants. Large positive values mean strong bonding, with π-bonding from metal to olefin dominating. Negative values are found for olefins less electronegative than Ni. Evidently σ-bonding is less effective than π-bonding.

    Further studies, including calorimetric ones, show that the variations in K$_{eq}$ for reaction 4 are due to variations in ΔH ($\underline{8}$). The value of the NiL$_2$-C$_2$H$_4$ bond strength is about 40 kcal/mol.

    According to Table II, ethylene has more π-bonding than σ-bonding. This agrees with accurate ab initio calculations ($\underline{9}$). For a more electrophilic metal center, such as Ag$^+$, we would expect ethylene to be a net electron donor, by way of σ-bonding. Unfortunately we cannot use equation 2 to calculate ΔN for Ag$^+$-C$_2$H$_4$. The reason is that the positive potential of the ion lowers the energy of the electrons in ethylene, even at a distance. Thus ethylene in the presence of a cation becomes more electronegative.

    Consider the series Cu$^+$, Ag$^+$ and Au$^+$, with χ = 14.0, 14.5 and 14.9 eV, respectively. Reacting with ethylene, we would expect only σ-bonding with Au$^+$ forming the strongest bond. In fact Cu$^+$ probably forms the strongest bond because it has the greatest amount of π-bonding ($\underline{10}$). This would not be possible if the cation did not enhance the electronegativity of the olefin. Note that π-bonding is more efficient than the same amount of σ-bonding, probably because of better overlap.

    Because of the large amount of data available, carbon monoxide is a good ligand to use to rate the zero-valent metal atoms in bonding ability. Even in the ground state, CO is more electronegative than any of the transition metals (Table I). The low spin metal atoms, characteristic of metal carbonyls have even lower χ values. Therefore π-bonding dominates.

Table III.   Bonding in Metal Carbonyle
[M + CO = M - CO]

M	ΔN	ΔH^+, Kcal[a]
V	0.211	$V(CO)_6$ stable[b]
Cr	.192	40
Mn	.149	37
Fe	.188	42
Co	.091	22
Ni	.128	22
Cu	.010	-
Mo	.148	40
Ru	.125	28
Pd	.070	(11)[c]
Ag	-.029	-
Pt	.037	(14)[c]
Au	-.024	-

[a] Activation enthalpy for loss of first CO in known carbonyls
[b] Does not dissociate readily
[c] Theoretical values. T. Ziegler, V. Tschinke, C. Ursenbach, J. Am. Chem. Soc. 1987, 109, 4825.

Table III contains, for a number of metal atoms in their low spin condition (11), the values of ΔN calculated for their reaction with CO using equation 2. These are to be compared with the numbers in the last column, which are the enthalpies of activation for the loss of the first CO ligand in known carbonyls such as $Ni(CO)_4$.

The correlation between ΔN and the first M-CO bond strength is seen to be very good. Large values of ΔN correspond to strong bonds. The smallest values of ΔN are calculated for metals where no stable carbonyls are known. The results for Cu, Ag and Au would be even less favorable for M-CO bonding if the ionization potential for removing an (n-1) $\underline{d}$ electron, rather than an $\underline{ns}$ electron, had been used.

The successful use of Equation 2 is dependent on keeping the interacting orbitals of M and L relatively constant. Failure to do so creates problems, if we examine the bond energies of a series of diverse ligands with a given metal atom.

For example, rating a series of common ligands in order of decreasing π-bonding would give the following:

$$CS > CO \sim PF_3 > N_2 > PCl_3 > C_2H_4 > PR_3 \sim$$

$$AsR_3 > R_2S > CH_3CN > pyr > NH_3 > R_2O$$

Various criteria, such as ir spectra, are used to do the rating. Because π-bonding is so important with neutral metal atoms, the order is very roughly that of decreasing bond strengths as well. However the variation from one end to the other is surprisingly small, about 10 kcal/mol. This results from the best σ-donors being at the end of the list.

If we use Equation 2 for reaction with a metal atom such as nickel, we find the following order of decreasing ΔN:

$$CS > N_2 \sim PCl_3 > CO > PF_3 \gg pyr \sim CH_3 CN >$$

$$C_2H_4 \gg As(CH_3)_3 > P(CH_3)_3 \sim NH_3 > (CH_3)_2S> (CH_3)_2O$$

The last five have ΔN negative, meaning they are net σ-donors.

The ordering given by ΔN is similar to that given by experiment, but there are discrepancies. These are due, for the most part, to orbital effects. For example, Fig. 2 shows the π^* orbitals of CO and N_2, oriented to overlap with a filled metal d orbital. The π^* orbital of CO is concentrated on carbon, just as the π orbital is concentrated on oxygen. Clearly it will give better overlap with the metal $\underline{d}$ orbital, compared to the π^* orbital of N_2. The π^* orbitals of CH_3CN and pyridine are also poorly suited to give good overlap (12).

BONDING OF ANIONIC LIGANDS

For anions and polyatomic cations (I+A) and (I-A) are no longer useful, or accessible, approximations to χ and η. Since electron transfer between M^+ and X^- would be so one-sided, better approximations to χ are I for X^- and A for M^+. For the hardness values, no numbers are available, but rank ordering is possible (13).

In principle, one could avoid the problems of considering the reaction of, say, Ag^+ with Br^-, by simply taking the atoms as a basis. For polyatomic ions, one could use the corresponding radicals. Thus one could apply Equation 2 to the reaction of Ag with Br.

But there is little to be learned by doing this. Bond energies would not be simply related to ΔN, for a series of atoms or radicals reacting with Ag. The value of ΔN for maximum covalency would be zero, and for maximum ionic bonding ΔN would equal one. At best, we could gain some insight into the polarity of the several bonds. Since the effects of covalency and ionic bonding are not included in Equation 2, the values of ΔN do not give reliable values for the final ionicity (14).

The most useful way to handle ionic reactants is to rank order them by means means of the HSAB principle, the original method for relating bond energies to hardness (13). The principle may be stated as follows:

$$sh + hs = ss + hh \quad 0 > \Delta H \tag{5}$$

where h and s are read as the harder, or the softer, acid or base.

For Equation 5 to be valid, all other bond determining factors should be held constant. These include the electronegativities of the bonded groups, their charges and sizes, orbital overlaps and steric repulsions. While it is impractical to keep all of these constant, at least one should compare acids, or bases, of the same charge, and with similar steric requirements. Also for a series of acids, or bases, the mean bond strengths to reference bases or acids should be about the same (13).

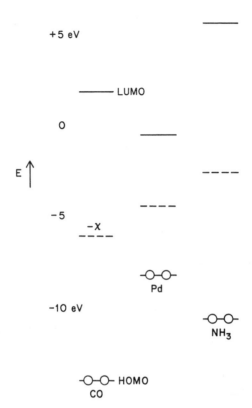

Fig. 1   Orbital energy diagram for CO, Pd and NH$_3$, showing HOMO, LUMO, and χ.

Fig. 2   The π^* orbitals of CO and N$_2$ showing greater overlap of CO with $\underline{d}$ orbitals on a transition metal atom, M.

The known values of I and A for Ag^+ and Na^+ show that the former is much softer ($\underline{4}$). Then thermodynamic data for the reaction

$$AgF_{(g)} + NaI_{(g)} = Ag\ I_{(g)} + NaF_{(g)} \quad \Delta H = -28 \text{ kcal/mol} \tag{6}$$

shows that I^- is much softer than F^-, as expected. The reaction

$$CH_3F_{(g)} + HI_{(g)} = CH_3I_{(g)} + HF_{(g)} \quad \Delta H = -12 \text{ kcal/mol} \tag{7}$$

shows that CH_3^+ is softer than H^+.

With these results, it is possible to rank order a large number of anionic bases ($\underline{13}$). The most convenient way is to take differences in D^o, the usual homolytic bond energies.

$$HX_{(g)} = H_{(g)} + X_{(g)} \qquad D_H^o \tag{8}$$

$$CH_3X_{(g)} = CH_3{(g)} + X_{(g)} \qquad D^oCH_3 \tag{9}$$

$$\Delta = D_H^o - D^oCH_3 \tag{10}$$

Table IV shows the values of Δ calculated for a number of common anions. A large value of Δ means a hard base.

The results are seen to be much as anticipated, based on earlier criteria, such as polarizability. The order cannot be considered as absolute, since a different choice of reference acids gives somewhat different orders ($\underline{15}$). These arise from specific interactions, between the various bases and the reference acids.

Table IV. Empirical Hardness Parameters for
Anionic Bases

X	Δ, Kcal	X	Δ
F^-	27	$CH_3CO_2^-$	23
Cl^-	19	$C_6H_5O^-$	23
Br^-	18	NO_3^-	22
I^-	15	CH_3O^-	21
OH^-	27	HO_2^-	19
SH^-	17	ONO^-	18
SeH^-	12	NCS^-	19
NH_2^-	22	$C_6H_5NH^-$	17
PH_2^-	12	$C_6H_5S^-$	14
AsH_2^-	12	$C_6H_5^-$	12
CH_3^-	15	CN^-	2^a
SiH_3^-	2	NC^-	12^b
GeH_3^-	4	H^-	-1

^{a} C-bonded
^{b} N-bonded

A good example is H^-, which is the softest base in Table IV. The hydride ion is not capable of π-bonding, unlike most other soft bases. Thus soft Lewis acids with filled d shells do not bond as strongly to H^- as might be expected. Instead the electronegativity of the acid is dominant (13).

Table V.   Bond Energies for $TiCl_3X$ and $Co(CO)_4X$,[a] kcal/mol

X^-	$TiCl_3X$	$Co(CO)_4X$	Δ'	Δ[b]
OH^-	108	56	52	27
NH_2-	87	35	52	22
SH^-	68	40	28	17
CH_3-	64	38	26	17
CN^-	98	73	25	2
SiH_3-	50	51	$^-1$	2
H^-	60	55	5	$^-1$

[a] Data from reference 16
[b] From Table 4

Table V lists some bond dissociation ion energies for the reactions

$$TiCl_3X_{(g)} = TiCl_{3(g)} + X_{(g)} \tag{11}$$

$$Co(CO)_4X_{(g)} = Co(CO)_4 + X_{(g)} \tag{12}$$

These are theoretical values, but calculated by a method that gives good agreement with experiment when tested (16). Also listed are Δ', the difference between the energies for reaction 11 and 12 for each X, and the corresponding values of Δ from Table 4.

The conclusion can be drawn that $TiCl_3^+$ is a much harder acid than $Co(CO)_4^+$, just as H^+ is harder than CH_3^+. It appears that CN^- is showing some specific effect with one of the reference acids. All four acids can be put in rank order by the reactions

$$MOH(g) = M(g) + OH(g) \tag{13}$$

$$MSH_{(g)} = M_{(g)} + SH_{(g)} \tag{14}$$

$$\Delta'' = D_{13}^{\,o} - D_{14}^{\,o} \tag{15}$$

	$TiCl_3^+$	H^+	CH_3^+	$Co(CO)_4^+$
Δ''	41 kcal/mol	28	18	15

Evidently $TiCl_3^+$ is the hardest of the four acids, and $Co(CO)_4^+$ is the softest. The two Lewis acids $Pt(dppe)CH_3^+$ and $Ru(C_5(CH_3)_5)(P(CH_3)_3)_2^+$ have Δ'' less than 18 kcal/mol, and are also quite soft as expected (17). Similar results would be found for the middle to late transition metals in low oxidation states.

The relative bond strengths of M-H and M-CH_3 are of great
importance in applications of transition metals to catalysis.
Table 4 shows that H^- is much softer than CH_3^-. The difference is
large enough to be surprising, since the earlier belief was that
carbanions were quite soft. Clearly the values of Δ for F^-, OH^-,
NH_2^- and CH_3^- form a well defined series, which correlate with the
extent to which their valence shell electrons can produce π-repul-
sions.

The hardness of CH_3^- compared to H^- accounts for several
important features regarding their relative bond strengths.
Increasing the positive oxidation state of the metal always
increases the relative bonding of CH_3^- over H^- (_18_). This follows
from the increased hardness of the metal ion. The early transition
metals favor methyl over hydride, compared to the late transition
metals (_19_). This is also a result of increased positive charge
for the early metals. Being less electronegative, the bonds are
more ionic, with the positive charge on the metal. Also there are
fewer d electrons for the early metals and less repulsion with
methyl.

Increasing the number of ligands on the metal favors H^- over
CH_3^-. This results from repulsive interactions between the valence
electrons of CH_3^- and filled orbitals on the ligands. This is just
a more sophisticated way to describe steric repulsions.

MORE ON M-H BONDS

To understand metal-hydrogen bond energies, it is necessary to
appreciate the important role that electronegativity plays in covalent bonding (_19_). Electronegative atoms hold on to all their
electrons tightly, including the electrons of the covalent bond.
This makes the bond stronger. Bonds to hydrogen range from 136
kcal/mol for F (χ - 10.41 eV), to only 42 kcal for Cs (χ = 2.18
eV).

The values of χ for the transition metals range from 3.51 eV
for Ti, to 5.77 eV for Au. This indicates a range of M - H bond
values from about 50 kcal/mol for Ti, to 75 kcal/mol for Au. This
agrees well with experiment (_21_). χ increases as we go from left
to right in any series, and as we go down in any triad.

But there is an opposing effect of electronegativity, if we
look at M-H bond breaking in another way.

$$HM^{(I)}L_n(g) = M^{(0)}L_n(g) + H_{(g)} \tag{16}$$

Loss of an H atom from a transition metal is a reductive-
elimination. This should occur _more_ readily for a metal with a
large value of χ. Eventually, as χ approaches that for the H atom
(χ = 7.17 eV), the bond will become non-polar, and Equation 16 will
lose its validity because HML_n does not contain H^-.

The net effect is that M-H bond energies increase in going down
any triad, and change very little in going from left to right in
any series (_18,20_).

The auxiliary ligands in a metal complex can also influence the
strength of the bond to hydrogen, though there has been little sys-

tematic study. The concept of hardness, however, can be used to make some definite predictions.

It is well known that increased positive charge increases the hardness of the central atom ($\underline{4}$). Considering Equation 16, we can make the prediction that hard ligands will stabilize the higher oxidation state, and soft ligands the lower. Thus hard ligands should increase the strength of M-H bonds. The only direct evidence seems to be the observation of Marks and Schock ($\underline{19}$) that $(C_5(CH_3)_3)_2$ Zr(OR)H has stronger bonds than $(C_5(CH_3)_5)_2$ ZrH_2 by 5 kcal/mol.

If we accept spectroscopic evidence, ir and nmr, for relative M-H bond strengths, then the prediction finds further support. In a series of $\underline{trans}$-Ni(PR_3)_2XH complexes, the M-H bond strength increases in the order ($\underline{22}$)

$$X = C_6H_5^- < CH_3^- < CN^- < C_6H_5S^- < I^-$$

$$SCN^- < Cl^- < HCO_2^- < CH_3CO_2^- < C_6H_5O^- < CF_3\ CO_2^-$$

Except for the somewhat anomalous position of CN^-, this is about as expected, with $CF_3CO_2^-$ being the hardest ligand.

Another kind of indirect evidence also exists. There is a correlation between M-H bond strengths and the Bronsted acidity of the hydride ($\underline{21}$). Stronger acids have weaker bonds and lose hydrogen more readily. The reason for this can readily be seen by writing the equation for acidity.

$$HM^{(I)}L_n = M^{(-I)}L_n + H^+ \tag{7}$$

This is also a reductive elimination, with the metal atom changing its oxidation state by two units. Hard ligands will stabilize HML_n, and reduce its acidity.

There is now considerable evidence for this rule. ($\underline{21,23}$). For example, the rhodium (III) complexes,

$$Rh(NH_3)_4(H_2O)H^{2+} \qquad pKa > 14$$

$$Rh(bipy)_2(H_2O)H^{2+} \qquad pKa = 9.5$$

$$Rh(CNR)_4(H_2O)H^{2+} \qquad pKa < 0$$

have pKa values as predicted. These changes are quite large. For reaction 16, the expected effect would be only about half as great.

ACKNOWLEDGMENT

The support of the Department of Energy over many years is gratefully acknowledged.

LITERATURE CITED

1. Parr, R.G.; Donnelly, R.A.; Levy, M.; Palke, W.E. J. Chem. Phys. 1978, $\underline{68}$, 3801.

2. Parr, R.G.; Yang, W. Density Functional Theory for Atoms and Molecular, Oxford Press, New York, 1989.
3. Parr, R.G.; Pearson, R.G. J. Am. Chem. foc. 1983, 105, 1503.
4. Pearson, R.G. Inorg. Chem. 1988, 27, 734
5. Pearson, R.G. J. Org. Chem. 1989, 83, 8440.
6. Pearson, R.G. Proc. Natl. Acad. Sci. U.S.A. 1986, 83, 8440.
7. Tolman, C.A. J. Am. Chem. Soc. 1974, 96, 2280.
8. Tolman, C.A.; Seidel, W.C.; Gosser, L.W. Organomet 1983, 2, 1391.
9. Ziegler, T. Inorg. Chem. 1985, 24, 1547.
10. Ziegler, T.; Rauk, A. Inorg. Chem. 1979, 18, 1558.
11. Pearson, R.G. Inorg. Chem. 1984, 23, 4675.
12. Jorgensen, W.L.; Salem, L. The Organic Chemists Book of Orbitals, Academic Press, new York, 1973.
13. Pearson, R.G. J. Am. Chem. Soc. 1988, 110, 7684.
14. Pearson, R.G. J. Am. Chem. Soc. 1985, 107, 6801.
15. Bochkov, A.F. Zhur. Org. Khim. 1986, 22, 2041.
16. Ziegler, T.; Tschinke, V.; Versluis, L.; Baerends, E.J.; Ravenek, W. Polyhedron, 1988, 7, 1625.
17. Bryndza, H.S.; Fong, L.K.; Bercaw, J.E. J. Am. Chem. Soc. 1987, 109, 1444
18. Ziegler, T.; Tschinke, V.; Becke, A. J. Am. Chem. Soc. 1987, 109, 1351.
19. Schock, L.E.; Marks, T.J. J. Am. Chem. Soc. 1988, 110, 7701.
20. Kutzelnigg, W. Angew. Chem. Int. Ed. Engl. 1986, 23, 272.
21. Pearson, R.G. Chem. Revs. 1985, 85, 41.
22. Darensbourg, M.; Ludwig, M.; Riordan, C.Y. Inorg. Chem. 1988, 28, 1630.
23. Spillett, C. Ph.D. thesis, University California, Santa Barbara, 1989.

RECEIVED December 28, 1989

Chapter 18

Gas-Phase Chemistry of First-Row Transition Metal Ions with Nitrogen-Containing Compounds

Theoretical and Experimental Investigations

A. Mavridis[1], K. Kunze[2], J. F. Harrison[2], and J. Allison[2]

[1]Chemistry Department, University of Athens, 13a Navarinou Street, Athens 10680 Greece
[2]Department of Chemistry, Michigan State University, East Lansing, MI 48824

The rich gas phase chemistry of first row transition metal (+1) ions with ammonia, and organic compounds (R-X where X=NH_2, CN, NO_2) is discussed. Ongoing theoretical investigations into the Sc^+/NH_3 system are presented, which provide some insights into the bonding and energetics of a variety of MNH_x^+ complexes.

There are a number of mass spectrometric techniques that have been developed for the study of the low pressure gas phase reactions of ionic species with organic molecules. The earliest experiments involving ion/molecule reactions involved chemical ionization mass spectrometry, and the most recent utilize ion beam and Fourier transform ion cyclotron resonance methods. The earliest studies were predominantly organic in nature. More recently, these methods have been used to study organometallic chemistry in the gas phase. The various ionization techniques available in mass spectrometry allow for the generation of unique gas phase species including bare transition metal ions (such as Co^+, Ni^+) and ligated species (such as $CoCO^+$, $CoNO^+$, $NiPF_3^+$, $NiC_5H_5^+$). Their chemistry with small molecules and a variety of larger molecules containing the functional groups of organic chemistry has been extensively studied in the past 15 years(1). The earlier studies were largely mechanistic in nature, to gain an understanding of how product ions were formed. Recently, combined experimental and theoretical efforts have provided important insights into how these reactions occur. These insights will become, we believe, the basis for a fresh evaluation of the reactivity of inorganic and organometallic species that are studied and utilized in condensed phases.

0097–6156/90/0428–0263$06.00/0
© 1990 American Chemical Society

Both polar and nonpolar organic compounds exhibit a rich chemistry with bare transition metal ions($\underline{1}$). Small polar compounds react with ions such as Fe^+ and Co^+, in a single, bimolecular step to form a metal-olefin complex, reaction ($\underline{1}$).

$$Co^+ + C_3H_7X \quad \rightarrow CoC_3H_6^+ + HX \qquad (1)$$

Such reactions have been reported for X=Cl, Br, OH, SH, OR and even for X=H and R (R=alkyl substituent). The mechanism by which these products are formed was proposed by Allison and Ridge in 1979($\underline{2}$), and is shown in reaction (2). The reactants first form a complex (a), which may be simply electrostatically bound. In (b), the transition metal ion inserts into the polar C-X bond (formal "oxidative addition"), followed by the shift of a H

$$Co^+ + C_3H_7X \rightarrow \quad C_3H_7X \cdot \quad \cdot \cdot Co^+ \rightarrow C_3H_7 - Co^+ - X$$

$$\qquad\qquad (a) \qquad\qquad\qquad (b) \qquad (c)\downarrow$$

$$\qquad\qquad\qquad\qquad (C_3H_6)Co^+(HX) \qquad (2)$$

$$\qquad\qquad\qquad\qquad\qquad (d)\downarrow$$

$$\qquad\qquad\qquad\qquad\qquad CoC_3H_6^+ + HX$$

atom that is on a C which is β to the metal (β-H shift), across the metal onto X, resulting in the degradation of the molecule into two smaller, stable species that reside as ligands on the metal. The last step (d) is a competitive ligand loss. In this process, it appears that the ligand that is more weakly bound to the metal is preferentially lost($\underline{3}$). In some cases, two products result from this dissociation. For example, Co^+ reacts with propanol to form both CoH_2O^+ and $CoC_3H_6^+(\underline{2})$. While studies of series of, e.g., alcohols, and labeling studies provided some insights into the mechanisms that are operative in the chemistry, many important questions are difficult to approach experimentally. To understand the chemistry, it is certainly important to understand the bonding of the molecules and fragments involved to the transition metal and the strengths of these bonds. Many features such as the distribution of the charge in the various intermediates may also be important in controlling the types of products that are formed. Thus, theoretical studies are necessary to understand the "driving forces" that dominate these reactions and lead to the rich chemistry.

The chemistry of polar compounds containing many types of functional groups and multifunctional organic compounds, with bare transition metal ions has been reported. The richest chemistry certainly involves those functional groups that contain nitrogen, and a few examples will be provided here.

Stepnowski and Allison(4) reported the chemistry of a variety of transition metal ions with a series of alkyl cyanides. While ions such as Fe^+ and Co^+ insert into the C-X bond in many polar compounds, they do not appear to do so for -X = -CN. Instead, C-C bonds are cleaved, as shown for the Co^+/n-propyl nitrile system in reactions (3) and (4).

$$Co^+ + n-C_3H_7CN \rightarrow CoCH_3CN^+ + C_2H_4 \qquad (3)$$

$$\rightarrow CoC_2H_4^+ + CH_3CN \qquad (4)$$

The reactivity of nitriles may be dominated by a number of aspects of the chemistry including the fact that C-CN bond dissociation energy (BDE) is substantially larger than the BDE's for C-Cl and C-OH bonds(4). Also, there is evidence that the geometry of the initial complex has a dramatic effect - an "end-on" interaction of the metal ions with the nitrile group may make insertion into the C-CN bond geometrically inaccessible(5).

Another interesting functional group containing nitrogen is the nitro group. The chemistry of Co^+ with a series of nitroalkanes has been reported by Cassady et al.(6). Consider the Co^+/n-$C_3H_7NO_2$ system. Fourteen different products were reported. Product ions such as $Co(C_3H_6)^+$ and $Co(HNO_2)^+$ suggested that Co^+ inserted into the C-NO_2 bond. Other product ions such as $CoC_3H_7O^+$ and $CoNO^+$ suggest that RNO_2 may be converted, in part, to RO- and NO- on the metal center. Thus the -NO_2 group actively interacts with the metal. The chemistry of organometallic anions with nitroalkanes has also been reported(7).

The chemistry of NO with transition metal ions has also received some attention, and deserves comment here. One of the first ways in which Co^+ ions were generated for mass spectrometric study was by electron impact ionization of the volatile compound $Co(CO)_3NO$. In addition to Co^+, ions such as $CoCO^+$ and $CoNO^+$ were formed. While Co^+ and $CoCO^+$ ions frequently react with organic molecules in the gas phase, $CoNO^+$ is unreactive(8). The reason for this change in chemistry upon change of ligand was unclear, since there was only "traditional" bonding-schemes to consider at the time. However, theoretical studies show that the CO ligand is not bonded to first row transition metal ions via the Dewar-Chatt model, but is electrostatically bound(8,9) - thus it's presence does not change the electronic structure of the metal. Presumably NO acts as a 1 or 3-electron donor with Co^+ (which has a $3d^8$ ground state configuration), so the $CoNO^+$ ion has a single unpaired electron on the metal, and should not participate in reactions that require an insertion step. We also note that the NO ligand does appear to be reactive when bound to Co_2^+. Jacobson($\underline{10}$) reported the observation of reaction (5) involving CO.

$$Co_2NO^+ + CO \rightarrow Co_2N^+ + CO_2 \qquad (5)$$

There have been a number of studies reported involving metal ion reactions with amines and ammonia. The first of these involved Co^+ with a series of amines, reported by Radecki and Allison in 1984($\underline{11}$). Reactions (6-9) were observed for the $Co^+/n-C_3H_7NH_2$ system.

$$Co^+ + n-C_3H_7NH_2 \rightarrow CoC_3H_7N^+ + H_2 \qquad (6)$$

$$\rightarrow CoC_3H_5N^+ + 2\ H_2 \qquad (7)$$

$$\rightarrow CoCH_5N^+ + C_2H_4 \qquad (8)$$

$$\rightarrow C_3H_8N^+ + CoH \qquad (9)$$

These results were certainly surprising since insertion into the $C-NH_2$ bond was not observed, although the reaction (10)

$$Co^+ + C_3H_7NH_2 \rightarrow CoC_3H_6^+ + NH_3 \qquad (10)$$

should be exothermic. It was proposed that H_2 elimination occurred by initial insertion into the $N-H$ bond. While Co^+ did not insert into the $C-NH_2$ bond of primary amines, it does appear to insert into the $C-N$ bond in the tertiary amine, $(C_2H_5)_3N$, and in allyl amine. To explain the failure to observe reaction (10) it was proposed that there was a barrier along the reaction channel - that the insertion intermediate was energetically inaccessible for thermal energy reactions, which could only be due to a weak Co^+-NH_2 bond. The BDE would have to be unusually small, < 20 kcal/mol. It has since been shown, as discussed below, that this is not the case. Another explanation may be that Co^+ does react with, e.g., propyl amine, to form propene and ammonia, but there is a barrier to the dissociation step of $(C_3H_6)Co^+(NH_3)$ - which cannot be distinguished in the mass spectrum from the complex containing the intact molecule, $Co^+(C_3H_7NH_2)$. The gas phase chemistry of a number of metal ions with amines has been studied. Babinec and Allison($\underline{12}$) reported the chemistry of Cr^+, Mn^+, Fe^+, Ni^+, Cu^+ and Zn^+ with n-propyl amine. Only Fe^+ clearly inserted into the $C-N$ bond, as evidenced by the observation of the two products $FeNH_3^+$ and $FeC_3H_6^+$. Mn^+ (with a half-filled d-shell, $3d^5 4s^1$ ground state) and Zn^+ (with a filled d-shell, $3d^{10}4s^1$ ground state) were unreactive with propyl amine. In contrast, Cr^+($3d^5$ ground state) and Cu^+($3d^{10}$ ground state) were both observed to induce H_2 elimination only. Sigsworth and Castleman([13]) studied the reactions of Ag^+ and Cu^+ with methyl amine, dimethyl amine and trimethyl amine. In these studies, hydride abstraction was observed to yield the (amine-H)$^+$ ion and the MH neutral product.

In 1987, Buckner and Freiser([14]) showed that Co^+ does form a strong bond to NH_2, and did so as follows. In an FTMS instrument they formed $CoOH^+$, which reacts with ammonia, reaction (11).

$$CoOH^+ + NH_3 \rightarrow CoNH_2^+ + H_2O \qquad (11)$$

They also found that the NH_2 group can be displaced from Co^+ by benzene, but not by acetonitrile. With this information, they concluded that the BDE for Co^+-NH_2 is 65 ± 8 kcal/mol. In this same work they reported the chemistry of MNH_2^+ with some hydrocarbons. They also found that FeO^+ reacts with ammonia to form $FeNH^+$, and reported some chemistry for this ionic species. Their work suggested that the BDE for $FeNH^+$ was > 41 kcal/mol and < 81 kcal/mol. This work was extended in a report by Buckner, Gord and Freiser in 1988([15]). They found that a variety of metal ions including Sc^+, Ti^+ and V^+ react with ammonia in an exothermic process by reaction (12), which suggests that these M^+-NH bonds are stronger than 93 kcal/mol. The BDE for V^+-NH was determined to be 101 ±7 kcal/mol, and the BDE for Fe^+-NH to be lower, 54 ± 14 kcal/mol.

$$M^+ + NH_3 \rightarrow MNH^+ + H_2 \qquad (12)$$

They also reported a number of reactions for MNH^+ with neutral molecules. A variety of mechanisms for reaction (12) was discussed, including dehydrogenation via direct decomposition of the intermediates $(H)_2M^+$=NH and/or H-M^+-NH$_2$.

The Reaction of Sc$^+$ with NH$_3$
Our intent is to gain theoretical insights into the initial interaction of M^+ with a polar molecule, the insertion step and reductive elimination by a systematic study of the interaction of M^+ with NH_3. We begin here with Sc^+ because it has only two valence electrons and thus its theoretical description and its calculated interaction with NH_3 and its fragments will be more reliable than with the latter transition elements. We will first describe the ScN$^+$ core and then build up the various intermediates by adding hydrogens. Specific fragments to be studied include ScNH$^+$, ScNH$_2^+$, HScNH$_2^+$ and ScNH$_3^+$. The calculations will include all electrons and will use both MCSCF and

CI techniques.   The basis sets used and the general approach have been described previously($\underline{16}$).

ScN$^+$

The relative energies of the low lying states of Sc$^+$, N, NH and NH$_2$ are collected in Figure 1. Because the excited states of N are much higher in energy than those of Sc$^+$ we will consider the states of ScN$^+$ which correlate to the ground state of N and the sd and d^2 configuration of Sc$^+$($\underline{16}$).   As Sc$^+$ in an sd configuration approaches N we can imagine forming a σ bond between the 4s electron on Sc$^+$ and a 2p σ electron on N and a π bond between the 3d$_{xz}$ and 2p$_x$ electrons. This leaves an unpaired electron on N in the 2p$_y$ orbital and results in a $^2\Pi$ state represented by

$$\left[\overset{\underset{\displaystyle \cdot \overline{\underset{\sigma}{}} \cdot}{\pi}}{\text{Sc} \cdot \overline{} \cdot \text{N}} _{\diagdown \text{p}_y} \right]^+ \longrightarrow \text{Sc}=\text{N}^+ \quad ^2\Pi$$

It is only at large ScN separations that the Sc electron in the σ bond is pure 4s. As the bond forms other orbitals on Sc of σ symmetry mix into the bonding orbital. This is seen most vividly in Figure 2a which shows the occupancies of various atomic orbitals in the $^2\Pi$ state of ScN$^+$ as a function of separation.   At equilibrium the atomic orbitals in the σ bond have the occupancy

$$(4s^{0.0} \; 4p_\sigma{}^{0.0} \; 3d_\sigma{}^{0.66})_{Sc} (2p_\sigma{}^{1.28})_N$$

and the 4s orbital has gone from an occupancy of 1 to 0. Interestingly the π bond occupancy is equal to that of the σ bond

$$(3d_{xz}{}^{0.66})_{Sc} (2p_x{}^{1.28})_N$$

while the unpaired electron remains essentially localized on N.

$$(3d_{yz}{}^{0.07})_{Sc} (2p_y{}^{0.91})_N$$

The net result is that Sc$^+$ looses 0.49 electrons to N resulting in the charge distribution

$$^{+1.51}\text{Sc} \; \text{N}^{-0.49}.$$

If we break the π bond in this state and place the high spin electron on Sc in a d$\delta_\pm$ orbital we will form a $^4\Delta_\pm$ state.

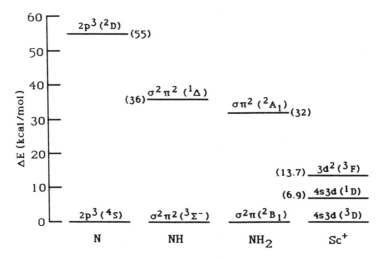

Figure 1. Relative Energy Levels (kcal/mol^{-1}) of N($\underline{19}$), NH($\underline{20}$), NH$_2$ ($\underline{21}$) and Sc$^+$($\underline{19}$).

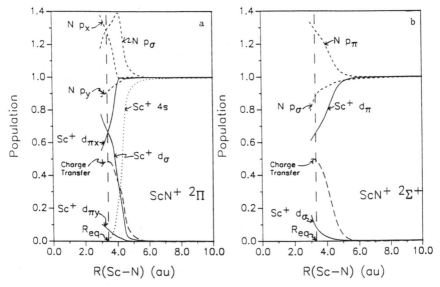

Figure 2. Orbital occupancy as a function of internuclear distance for ScN$^+$ in the a) $^2\Pi$ and b) $^2\Sigma^+$ states.

$$\left[\overset{\bullet}{S}c\cdot\underset{\sigma}{\bullet}\overset{\bullet}{\underset{\bullet}{N}}_{\diagdown P_y} \right]^+ \longrightarrow \overset{\bullet}{S}c-\overset{+}{\underset{\bullet}{N}} \quad {}^4\Delta_{\pm}$$

In a similar way we can consider states of ScN$^+$ which correlate to the d^2 (excited) state of Sc$^+$. For example the $^2\Sigma^+$ state has a π, π double bond and an unpaired electron on N in the σ system

$$\left[\overset{\bullet\overline{\pi}\bullet}{\underset{\bullet\underline{\pi}\bullet}{Sc}} \overset{}{\bullet N} \right]^+ \longrightarrow Sc=\overset{+}{N}\overset{\bullet}{\diagdown}_{P_\sigma} \quad {}^2\Sigma^+$$

If we break a π bond in this structure and place the then uncoupled $d\pi$ electron in a $d\delta_{\pm}$ orbital we have ScN$^+$ in a $^4\Phi$ state held together by a single π bond

$$\left[\overset{\delta_{\pm}\diagup\overset{\bullet}{S}c}{\underset{\underline{\pi}}{}}\overset{\bullet\overset{+}{N}}{\underset{\bullet}{}}\diagdown P_x \right]^+ \longrightarrow \overset{\bullet}{S}c-\overset{\bullet\overset{+}{N}}{}\overset{}{\diagdown}_{P_\sigma} \quad {}^4\Phi$$

Configuration interaction calculations on these four states result in the potential curves shown in Figure 3. The orbital populations in the $^2\Sigma^+$ state are shown in Figure 2b and the properties of all four states are summarized in Table I.

Table I.   Calculated Properties of Several States of ScN$^+$

State	Lewis	D_e(kcal/mol)	R_e(Å)	ω_e(cm^{-1})	Q(Sc)
$^2\Sigma^+$(d^2)	$\overset{+}{Sc}\underset{\pi}{\overset{\pi}{\equiv}}N\cdot^{\diagup P_\sigma}$	63.0	1.738	871	+1.51
$^2\Pi_x$(sd)	$\overset{+}{Sc}\underset{\sigma}{\overset{\pi}{\equiv}}\overset{}{N}^{\diagup P_x}$	55.3	1.804	811	+1.51
$^4\Delta$(sd)	$\delta^{\diagup}\overset{+}{\cdot Sc}\overset{\sigma}{-}N\overset{\bullet^{\diagup P_x}}{\underset{\bullet-P_y}{}}$	26.8	2.101	534	+1.46
$^4\Phi$(d^2)	$\delta^{\diagup}\overset{+}{\cdot Sc}\overset{\pi}{\underline{\equiv}}N\overset{\bullet^{\diagup P_\sigma}}{\underset{\bullet-P_y}{}}$	21.1	1.979	674	+1.47

Note that all D_e's are reported relative to the ground states of Sc$^+$(^3D) and N(^4S). Most interestingly, the ground state is the $^2\Sigma^+$ which contains two π bonds and which correlates with the excited d^2 configuration of

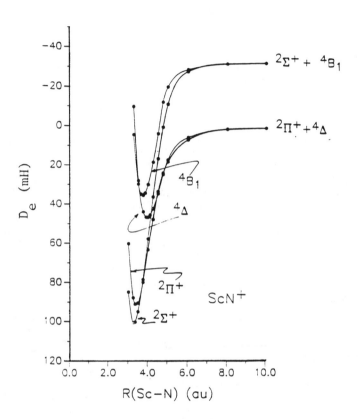

Figure 3. Potential Energy Curves for 4 States of ScN⁺.

Sc^+. It is also noteworthy that the charge on Sc^+, Q, is independent of the number and type of formal bonds in the molecule.

ScNH+

We can imagine $ScNH^+$ being formed by addition of a H to either the $^2\Sigma^+$ or $^2\Pi$ state of ScN^+.

$$Sc = N \overset{+}{\cdot} [\,^2\Sigma^+\,] \; + \; \cdot H\,[\,^2S\,] \; \longrightarrow \; Sc = N - H^+ \; [\,^1\Sigma^+\,]$$

or

$$Sc = \overset{..}{N}{}^+ [\,^2\Pi\,] \; + \; \cdot H\,[\,^2S\,] \; \longrightarrow \; Sc = \overset{H}{\overset{|}{N}}{}^+ [\,^1A'\,]$$

Optimizing the structure of this system at the CI level results in the linear structure with a $^+$ScN-H dissociation energy of 118.5 kcal/mol (Kunze, K.L. and Harrison, J.F., MSU, unpublished results). This is a remarkably strong NH bond (free NH has a D_e of 80 kcal/mol and we require 108 kcal/mol to break the first N-H bond in NH_3).

The reason for this enhanced NH bond strength is apparent when one analyzes the electron distribution in the $ScNH^+$ molecule and compares it with ScN^+ and NH. This analysis shows that when H bonds to ScN^+ the N 2s orbital loses 0.45 electrons, the H 1s loses 0.17e and the N $2p_\sigma$ orbital gains 0.30 electrons. If, for bookkeeping purposes, we assume all of the electrons lost by H go into the N $2p_\sigma$ and that the remaining gain in the N $2p_\sigma$ comes from the N 2s, then we can allot the remaining 0.32 electrons lost by the N 2s to the σ orbitals on Sc^+. This accounts nicely for the increase of 0.33 electrons in the Sc^+ σ system and suggests that the enhanced N-H bond strength in $ScNH^+$ is due to a significant dative bond formed between Sc and the N 2s electron pair. Alternatively, we may imagine forming $ScNH^+$ via the lowest state (^3F) of the excited d^2 configuration.

$$Sc^+ \; (^3F) + NH \; (^3\Sigma^-) \rightarrow ScNH^+ \; (^1\Sigma^+)$$

The Sc^+ - NH bond strength is calculated to be 105.9 kcal/mol relative to the ground state fragments (Kunze, K.L. and Harrison, J.F., MSU, unpublished results), which should be compared with recent results by Armentrout et al. (Armentrout, P., Univ. of Utah, private communication, 1989) suggesting a Sc^+-NH bond strength of 118 kcal/mol. This Sc^+-NH calculated bond strength is 43 kcal/mol stronger than the 63 kcal/mol calculated for the ScN^+ bond in the $^2\Sigma^+$ state. Note that the calculated N-H bond strength in ScN^+-H is also 43 kcal/mol stronger than the 75.5 kcal/mol calculated bond strength in NH. This synergistic relationship between the N-H and Sc-N bonds in $ScNH^+$ is a consequence of the freedom given to the σ system by the unusual π, π double bond between the metal and N. As H bonds to the $p\sigma$ electron on N considerable 2s character is mixed into the NH bonding orbital. This hybridization not only strengthens the N-H bond but the companion orbital which points toward the Sc simultaneously stabilizes the ScN bond by a dative interaction. The $^2\Sigma^+$ state represented above is more accurately written as

$$^+Sc \leftarrow N\text{-}H \; (^1\Sigma^+)$$

$ScNH_2^+$

The ground state of NH_2 is of $^2B_1$ symmetry and is characterized by a $\sigma^2\pi^1$ electronic configuration and an NH_2 angle of 104°, whereas, the first excited state is of $^2A_1$ symmetry and is 32 kcal/mol above the ground $^2B_1$. It is characterized by a $\sigma^1\pi^2$ electronic configuration and an angle of 144°

We may imagine the ground state of NH_2 reacting with either the ground or excited state of Sc^+ to form a doublet with a nonplanar structure in which the bonding electron on Sc^+ is formally either 4s or 3dσ.

Alternatively we can imagine the ground state of NH_2 reacting with the $3d^2$ configuration of Sc^+ to form the planar molecule having a π bond and a (dative) σ bond. The unpaired electron on Sc would most likely be in a $\delta_\pm$ orbital to optimize the ability to form a π and dative σ bond.

All electron multireference single and double CI calculations (Mavridis, A.; Herrera, F.; and Harrison, J.F., MSU, unpublished work) predict that the ground state of $ScNH_2^+$ is of $^2A_2$ symmetry and that it is planar with the geometry

Only small energies are required to deviate from the planar structure. Our calculations suggest that 0.3 kcal/mol is required to move Sc 10° out of plane and 0.9 kcal/mol to move it 20°.
The electron distribution at equilibrium suggests that Sc has a total charge of +1.51 having lost 0.49 electrons to the NH_2 group. The electrons remaining on Sc are distributed as follows

$$4s^{0.06}4p_\sigma^{0.0}4p_x^{0.07}3d_\sigma^{0.16}3d\pi_x^{0.22}3d_\delta^{1.00}$$

We calculate the $^+Sc=NH_2$ bond dissociation energy for the $^2A_2$ state to be 79 kcal/mol which compares favorably with the 85 kcal/mol recently determined by Armentrout et al. (Armentrout, P.B., Univ. of Utah, private communication, 1989). The lowest $^2A_1$ state has the unpaired electron in a δ_+ orbital and is essentially degenerate with the ground $^2A_2$ state. The lowest $^2B_2$ state has a D_e of 70 kcal/mol and results when the unpaired electron is in a $d\pi_y$ orbital. Placing the unpaired electron in a $d\pi_x$ orbital results in the $^2B_1$ state which is 22 kcal/mol above the $^2A_2$ ground state. We have also investigated several quartet states which are bound by ~30 kcal/mol, primarily by charge-dipole interactions.

$ScNH_3^+$
Three isomers with this formula are relevant. The first is a charge-dipole complex in which NH_3 is intact. This complex is bound relative to ground state fragments by 36.8 kcal/mol.
The second isomer is the insertion product $H-Sc-NH_2^+$. Generalized Valence Bond calculations on this system suggest that it is bound relative to Sc^+ + NH_3 by 16 kcal/mol.
The geometry is

$$\left[\begin{array}{c} H \diagdown^{111°} H \\ 1.79 \diagup Sc=N \diagdown^{} \Big)107° \\ 1.92^{} 1.01^{} H \end{array} \right]^+$$

The third isomer is the electrostatic complex $(H_2)ScNH^+$ that is bound relative to Sc^+ + NH_3 by 14 kcal/mol. Interestingly the H_2 molecule is bound to the $ScNH^+$ ion by 5 kcal/mol.

Summary of Energetics
The relative energies of the various products of the reaction of Sc^+ (^3D) with NH_3 are collected in Figure 4. This figure was constructed using the energies discussed in the previous section as well as binding energies of ScH^+ and ScH_2^+ (18) reported earlier. Energies determined experimentally are shown in parentheses.

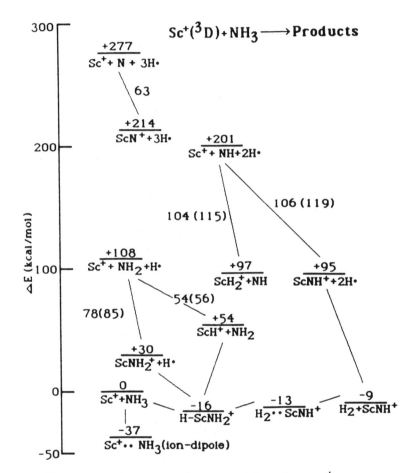

Figure 4. Summary of Energetics for $Sc^+ + NH_3$.

These calculations suggest that both the insertion product $H-Sc-NH_2^+$ and the complex $H_2 \cdots ScNH^+$ are exothermic products of the reaction of Sc^+ with NH_3. The insertion product is formed exothermically because the $Sc-NH_2$ bond strength and the $Sc-H$ bond strength are large and able to overcome the 108 kcal/mol required to break a $N-H$ bond in NH_3. Our calculations suggest that the $^+Sc-NH_2$ bond is strong because:

1. The NH_2 group donates charge to the transition metal ion via the σ^2 pair on N. This dative interaction is optimized when there are no 4s or 3dσ electrons on the metal.

2. The metal donates charge to the π orbital of the NH_2 group. This transfer is optimized when there is one dπ electron in the π system being coupled with the NH_2 π orbital.

   Interfering with either of these mechanisms will reduce the metal $-$ NH_2 bond strength and consequently the potential for an exothermic $H-M-NH_2^+$ insertion product. Likewise reducing the metal-H bond strength would also reduce the possibility of an exothermic insertion product.

   For example, Co^+ is a d^8 system and regardless of the configuration of the electrons in the ground ^3F state one will always have at least one electron, in a dσ orbital. Consequently the planar configuration is not obviously better than the singly bonded non planar structure

$$^+Co\cdot$$

$$\sigma \diagdown \overset{|}{\underset{\bullet\bullet}{\bigcirc}} \overset{\text{\tiny\\\\\\}}{\diagdown} H$$

$$\underset{\bigcirc}{\circledcirc} N \diagdown_{\blacktriangleright H}$$

and, as a result, the $M-NH_2$ bond strength will be smaller for Co than Sc. So while Sc^+ and Co^+ both have two singly occupied d orbitals Co^+ is less energetically favored to react with NH_3 to form

$$H - Co - NH_2^+.$$

Similar considerations apply to the exothermic products $H_2 + ScNH^+$. In this case the exothermicity results from the strong M-NH bond which arises from:

1. Each $2p_\pi$ electron on N bonding to a singly occupied orbital on Sc^+ and

2. The lone pair on N bonding to the empty 4s, 3d orbitals on Sc.

It is not possible for Co^+ to satisfy both of these conditions simultaneously and consequently, $CoNH^+$ should have a relatively weak Metal-NH bond strength. The process

$$Co^+ + NH_3 \rightarrow Co=NH^+ + H_2$$

will be less exothermic if the $Co=NH^+$ bond strength is less than 90 kcal/mol.

Acknowledgments

This work was partially supported under NSF Grants CHE8519752 (JFH) and CHE8722111 (JA), as well as a NATO Grant CRG890502 to A. Mavridis and J. F. Harrison. The Electron Structure codes provided by the Argonne Theory group have been indispensable to this work.

Literature Cited

1. Allison, J. *Prog. Inorg. Chem.* vol. 34, S.J. Lippard, Ed. **1986**, p. 627.
2. Allison, J.; Ridge, D.P. *J. Am. Chem. Soc.* **1979**, *101*, 1332.
3. Tsarbopoulos, A.; Allison, J. *Organometallics* **1984**, *3*, 86.
4. Stepnowski, R.M.; Allison, J. *Organometallics* **1988**, *7*, 2097.
5. An end-on interaction of the metal ion with the CN group results in reactions that are remote from the site of complexation. Such reactions are also discussed in: Tsarbopoulos, A.; Allison, J. *J. Am. Chem. Soc.* **1985**, *107*, 5085.
6. Cassady, C.J.; Freiser, B.S.; McElvany, S.W.; Allison, J. *J. Am. Chem. Soc.* **1984**, *106*, 6125.
7. McElvany, S.W.; Allison, J. *Organometallics* **1986**, *5*, 1219.
8. Allison, J.; Mavridis, A.; Harrison, J.F. *Polyhedron,* **1988**, *16/17*, 1559.
9. Mavridis, A.; Harrison, J.F.; Allison, J. *J. Am. Chem. Soc.* **1989**, *111*, 2482.
10. Jacobson, D.B.; *J. Am. Chem. Soc.* **1987**, *109*, 6851.
11. Radecki, B.D.; Allison, J. *J. Am. Chem. Soc.* **1984**, *106*, 946.
12. Babinec, S.J.; Allison, J. *J. Am. Chem. Soc.* **1984**, *106*, 7718.
13. Sigsworth, S.W.; Castleman, Jr., A.W. *J. Am. Chem. Soc.* **1989**, *111*, 3566.
14. Buckner, S.W.; Freiser, B.S. *J. Am. Chem. Soc.* **1987**, *109*, 4715.
15. Buckner, S.W.; Gord, J.R.; Freiser, B.S. *J. Am. Chem. Soc.* **1988**, *110*, 6606.
16. Kunze, K.L. and Harrison, J.F. *J. Phys. Chem.* **1989**, *93*, 2983.
17. Alvarado Swaisgood, A.E. and Harrison, J.F. *J. Phys. Chem.* **1985**, *89*, 5198.

18. Sunderlin, L.; Aristov, N and Armentrout P.B. _J. Am. Chem. Soc._ **1987**, 109, 78
19. Moore, C. E., "Atomic Energy Levels; Nat. Stand. Ref. Data Ser., Nat. Bur. Stand. Circular 35, Washington, DC, **1971**, Vol. I and II.
20. Huber, K.P. and Herzberg, G., "Moleculear Spectra and Molecular Structure", Van Nostrand Reinhold Co., New York, **1979**.
21. Johns, J.W.C.; Ramsay, D.A. and Ross, S.C., _Can. J. Phys._, **1976**, 54, 1804.

RECEIVED December 19, 1989

Chapter 19

Periodic Trends in the Bond Energies of Transition Metal Complexes

Density Functional Theory

Tom Ziegler and Vincenzo Tschinke

Department of Chemistry, University of Calgary, Calgary, T2N 1H4, Canada

Accurate Density Functional calculations make possible today the systematic investigation of periodic trends in the bond energies of transition metal complexes. Computational results are presented for metal-metal bonds in dimers of the group 6 transition metals, metal-ligand bonds in early and late transition metal complexes, and metal-carbonyl bonds in hexa- penta- and tetra-carbonyl complexes.

The dearth of reliable experimental data on bond dissociation energies is felt throughout the field of organometallic chemistry. Accurate theoretical studies should afford a much needed supplement to the sparse available experimental data on metal-ligand bond energies, necessary for a rational approach to the synthesis of new transition metal complexes.

Recently, Density Functional investigations of molecular bond energies have gained novel impetus due to the introduction by Becke (1) of a gradient correction to the Hartree-Fock-Slater local exchange expression,

$$E_X^{LSD/NL} = E_X^{HFS} - \sum_\gamma \beta_B \int \frac{|\nabla_1 \, \rho_1^\gamma(\vec{r}_1)|^2}{[\rho_1^\gamma(\vec{r}_1)]^{7/3}} \left\{ 1 + \gamma_B \frac{|\nabla_1 \, \rho_1^\gamma(\vec{r}_1)|^2}{[\rho_1^\gamma(\vec{r}_1)]^{8/3}} \right\}^{-1} \delta \vec{r}_1 \quad (1),$$

where ρ_1^γ is a spin density and β_B and γ_B are parameters. In conjunction with appropriate approximations for antiparallel spin correlations, the expression of Eq.(1) provides near-quantitative estimates (1) of bond energies in main-group compounds.

In this contribution we shall present several applications of the new method, which we shall refer to as LSD/NL, to the calculation of bond energies of transition metal complexes. We shall focus on trends along a transition period and/or down a transition triad. The following subjects will be discussed: a) metal-metal bonds in dimers of the group 6 transition metals; b) metal-ligand bonds in early and late transition metal complexes; c) the relative strength of metal-hydrogen and metal-methyl bond in transition metal complexes; d) the metal-carbonyl bond in hexa- penta- and tetra-carbonyl complexes.

0097–6156/90/0428–0279$06.00/0
© 1990 American Chemical Society

Table 1   Calculated bond energies [D(M-M)] (eV) and metal-metal bond distances (R_{M-M}) (Å) for Cr_2, Mo_2 and W_2

	D(M-M)		R_{M-M}	
	Calc.	Exp.[11]	Calc.	Exp.[11]
Cr_2	1.75	1.56±0.2	1.65	1.69
Mo_2	4.03	4.18±0.2	1.95	1.93
W_2	4.41(3.54)[a]	–	2.03(2.07)[a]	–

[a] Non-relativistic results.

Computational Details

In the present set of calculations we have used the functional proposed by Becke (1), which adopts a non-local correction to the HFS exchange, and treats correlation between electrons of different spins at the local density functional level. All calculations presented here were based on the LCAO-HFS program system due to Baerends et al. (2) or its relativistic extension due to Snijders et al.(3), with minor modifications to allow for Becke's non-local exchange correction as well as the correlation between electrons of different spins in the formulation by Stoll et al. (4) based on Vosko's parametrization (5) from homogeneous electron gas data.   Bond energies were evaluated by the Generalized Transition State method (6), or its relativistic extensions (7).

A double ζ-STO basis (8) was employed for the ns and np shells of the main group elements augmented with a single 3d STO function, except for Hydrogen where a 2p STO was used as polarization.   The ns, np, nd, $(n + 1)s$ and $(n + 1)p$ shells of the transition metals were represented by a triple ζ-STO basis (3).   Electrons in shells of lower energy were considered as core and treated according to the procedure due to Baerends et al. (2).   The total molecular electron density was fitted in each SCF-iteration by an auxiliary basis (9) of s, p, d, f and g STOs, centred on the different atoms, in order to represent the Coulomb and exchange potentials accurately.

Metal-Metal Bond Strength of the Dimers Cr_2, Mo_2 and W_2 (10)

The bond energies (11) of the metal dimers Cr_2 and Mo_2 are accurately known experimentally and we note that several theoretical accounts of the bonding in these systems have already appeared, based on ab initio (11a) and Density Functional Theory (12). However, no calculation has been reported for W_2, nor are there any experimental data available.

The bond energies in Table 1 were calculated by evaluating the energy difference

$$\Delta E = 2E(^7S) - E(M_2) \qquad (2)$$

between two metal atoms in the ^7S state corresponding to the $nd^5(n + 1)s^1$ configuration, and M_2. For Cr_2 and Mo_2, ΔE represents the bond energies D(Cr-Cr) and D(Mo-Mo), respectively, since Cr and Mo have a spherical ^7S ground-state. However, the W-atom has a ^5D ground-state with the configuration $5d^46s^2$, thus we have subtracted for W_2 the experimental energy difference (.37 eV) (13) between the ^5D and the ^7S states twice to arrive at D(W-W) of Table 1.

Table 2   Calculated [D(M-L)] bond energies (kJ mol^{-1}) and optimized (R_{M-L}) bond distances (Å) in Cl_3ML

| | D(M-L) | | | R_{M-L}[a] | | |
| | M | | | M | | |
L	Ti	Zr	Hf	Ti	Zr	Hf
H	250.7	297.2	313.5	1.70	1.82	1.80
CH_3	267.5	309.5	326.6	2.13	2.26	2.25
SiH_3	210.9	239.5	272.2	2.63	2.78	2.79
OH	453.2	527.2	535.9	1.83	1.95	1.97
OCH_3	426.9	484.5	506.6	1.86	1.99	2.01
SH	293.3	347.9	360.1	2.28	2.47	2.47
NH_2	364.7	420.6	439.1	1.87	2.01	2.04
PH_2	190.6	225.6	233.9	2.24	2.48	2.47
CN	410.4	457.6	477.9	2.06	2.21	2.23

[a] Optimized from a quadratic fit through three energy points corresponding to three different M-L distances.

The calculated bond energies and equilibrium bond distances R_{M-M} for Cr_2 and Mo_2 are in good accord with experimental values, as can be seen from Table 1. In contrast to other calculations based on DFT, we have employed in the present work (n + 1)f polarization functions. Their contribution to the bond energies are modest, 0.2 - 0.4 eV. On the other hand, the contributions to D(M-M) from the non-local correction to the exchange are -1.8 and -2.4 eV for Mo_2 and Cr_2, respectively, and are thus important in determining the agreement with experiment.

We predict that W_2, after the inclusion of relativistic effects, should have a stronger metal-metal bond than Mo_2. Even in the non-relativistic case, the bonding interaction is stronger in W_2 than in Mo_2 if the two metal atoms are referred to the same ^7S reference state.

Metal-ligand Bond Strengths in the Early Transition Metal Systems Cl_3ML and Late Transition Metal Systems $LCo(CO)_4$ (*14*)

The way in which metal-ligand bond energies of early transition metals and f-block elements differ from those of middle to late transition metals, or metal-ligand bond energies of 3d and 4f elements differ from those of their heavier congeners, has been the subject of many experimental (*15*) as well as a few theoretical studies (*16*) over the past decade.

We shall present here calculations on the D(M-L) bond strength in the Cl_3ML (**1**) model systems of the early transition metals M = Ti, Zr and Hf, as well as the $LM(CO)_4$ model system with the late transition metal M = Co, for a number of rudimentary ligands, L = H, CH_3, SiH_3, OH, SH, OCH_3, NH_2, PH_2 and CN. The calculated bond energies D(M-L) are displayed in Table 2. The ligands L = OH, OCH_3, with the coordinating atoms of the highest electronegativity and the most polar Cl_3M-L bond have the largest D(M-L) bond energies. The ligand L = SiH_3, with the coordinating atom of the lowest electronegativity and the least polar M-L bond, has a modest D(M-L) bond energy. For the series of ligands NH_2, SH, CH_3

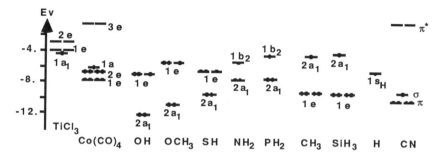

Fig. 1 Energies for the frontier orbitals of the two metal fragments $TiCl_3$ and $Co(CO)_4$ as well as the ligands L = OH, OCH_3, SH, NH_2, PH_2, CH_3, SiH_3, CN and H.

and H one finds in Table 2 that D(M-L) follows the order of the electronegativity of the coordinating atom, with NH_2 > SH > CH_3 > H.

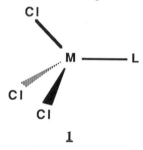

1

It is clear from Table 2 that zirconium, and even more so hafnium, form stronger M-L bonds to the ligands under investigation than titanium in the Cl_3M-L systems. We have found that the calculated increase in D(M-L) down the triad is primarily caused by a corresponding increased overlap between the singly occupied $1a_1$-orbital of Cl_3M and the singly occupied orbital on the ligand L, see Figure 1, which is in turn responsible for an increased σ-bonding interaction between the metal centre and the ligand (*17*).

There are few thermochemical data available for M-L bond involving group 4 metals. To our knowledge, of the ligands under consideration data (*15a*) are only available for L = CR_3, NR_2 and OR (R = alkyl), for the homoleptic $M(CR_3)_4$, $M(NR_2)_4$, and $M(OR)_4$ systems with M = Ti, Zr and Hf. The M-L bonds in these systems follow the same trend as observed here for the Cl_3M-L systems, with the bond energy increasing down the triad as well as with the increasing electronegativity of the corresponding atom on the ligand (O > N > C). Group 4 metals are known (*18*) to form several complexes involving M-L bonds with L = SiR_3, PR_2, and SR ligands. However, the corresponding D(M-L) bond energies have not been determined experimentally. Perhaps not surprisingly, we find that the M-L bonds of L = SiH_3, PH_2, and SH are weaker and less polar than the M-L bonds of the homologous ligands L = CH_3, NH_2, and OH (Table 2).

Comparative experimental data on M-H and M-CH_3 bond energies of early transition metals are not available. However, it has been asserted (*19*) that the M-H and M-CH_3 bond strengths of early transition metal complexes are quite similar. Indeed, we find D(M-H) and D(M-CH_3) to be quite similar in the Cl_3M-H and Cl_3M-CH_3 systems, respectively. We shall dedicate the next section to the relative strengths

Table 3 Calculated [D(Co-L)] bond energies (kJ mol^{-1}) and optimized (R$_{Co-L}$) bond distances (Å) in LCo(CO)$_4$

L	D(Co-L)	R$_{Co-L}$	L	D(Co-L)	R$_{Co-L}$
H	230.	1.55	SH	168.9	2.49
CH$_3$	160.	2.11	NH$_2$	145.5	2.09
SiH$_3$	211.6	2.73	PH$_2$	145.5	2.43
OH	232.4	2.09	CN	304.3	2.04

of the M-H and M-CH$_3$ bonds in systems ranging from early transition metal and f-block element complexes to middle and late transition metal complexes.

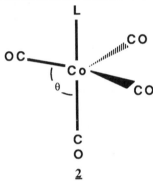

2

We shall now consider the M-L bond energies in the late transition metal complexes LCo(CO)$_4$ (**2**). The σ-bond in LCo(CO)$_4$ is considerably less polar than in Cl$_3$ML. This is primarily so because cobalt is more electronegative than the group 4 metals Ti, Zr and Hf, and as a consequence the 1a$_1$ metal based frontier orbital, involved in the σ-bond, is of lower energy than the frontier orbital 1a$_1$ of Cl$_3$M (see Figure 1).

For ligands other than H we had in the Cl$_3$ML systems favourable donor-acceptor interactions from occupied ligand orbitals to the two empty e-sets, see Figure 1. In Co(CO)$_4$ the metal based d-orbitals of e-symmetry are fully occupied and the corresponding interactions between occupied ligand orbitals and either 1e or 2e are as a consequence repulsive. As a consequence, we find in accord with available experimental evidence (*20*) that all ligand except hydrogen form stronger bonds to the early transition metal Ti than to the late transition metal Co, see Table 2 and 3.

The Relative Strengths of the Metal-Hydrogen and the Metal-Methyl Bonds in Transition Metal Complexes (*14,21,22*).

The breaking or formation of metal-hydrogen and metal-alkyl bonds is an integral part of most elementary reaction steps in organometallic chemistry. As a consequence, considerable efforts have been directed toward the determination of M-H (*15b*) and M-alkyl bond strength (*23*) as a prerequisite for a full characterization of the reaction enthalpies of elementary steps in organometallic chemistry.

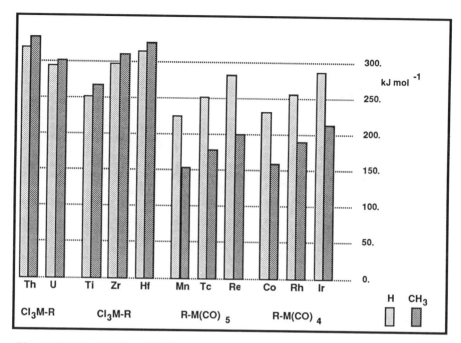

Fig. 2 Calculated M-H and M-CH₃ bond energies for several actinide and transition metal complexes.

As already mentioned, the strengths of M-H and M-Alkyl bonds are comparable for early transition metals (19). According to sparse experimental data, the same trend is observed in actinide complexes (24). By contrast, data for alkyl (20a,15e,25,26) and hydride (20e,26,27) complexes of middle to late transition metals indicate that the M-H bond is stronger than the M-Alkyl bond by some 40-80 kJ mol⁻¹. This difference in strength has implications for the relative ease by which ligands can insert into the M-H and M-Alkyl bonds (28). Also, it is one of the thermodynamic factors, along with the relative order of the bond energies H₂ < H–Alky < Alkyl–Alkyl, which favour the oxidative addition (19,29) to metal centres of H₂ compared to H–Alkyl and Alkyl–Alkyl bonds.

In the preceding section we have presented results on the relative bond strengths of the M-H and M-CH₃ bonds of model complexes Cl₃M-R (R = H, CH₃) involving the early transition metals Ti, Zr and Hf, as well as the late transition metal complex R-Co(CO)₄. Here, we shall present additional results on actinide metal complexes as well as on middle and late transition metal complexes, in an attempt to supplement the rather sparse experimental data available.

We have conducted calculations on the bond energy D(M-R) (R = H, CH₃) of the model actinide complexes Cl₃M-R (M = Th and U) as well as the middle transition metal complexes R-M(CO)₅ (M = Mn, Tc and Re) and the late transition metal complexes R-M(CO)₄ (M = Co, Rh and Ir). These results are depicted in Figure 2, along with the corresponding results presented in the preceeding section.

The results depicted in Figure 2 seem to indicate that the trend which assigns comparable M-H and M-CH₃ bond strengths in early transition metal and actinide complexes but a stronger M-H bond in middle and late transition metal complexes, is of general validity.   It also appear that both the M-H and the M-CH$_3$ bond strengths increase down a triad for transition metal complexes.

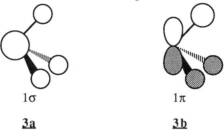

1σ   1π

3a   **3b**

The reduced strength of the M-CH$_3$ bonds in middle and late transition metals can be readily explained in terms of destabilizing three- and four-electron two-orbital interactions which occur between the fully occupied 1σ (**3a**) and 1π (**3b**) methyl orbitals and the fully or singly occupied d-orbitals of matching symmetries present on the metal centres.   By contrast, in early transition metal and actinide complexes, the metal orbitals of π-symmetry are vacant and are involved in stabilizing interactions with **3b**.   A destabilizing three-electron two-orbital interaction between **3a** and the single-occupied metal orbital 1a$_1$ (Figure 1) is still present.   However, the M-CH$_3$ bond is more polar in early transition metal complexes than in middle and late transition metal complexes and as a consequence electronic charge is transferred from the metal orbital 1a$_1$ to the methyl ligand, thereby relieving the destabilizing interaction between 1a$_1$ itself and **3a**.

The comparable strengths of the M-H bonds in the complexes studied is perhaps not too surprising, since H is a simple one-orbital ligand without additional occupied orbitals involved in four-electron two-orbital interactions or π-donor-acceptor interactions.   Finally, the increase in strength of both the M-H and M-CH$_3$ bonds down a triad, see Figure 2, is primarily related to an increase in the overlap between the metal 1a$_1$ orbitals and the matching ligand orbitals, which leads to a stronger σ– interaction.   Such an increase in overlap occurs as the metal d-orbitals become more diffuse down the triad.

A comparison between our calculated results for D(M-R) (R = H, CH$_3$) and the few available experimental data is presented in Table 4.   We find in general a good agreement with the experimental bond energies.   Also, the stability order D(M-L) > D(M-CH$_3$) in middle and late transition metal complexes supported by our theoretical study is consistent with data on organometallic reactions in which M-L and M-CH$_3$ bonds are formed or broken.   Thus, CO will readily insert into a M-CH$_3$ bond whereas the corresponding insertions into M-H bonds are virtually unknown (*28*), and methyl has likewise a larger migratory aptitude toward most other ligands than hydride.   The H$_2$ molecule is known to add oxidatively and exothermically to several metal fragments where the corresponding oxidative additions of the H-Alkyl and Alkyl-Alkyl bonds are unknown and probably endothermic as a consequence of the weak M-R bond (*19*).

At this stage, it is important to comment on the role of the non-local correction to the exchange in the calculated bond energies.   For middle and late transition metal complexes, the non-local correction reduces significantly the D(M-CH$_3$) values (by 105 kJ mol^{-1} for CH$_3$-Mn(CO)$_5$) whereas the corresponding D(M-H) bond energies are decreased to a lesser extent (by 13 kJ mol^{-1} for H-Mn(CO)$_5$).   Thus, it is apparent that Becke's non-local correction to the exchange is essential to assure the good agreement of the LSD/NL results with experiment, whereas the HFS and the LSD methods not only tend to give too large bond energies, but in some cases predict the wrong order for the M-H and M-CH$_3$ bond strengths.

Table 4 Calculated and experimental values for the bond energies D(M-R) (R = H, CH$_3$)
(kJ mol^{-1})

M-H	LSD/NL	Exp.	M-CH$_3$	LSD/NL	Exp.
Cl$_3$Th-H	318.0	~335.[a]	Cl$_3$Th-CH$_3$	333.9	~335.[a]
Cl$_3$U-H	293.3	319.7[a]	Cl$_3$U-CH$_3$	302.1	302.9[a]
Cl$_3$Ti-H	250.7	—	Cl$_3$Ti-CH$_3$	267.5	—
Cl$_3$Zr-H	297.2	—	Cl$_3$Zr-CH$_3$	309.5	—
Cl$_3$Hf-H	313.5	—	Cl$_3$Hf-CH$_3$	326.6	—
H-Mn(CO)$_5$	225	213[b]	CH$_3$-Mn(CO)$_5$	153	153[b]
H-Tc(CO)$_5$	252	—	CH$_3$-Tc(CO)$_5$	178	—
H-Re(CO)$_5$	282	—	CH$_3$-Re(CO)$_5$	200	—
H-Co(CO)$_4$	230	238[c]	CH$_3$-Co(CO)$_4$	160	—
H-Rh(CO)$_4$	255	—	CH$_3$-Rh(CO)$_4$	190	—
H-Ir(CO)$_4$	286	—	CH$_3$-Ir(CO)$_4$	212	—

[a] Experimental bond energies from Ref. 24 correspond to Cp$_2$ MCl-R systems. [b] Ref. 20c.
[c] Ref. 27e.

Thermal Stability and Kinetic Lability of the Metal-Carbonyl Bond (*30*)

The extensive use of coordinatively saturated mono-nuclear carbonyls as starting materials in organometallic chemistry, along with their volatility and high molecular symmetry, has prompted numerous experimental (*15a,31,32,33*) and theoretical (*34,35*) studies on their structure and reactivity. Special attention has been given to the degree of σ-donation and π-back-donation (*34b-g,35a,35e*) in the synergic (34k) M-CO bond.

However, in spite of many experimental (*32*) investigations, there is still a lack of basic data on the thermal stability and kinetic lability of the M-CO bond in essential metal carbonyls such as M(CO)$_6$ (M = Cr, Mo, W), M(CO)$_5$ (M = Fe, Ru, Os) and M(CO)$_4$ (M = Ni, Pd, Pt), particularly with respect to the carbonyls of the second- and third-row metals.

Theoretical methods have begun to play a role in determining the energetics of organometallics (*35g*) and *ab initio* type methods have recently been applied to calculation on the M-CO bond strength of Cr(CO)$_6$ (*35d-e*), Fe(CO)$_5$ (*35a-c,f*), and Ni(CO)$_4$ (*35a,f*), but not yet to M-CO bond strength of their second- and third-row homologues. Here, we shall present LSD/NL calculations on the intrinsic mean bond energy D(M-CO) and first CO dissociation energy ΔH of Cr(CO)$_6$, Fe(CO)$_5$, and Ni(CO)$_4$ as well as their second- and third-row homologues.

We shall here be concerned with periodic trends in the strength of the M-CO bonding interaction within the triads M = Cr, Mo, W; M = Fe, Ru, Os; and M = Ni, Pd, Pt. As measures for the M-CO bonding interaction in the hexacarbonyls (**4a**), pentacarbonyls (**4b**) and tetracarbonyls (**4c**), we will consider the intrinsic mean bond energy D(M-CO) between M (in its low-spin valence state) and the n CO ligands, as well as the bond energy ΔH between M(CO)$_{n-1}$ and CO.

Table 5   Comparison between calculated and experimental values for the mean bond energy E of several carbonyl complexes.   Calculated values for D(M-CO) and ΔE_{prep} are also given.   All values in kJ mol^{-1}

M(CO)$_n$ E(Exp.)^c	D(M-CO)	$1/n \cdot \Delta E_{prep}$	E^c	
Cr(CO)$_6$	211	100.7	110	110[a]
Mo(CO)$_6$	178	51.6	126	151[a]
W(CO)$_6$	210	54.4	156	179[a]
Fe(CO)$_5$	216.8	98.42	117	118.4[a]
Ni(CO)$_4$	178.9	—	—	191[b]

[a] Ref. 15a.   [b] Experimental intrinsic mean dissociation energy D(M-CO), Ref. 15a.   [c] Mean bond dissociation energy.

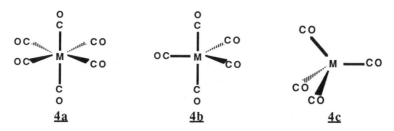

There are two sets of experimental data with a bearing on the M-CO bond strength in M(CO)$_n$, namely, the mean bond energy E corresponding to the process

$$M(CO)_n (g) \rightarrow M(g) + nCO(g) - nE \qquad (2a)$$

and the first bond dissociation energy ΔH corresponding to the process

$$M(CO)_n \rightarrow M(CO)_{n-1} + CO - \Delta H \qquad (2b).$$

It is important to note that E is given by

$$E = D(M-CO) - 1/n \Delta E_{prep} \qquad (3),$$

where ΔE_{prep} is the energy required to promote the metal atom from its high-spin electronic ground state to the low-spin valence configuration.   As a consequence, one can not conclude that the order of E will correspond to the order of D(M-CO) down a triad, since ΔE_{prep} might differ significantly for the three elements.

The first bond dissociation energy ΔH is on the other hand a direct measure for the strength of the M-CO bond interaction.   It is further an extremely important kinetic parameter, since the dissociation process of Eq.(2b) is assumed to be a key step in the large volume of kinetically useful substitution reactions (*36*)

$$M(CO)_n + L \rightarrow M(CO)_{n-1}L + CO \qquad (4)$$

where L is introduced into the coordination sphere of M by replacing one carbonyl ligand.

Our computational results for D(M-CO) and ΔH are depicted in Figure 3.   It appears from the values of D(M-CO) (Figure 3a) that strength of the M-CO bond

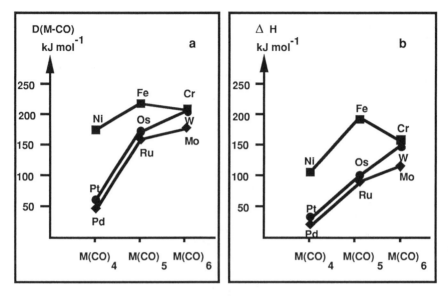

Fig. 3 Calculated bond energies for $M(CO)_6$ (M = Cr, Mo, W), $M(CO)_5$ (M = Fe, Ru, Os), and $M(CO)_4$ (M = Ni, Pd, Pt): a) intrinsic bond energies D(M-CO); b) first CO dissociation energies ΔH.

decreases going from the 3d to the 4d metal of a triad, to increase again for the complex of the 5d metal. The destabilization of the M-CO bond of the 4d and 5d elements, compared to their 3d counterparts increases in going from the Cr to the Ni group; in fact within the group 6 carbonyls, the M-CO bond in the W carbonyl is at a par in strength with the bond in the Cr complex. Among the systems studied, the strongest M-CO bond is assigned to the $Fe(CO)_5$ complex, whereas the weakest bond is found in the $Pd(CO)_4$ complex. It is important to note that the first dissociation bond energies ΔH (Figure 3b) follow closely the same trends observed for the intrinsic bond energies D(M-CO) (Figure 3a).

$$\pi_{CO}^*$$

$$\sigma_{CO}$$

5a   **5b**

The periodic trends discussed above can be readily rationalized in terms of the stabilizing electronic interactions and the destabilizing steric interactions which determine the strength of the M-CO bond. The electronic terms are represented by π-back-donation from occupied nd-orbitals on the metal centre to the empty π_{CO}^* orbital (**5a**), as well as σ-donation from the doubly-occupied σ_{CO} orbital (**5b**) to vacant nd orbitals. The steric terms are dominated by the repulsive four-electron two-orbital interactions between σ_{CO} and occupied nd orbitals on the metal centre. Our results show that electronic factors are most favourable for the pentacarbonyls where both π-

Table 6   Comparison between calculated and experimental values for the first CO dissociation energy ΔH, values in kJ mol^{-1}.   Calculated values do not include geometry relaxation of the fragments $M(CO)_{n-1}$

$M(CO)_n$	LSD/NL	Exp.	$M(CO)_n$	LSD/NL	Exp.
$Cr(CO)_6$	147	162[a]	$Fe(CO)_5$[b]	185[c]	176[d]
$Mo(CO)_6$	119	126[a]	$Ru(CO)_5$[b]	92	117[e]
$W(CO)_6$	142	166[a]	$Ni(CO)_4$	106	104[f]

[a] Ref. 37.   [b] Equatorial CO dissociation energy.   [c] The dissociation product is the fragment $Fe(CO)_4$ in its singlet state.   [d] Ref. 32a.   [e] Ref. 38.   [f] Ref. 39.

back-donation and σ-donation are important, whereas the steric interactions are most favourable for the M-CO bond among the hexacarbonyls, where all nd orbitals of σ-symmetry are empty and only mild repulsive interactions between σ_{CO} and the occupied ns and np metal orbitals are present.   For first-row transition metals, the repulsive interactions between occupied nd- and σ_{CO}-orbitals are still modest, since the nd-σ_{CO} overlap integrals are relatively small for the contracted 3d-orbitals compared to the more diffuse 4d and 5d orbitals.   Thus, electronic factors will make the intrinsic mean bond energy larger for $Fe(CO)_5$ than for $Cr(CO)_6$ (Figure 3a).   On the other hand, in carbonyls of 4d and 5d elements, where repulsive interactions between occupied nd- and σ_{CO}-orbitals are considerable, the steric factors cause the M-CO bonds in $Ru(CO)_5$ and $Os(CO)_5$ to be weaker than in $Mo(CO)_6$ and $W(CO)_6$, respectively.   The tetracarbonyls, in which all interactions between the nd- and σ_{CO}-orbitals are repulsive, have weaker M-CO bonds than the corresponding hexacarbonyls and pentacarbonyls in each of the transition series (Figure 3a).   Finally, relativistic effects will stabilize the 5d element carbonyl compared to the 4d metal carbonyl within a triad, see Figure 3a.   Such a stabilization is sufficient to bring the strength of the M-CO bond in $W(CO)_6$ at a par with the strength of the bond in $Cr(CO)_6$.   The rational given above for the variations in D(M-CO) can further be used to explain the trends in the first CO dissociation energy ΔH (Figure 3b).

We have calculated the promotion energy ΔE_{prep} needed to evaluate the mean bond dissociation energy E, according to Eq.(3).   In Table 5, we compare our calculated value for E with the available experimental mean bond dissociation energies.   We observe a good to fair agreement of our theoretical estimates with the experimental data; in particular, the experimental trend in the values of E down the only triad for which such data are available, the group 6 carbonyls, is correctly reproduced.   Here, see Table 5, the $Cr(CO)_6$ complex has the lowest average dissociation energy in spite of having the largest value of D(M-CO).   The different trends in E and D(M-CO) are related to variations in the promotion energies ΔE_{prep} (Table 5), as the promotion energy is seen to be much larger for Cr than for Mo or W.   This result is not unexpected since ΔE_{prep} depends on exchange integrals that in general are larger for the relatively contracted 3d-orbitals of chromium than for the more diffuse 4d- and 5d-orbitals of molybdenum and tungsten.

Our calculated values for the first CO dissociation energy ΔH are also in fair to good agreement with available experimental data, see Table 6.   Finally, the indication that the M-CO bonds are fairly weak in the $Pd(CO)_4$ and $Pt(CO)_4$ complexes, inferred by our theoretical estimates of both D(M-CO) and ΔH (Figure 3), finds support in the apparent instability of these species at room temperature.

The value of the non-local contribution to the exchange in the calculated bond energies for carbonyls ranges from 109 to 138 kJ mol^{-1}.   Thus, the HFS and LSD estimates (*16a,35e*) of bond energies are in some case twice as large as the correct experimental values, and the non-local correction to the exchange is once again found to be essential to achieve good agreement with experiment. For calculations on transition metal complexes, the LSD/NL method must also be considered more reliable than the HF-method itself, which tends to severely under-estimate the bond energies in such systems, as attested by the prediction, given by HF-results, that the $Fe(CO)_4$ and $Ni(CO)_4$ complexes be unstable with respect to the free metal atom and free CO molecules ($35a$). On the other hand, due to their slow convergency ($35b$), CI calculations are not at present widely applicable to transition metal complexes. These considerations point to the LSD/NL-approximation as one of today's methods of choice in the theoretical investigation of transition metal complexes.

Literature Cited

1. Becke, A.D. *J. Chem. Phys.* **1986**, *84*, 4524.
2. Baerends, E.J.; Ellis, D.E.; Ros, P. *Chem. Phys.* **1973**, *2*, 71.
3. Snijders, J.G.; Baerends, E.J.; D.E.; Ros, P. *Molec. Phys.* **1979**, *38*, 1909.
4. Stoll, H.; Golka, E.; Preuss, H. *Theor. Chim. Acta* **1980**, *55*, 29.
5. Vosko, S.H.; Wilk, L.; Nusair, M. *Can. J. Phys.* **1980**, *58*, 1200.
6. Ziegler, T.; Rauk, A. *Theor. Chim. Acta* **1977**, *46*, 1.
7. Ziegler, T.; Snijders, J.G.; Baerends, E.J. *J. Chem. Phys.* **1981**, *74*, 1271.
8. (a) Snijders, J.G.; Baerends, E.J.; Versnoijs, P. *At. Nucl. Data Tables* **1982**, *26*, 483.
   (b) Versnoijs, P.; Snijders, J.G.; Baerends, E.J. *Slater Type Basis Functions for the Whole Periodic System.* Internal Report; Free University: Amsterdam, 1981.
9. Krijn, J.; Baerends, E.J. *Fit Functions in the HFS-method.* Internal Report (in Dutch); Free University: Amsterdam, 1981.
10. Ziegler, T.; Tschinke, V.; Becke, A. *Polyhedron* **1987**, *6*, 685.
11. (a) Shim, I. K. *Dan. Videmsk. Selsk. Mat.-Fys. Medd.* **1985**, *41*, 47; and references therein.
    (b) Weltner, W.; Van Zee, R.J. *Annu. Rev. Phys. Chem.* **1984**, *35*, 291.
12. Baycara, N.A.; McMaster, B.N.; Salahub, D.R. *Molec. Phys.* **1984**, *52*, 891; and references therein.
13. Moore, C.E. In *Atomic Energy Levels, Nat. Bur. Stand. (U.S.)* C**1958**, *467*; Vol. 3.
14. Ziegler, T.; Tschinke, V.; Versluis, L.; Baerends, E.J.; Ravenek, W. *Polyhedron* **1988**, *7*, 1625.
15. (a) Connors, J.A. *Top. Curr. Chem.* **1977**, *71*, 71.
    (b) Pearson, R.G. *Chem. Rev.* **1985**, *85*, 41.
    (c) Beauchamp, J.L. *Chem. Rev.* in press.
    (d) Bryndza, H.E.; Fong, L.K.; Peciello, R.A.; Tam, W.; Bercaw, J.E. *J. Am. Chem. Soc.* **1987**, *109*, 1444.
    (e) Bruno, J.W.; Marks, T.J.; Morss, L.R. *J. Am. Chem. Soc.* **1983**, *105*, 6824.
16. (a) Ziegler, T.; Tschinke, V.; Versluis, L. *NATO ASI* **1986**, *Series C 176*, 189; and references therein.
    (b) Hay, P.J.; Rohlfing, C.M. *NATO ASI* **1986**, *Series C 176*, 135.
17. For a full account on the energy decomposition of the bond energies of Cl_3M-L and $L-Co(CO)_4$, see Ref. 14.
18. (a) Roddick, D.M.; Santasiero, B.D.; Bercaw, J. *J. Am. Chem. Soc.* **1985**, *107*, 4670.

(b) Cardin, D.J.; Lappert, C.L.; Raston, C.L.; Riley, P.L. In *Comprehensive Organometallic Chemistry*, Wilkinson, G., Ed.; Pergamon Press: New York, 1982.
19. Crabtree, R.H. *Chem. Rev.* **1985**, *85*, 245.
20. (a) Mandich, M.L.; Halle, L.F.; Beauchamp, J.L. *J. Am. Chem. Soc.* **1984**, *106*, 4403.
(b) Armentrout, P.B.; Halle, L.F.; Beauchamp, J.L. *J. Am. Chem. Soc.* **1981**, *103*, 6501.
(c) Connor, J.A.; Zafarani-Moattar, M.T.; Bickerton, J.; Saied, N.I.; Suradi, S.; Carson, R.; Al Tackhin, G.A.; Skinner, H.A. *Organometallics* **1982**, *1*, 1166.
21. (a) Ziegler, T.; Tschinke, V.; Baerends, E.J.; Snijders, J.G. *J. Phys. Chem.* in press.
(b) Ziegler, T.; Tschinke, V.; Becke, A. *J. Am. Chem. Soc.* **1987**, *109*, 1351.
22. Ziegler, T.; Wendan, C.; Baerends, E.J.; Ravenek, W. *Inorg. Chem.* **1988**, *27*, 3458.
23. Halpern, J. *Acc. Chem. Res.* **1982**, *15*, 238.
24. Bruno, J.W.; Stecher, H.A.; Mors, L.R.; Sonnenberg, D.C.; Marks, T.J. *J. Am. Chem. Soc.* **1986**, *108*, 7275.
25. (a) Georgiadis, R.; Armertrout, P.B. *J. Am. Chem. Soc.* **1986**, *108*, 2119.
(b) Aristov, N.; Armertrout, P.B. *J. Am. Chem. Soc.* **1986**, *108*, 1806.
26. Halle, L.F.; Armentrout, P.B.; Beauchamp, J.L. *Organometallics* **1982**, *1*, 963.
27. (a) Squires, R.R. *J. Am. Chem. Soc.* **1985**, *107*, 4385.
(b) Sallans, L.; Lane, K.R.; Squires, R.R.; Freiser, B.S. *J. Am. Chem. Soc.* **1985**, *107*, 4379.
(c) Schilling, J.B.; Goddard, W.A. III; Beauchamp, J.L. *J. Am. Chem. Soc.* in press.
(d) Girling, R.B.; Grebenik, P.; Perutz, R.N. *Inorg. Chem.* **1986**, *24*, 31.
(e) Ungvary, F. *Organomet. Chem.* **1972**, *36*, 363.
28. Ziegler, T.; Versluis, L.; Tschinke, V. *J. Am. Chem. Soc.* **1986**, *108*, 612.
29. Low, J.J; Goddard, W.A. III *Organometallics* **1986**, *5*, 609.
30. Ziegler, T.; Tschinke, V.; Ursenbach, C. *J. Am. Chem. Soc.* **1987**, *109*, 4825.
31. (a) Kettle, S.F.A. *Curr. Top. Chem.* **1977**, *71*, 111.
(b) Braterman, P.S. In *Metal Carbonyl Spectra*; Academic Press: London, 1975.
(c) Mingos, D.P.M. In *Comprehensive Organometallic Chemistry* ; Wilkinson, G.; Stone, F.G.A.; Abel, E.W., Eds.; Pergamon: New York;1982; Vol. 3, p1.
32. (a) Lewis, K.E.; Golden, D.M.; Smith, G.P. *J. Am. Chem. Soc.* **1984**, *106*, 3906.
(b) Bernstein, M.; Simon, J.D.; Peters, J.D. *Chem. Phys. Lett.* **1983**, *100*, 241.
33. (a) Rees, B.; Mitschler, A. *J. Am. Chem. Soc.* **1976**, *98*, 7918.
(b) Beagley, B.; Schmidling, D.G. *J. Mol. Struct.* **1974**, *22*, 466.
(c) Hedberg, L.; Lijima, T.; Hedberg,K. *J. Chem. Phys.* **1979**, *70*, 3224.
(d) Jones, L.J.; McDowell, R.S.; Boldblatt, M. *Inorg. Chem.* **1969**, *8*,2349.
34. (a) Guenzburger, D.; Saitovitch, E.M.B.; De Paoli, M.A.; Manela, J. *J. Chem. Phys.* **1984**, *80*, 735.
(b) Baerends, E.J.; Ros, P. *Mol. Phys.* **1975**, *30*, 1735.
(c) Heijser, W.; Baerends, E.J.; Ros, P. *J. Mol. Struct.* **1980**, *19*, 1805.
(d) Bursten, B.E.; Freier, D.G.; Fenske, R.F. *Inorg. Chem.* **1980**, *19*, 1804.
(e) Demuynk, J.; Veillard, A. *Theor. Chim. Acta* **1973**, *28*, 241.
(f) Caulton, K.G.; Fenske, R.F. *Inorg. Chem.* **1975**, *7*, 1273.
(g) Hubbard, J.L.; Lichtenberger, J. *J. Am. Chem. Soc.* **1982**, *104*, 2132.
(h) Elian, M.; Hoffmann, R. *Inorg. Chem.* **1975**, *14*, 1058.

(i) Saddei, D.; Freund, H.J.; Hohlneicher, G. *Chem. Phys.* **1981**, *55*, 1981.
(j) Johnson, J.B.; Klemperer, W.G. *J. Am. Chem. Soc.* **1977**, 99,7132.
(k) Chatt, J.; Duncanson, L.A. *J. Chem. Soc.* **1953**, 2939.
(l) Hillier, I.H.; Saunders, V.R. *Mol. Phys.* **1971**, *22*, 1025.
(m) Vanquickenborne, L.J.; Verhulst, J. *J. Am. Chem. Soc.* **1977**, *99*, 7132.
(n) Ford, P.C.; Hillier, I.H.; Pope, S.A.; Guest, M.F. *Chem. Phys. Lett.* **1983**, *102*, 555.
(o) Burdett, J.K. *J. CHem. Soc., Faraday Trans. 2* **1974**, *70*, 1599.
(p) Penzak, D.A.; McKinney, R.J. *Inorg. Chem.* **1979**, *18*, 3407.
(q) Osman, R.; Ewig, C.S.; Van Wazer, J.R. *Chem. Phys. Lett.* **1978**, *54*, 1341.
(r) Serafini, A.; Barthelat, J.C.; Durand, P. *Mol. Phys.* **1978**, *36*, 1341.
(s) Sakai, T.; Huzinaga, J. *Chem. Phys.* **1982**, *76*, 2552.
35. (a) Bauschlicher, C.W.; Bagus, P.S. *J. Chem. Phys.* **1984**, *81*, 5889.
(b) Lüthi, H.P.; Siegbahn, P.E.M.; Almlof, J. *J. Phys. Chem.* **1985**, *89*, 2156.
(c) Daniel, C.; Benard, M.; Dedieu, A.; Wiest, R.; Veillard, A. *J. Phys. Chem.* **1984**, *88*, 4805.
(d) Sherwood, D.E; Hall, M.B. *Inorg. Chem.* **1983**, *22*, 93.
(e) Baerends, E.J.; Rozendaal, A. *NATO ASI* **1986**, *Series C, 176*, 159.
(f) Rösch, N.; Jorg, H.; Dunlap, B. *NATO ASI* **1986**, *Series C, 176*, 179.
(g) Veillard, A., Ed. *NATO ASI* **1986**, *Series C, 176*.
(h) Rolfing, C.M.; Hay, P.J. *J. Che. Phys.* **1985**, *83*, 4641.
36. Kirtley, S.W. In *Comprehensive Organometallic Chemistry*; Wilkinson, G.; Stone, F.G.A.; Abel, E.W., Eds.; Pergamon: New York;1982; Vol. 3, p 783.
37. (a) Angelici, R.J. *Organomet. Chem. Rev. A* **1968**, *3*, 173.
(b) Covey, W.D.; Brown, T.L. *Inorg. Chem.* **1973**, *12*, 2820.
(c) Centini, G.; Gambino, O. *Atti Acc. Sci. Torino I* **1963**, *97*, 757.
(d) Werner, H. *Angew. Chem. Int. Ed. Engl.* **1968**, *7*, 930.
(e) Graham, J.R.; Angelici, R.J. *Inorg. Chem.* **1967**, *6*, 2082.
(f) Werner, H.; Prinz, R. *Chem. Ber.* **1960**, *99*, 3582.
(g) Werner, H.; Prinz, R. *J. Organomet. Chem.* **1966**, *5*, 79.
38. Huq, R.; Poe, A.J.; Chawla, S. *Inorg. Chem. Acta* **1979**, *38*, 121.
39. Turner, J.J.; Simpson, M.B.; Poliakoff, M.; Maier, W.B. *J. Am. Chem. Soc.* **1983**, *105*, 3998.

RECEIVED November 15, 1989

INDEXES

Author Index

Affiliation Index

Subject Index

Production: BethAnn Pratt-Dewey
Indexing: Deborah H. Steiner
Acquisition: Cheryl Shanks

Elements Typeset by Hot Type Ltd., Washington, DC
Printed and Bound by Maple Press, York, PA

Paper meets minimum requirements of American National Standard
for Information Sciences–Permanence of Paper for Printed Library
Materials, ANSI Z39.48–1984 ∞

Other ACS Books

Chemical Structure Software for Personal Computers
Edited by Daniel E. Meyer, Wendy A. Warr, and Richard A. Love
ACS Professional Reference Book; 107 pp;
clothbound, ISBN 0–8412–1538–3; paperback, ISBN 0–8412–1539–1

Personal Computers for Scientists: A Byte at a Time
By Glenn I. Ouchi
276 pp; clothbound, ISBN 0–8412–1000–4; paperback, ISBN 0–8412–1001–2

Biotechnology and Materials Science: Chemistry for the Future
Edited by Mary L. Good
160 pp; clothbound, ISBN 0–8412–1472–7; paperback, ISBN 0–8412–1473–5

Polymeric Materials: Chemistry for the Future
By Joseph Alper and Gordon L. Nelson
110 pp; clothbound, ISBN 0–8412–1622–3; paperback, ISBN 0–8412–1613–4

The Language of Biotechnology: A Dictionary of Terms
By John M. Walker and Michael Cox
ACS Professional Reference Book; 256 pp;
clothbound, ISBN 0–8412–1489–1; paperback, ISBN 0–8412–1490–5

Cancer: The Outlaw Cell, Second Edition
Edited by Richard E. LaFond
274 pp; clothbound, ISBN 0–8412–1419–0; paperback, ISBN 0–8412–1420–4

Practical Statistics for the Physical Sciences
By Larry L. Havlicek
ACS Professional Reference Book; 198 pp; clothbound; ISBN 0–8412–1453–0

The Basics of Technical Communicating
By B. Edward Cain
ACS Professional Reference Book; 198 pp;
clothbound, ISBN 0–8412–1451–4; paperback, ISBN 0–8412–1452–2

The ACS Style Guide: A Manual for Authors and Editors
Edited by Janet S. Dodd
264 pp; clothbound, ISBN 0–8412–0917–0; paperback, ISBN 0–8412–0943–X

Chemistry and Crime: From Sherlock Holmes to Today's Courtroom
Edited by Samuel M. Gerber
135 pp; clothbound, ISBN 0–8412–0784–4; paperback, ISBN 0–8412–0785–2

For further information and a free catalog of ACS books, contact:
American Chemical Society
Distribution Office, Department 225
1155 16th Street, NW, Washington, DC 20036
Telephone 800–227–5558